Springer Series in Statistics

Springer Series in Statistics

D. F. Andrews and A. M. Herzberg, Data: A Collection of Problems from Many Fields for the Student and Research Worker. xx, 442 pages, 1985.

F. J. Anscombe, Computing in Statistical Science through APL. xvi, 426 pages, 1981.

J. O. Berger, Statistical Decision Theory: Foundations, Concepts, and Methods, 2nd edition. xiv, 425 pages, 1985.

P. Brémaud, Point Processes and Queues: Martingale Dynamics. xviii, 354 pages, 1981.

K. Dzhaparidze, Parameter Estimation and Hypothesis Testing in Spectral Analysis of Stationary Time Series. xii, 300 pages, 1985.

R. H. Farrell, Multivariate Calculation. xvi, 367 pages, 1985.

L. A. Goodman and W. H. Kruskal, Measures of Association for Cross Classifications. x, 146 pages, 1979.

J. A. Hartigan, Bayes Theory. xii, 145 pages, 1983.

H. Heyer, Theory of Statistical Experiments. x, 289 pages, 1982.

H. Kres, Statistical Tables for Multivariate Analysis. xxii, 504 pages, 1983.

M. R. Leadbetter, G. Lindgren and H. Rootzén, Extremes and Related Properties of Random Sequences and Processes. xii, 336 pages, 1983.

R. G. Miller, Jr., Simultaneous Statistical Inference, 2nd edition. xvi, 299 pages, 1981.

F. Mosteller, D. S. Wallace, Applied Bayesian and Classical Inference: The Case of *The Federalist* Papers. xxxv, 301 pages, 1984.

D. Pollard, Convergence of Stochastic Processes. xiv, 215 pages, 1984.

J. W. Pratt and J. D. Gibbons, Concepts of Nonparametric Theory. xvi, 462 pages, 1981.

L. Sachs, Applied Statistics: A Handbook of Techniques. xxviii, 706 pages, 1982.

E. Seneta, Non-Negative Matrices and Markov Chains. xv, 279 pages, 1981.

D. Siegmund, Sequential Analysis: Tests and Confidence Intervals. xii, 272 pages, 1985.

V. Vapnik, Estimation of Dependences based on Empirical Data. xvi, 399 pages, 1982.

K. M. Wolter, Introduction to Variance Estimation. xii, 428 pages, 1985.

David Siegmund

Sequential Analysis

Tests and Confidence Intervals

With 13 Illustrations

Springer-Verlag
New York Berlin Heidelberg Tokyo

David Siegmund
Department of Statistics
Stanford University
Stanford, CA 94305
U.S.A.

AMS Classification: 62L10

Library of Congress Cataloging in Publication Data
Siegmund, David
Sequential analysis.
(Springer series in statistics)
Bibliography: p.
Includes index.
1. Sequential analysis. 2. Statistical hypothesis testing.
3. Confidence intervals. I. Title. II. Series.
QA279.7.S54 1985 519.2 85-7942

Typeset by Asco Trade Typesetting Ltd., Hong Kong.
Printed and bound by R. R. Donnelley and Sons, Harrisonburg, Virginia.
Printed in the United States of America.

9 8 7 6 5 4 3 2 1

ISBN 0-387-96134-8 Springer-Verlag New York Berlin Heidelberg Tokyo
ISBN 3-540-96134-8 Springer-Verlag Berlin Heidelberg New York Tokyo

Preface

The modern theory of Sequential Analysis came into existence simultaneously in the United States and Great Britain in response to demands for more efficient sampling inspection procedures during World War II. The developments were admirably summarized by their principal architect, A. Wald, in his book *Sequential Analysis* (1947).

In spite of the extraordinary accomplishments of this period, there remained some dissatisfaction with the sequential probability ratio test and Wald's analysis of it. (i) The open-ended continuation region with the concomitant possibility of taking an arbitrarily large number of observations seems intolerable in practice. (ii) Wald's elegant approximations based on "neglecting the excess" of the log likelihood ratio over the stopping boundaries are not especially accurate and do not allow one to study the effect of taking observations in groups rather than one at a time. (iii) The beautiful optimality property of the sequential probability ratio test applies only to the artificial problem of testing a simple hypothesis against a simple alternative.

In response to these issues and to new motivation from the direction of controlled clinical trials numerous modifications of the sequential probability ratio test were proposed and their properties studied—often by simulation or lengthy numerical computation. (A notable exception is Anderson, 1960; see III.7.) In the past decade it has become possible to give a more complete theoretical analysis of many of the proposals and hence to understand them better.

The primary goal of this book is to review these recent developments for the most part in the specific framework of Wald's book, i.e., sequential hypothesis testing in a non-Bayesian, non-decision-theoretic context. In contrast to the sequential probability ratio test, the emphasis is on closed (truncated) sequential tests defined by non-linear stopping boundaries and often applied to grouped data. In particular the repeated significance tests of Armitage (1975) and the group repeated significance tests of Pocock (1977) are given an extensive theoretical treatment. To the extent that there is a unifying theme to the book, it is an attempt to understand repeated significance tests theoretically, to criticize them, and to suggest some improvements as a response to this criticism.

A secondary goal is to redress to some extent the imbalance in the literature of sequential analysis between considerations of experimental design and considerations of statistical inference. Here the term "experimental design" refers to selection of an experiment, hence to selection of the data to be observed, from a set of possibilities. Since choosing a sequential test includes the problem of choosing a stopping rule, it should properly be considered an aspect of experimental design.

Inferential summaries of the data actually obtained in an experiment include attained significance levels (p-values) and confidence intervals, which play important roles in fixed sample statistics, but until recently have been almost totally ignored in discussions of sequential analysis. Chapters III and IV examine these concepts, which turn out to have certain implications for the selection of a stopping rule, hence for the experimental design itself.

Some additional subjects which can be studied by the methods developed to investigate truncated sequential tests are discussed briefly. Thus cusum tests are introduced in Chapter II and a simple approximation for their average run length is given in Chapter X. However, they are not systematically compared with other competing possibilities for detecting a change of distribution. Fixed precision confidence intervals receive a similar cursory treatment in Chapter VII.

I have attempted to make this book more widely accessible than most of the literature on which it is based by delaying as long as possible the introduction of "heavy" mathematics. Chapters I–VII emphasize the statistical ideas accompanied by enough of the mathematical development that the reader can regard himself as a participant rather than a spectator observing the opening of a black box. The proofs of many results are delayed until after their significance and application have been discussed; in fact most proofs fall short of the currently accepted mathematical standard. Woodroofe (1982) has given a more precise and extensive development of the mathematical theory.

The requirements of exposition have led to numerous compromises. The most significant is the use of Brownian motion in Chapters I–VI to provide basic qualitative insight. The corresponding discrete time results are stated so that they can be used for numerical calculations, but their generally more difficult justification is then deferred. Consequently the mathematical methods have been chosen to a certain extent because they provide a unified treatment for the range of problems considered, linear and non-linear in discrete and continuous time. In many cases this meant not using what appears to be the "best" method for a particular problem. For example, the methods pioneered by Woodroofe, which often seem to deliver the best result in discrete time, have not been developed completely; and the purely continuous time methods of Jennen and Lerche (1981, 1982) are not discussed at all.

Finally it should be noted that the book is primarily concerned with very simple models—especially those involving the normal distribution. There exists an extensive literature devoted to approximating more complex models by simple ones. A notable early contribution is due to Cox (1963). See also Hall and Loynes (1977) and Sen (1982). Digressions illustrating the nature of such approximations appear in III.9 and again in V.5, which is concerned with the log rank test of survival analysis.

I would like to thank a number of people who have directly or indirectly contributed to this project. Herbert Robbins introduced me to the subject of Sequential Analysis and has been a constant source of intellectual guidance and inspiration. T. L. Lai worked with me on two papers which form the mathematical foundation of the presentation given here. Michael Woodroofe's research, beginning with his brilliant 1976 paper, has been a rich source of ideas and a stimulus to me to improve mine. In addition I want to thank Michael Hogan, Steve Lalley, and Thomas Sellke for many helpful discussions and technical assistance during the past several years. Some of their specific contributions are mentioned in the bibliographical notes. I thank Peter Elderon and Inchi Hu for proofreading and helpful suggestions on exposition. Thanks are also due to Jerri Rudnick and Judi Davis for their superb typing and cheerful retyping. And finally I want to thank the Office of Naval Research and the National Science Foundation for their support of my research.

Stanford, California David Siegmund
May, 1985

Contents

CHAPTER I

Introduction and Examples

In very general terms there are two reasons for introducing sequential methods into statistical analysis. One is to solve more efficiently a problem which has a fixed sample solution. The other is to deal with problems for which no fixed sample solution exists. It is the first category which is the primary concern of this book, but we begin here with a few comments about the second.

Some problems are intrinsically sequential and cannot be discussed without considering their sequential aspects. An important example is a control system with unknown dynamics, about which something can be learned as the system operates. Dynamic programming is one method for dealing with problems of this sort. A beautiful recent summary is given by Whittle (1982, 1983).

Another intrinsically sequential problem is the fixed precision estimation of a parameter in the presence of an unknown nuisance parameter. It is almost obvious that one cannot give a confidence interval of prescribed length for the mean of a normal distribution based on a sample of some fixed size n if one does not know the variance of the distribution. (See Dantzig, 1940, for a formal proof.) However, by taking data sequentially one can use the data to estimate the variance and the estimated variance to determine a (random) sample size which will permit the mean to be estimated by a fixed length confidence interval. See Stein (1945) and Chapter VII. (In spite of its apparent omnipotence the method of dynamic programming appears not to have been applied to this problem.)

The principal subject of this book is sequential hypothesis testing and related problems of estimation. In contrast to the preceding examples, for most of the problems studied in detail there exist fixed sample solutions, and the reason for introducing sequential methods is to provide greater efficiency in some sense to be defined. Many of the problems might be attacked by dynamic programming. In fact, dynamic programming is a far reaching generalization of the method originally developed in the pioneering papers of Wald (1947b), Wald and Wolfowitz (1948), and perhaps most importantly Arrow *et al.* (1949) to find Bayes solutions to problems of sequential hypothesis testing. Nevertheless, because we shall be primarily concerned with problems having vaguely specified loss functions, for the most part we shall ignore the possibility of finding optimal solutions and concentrate instead on procedures which can be directly compared with and improve upon those used most often in practice, to wit fixed sample size procedures evaluated in the classical terms of significance level, power, and sample size.

The simplest sequential test is a so-called curtailed test. Suppose that a machine produces items which may be judged good or defective, and we wish to infer on the basis of a random sample whether the proportion of defectives in a large batch of items exceeds some value p_0. Assume that the inference will be based on the number S_m of defectives in a random sample of size m. If m is a small proportion of the batch size, then S_m has approximately a binomial distribution with mean mp, where p is the true proportion of defectives in the batch; and a reasonable rule to test the hypothesis H_0: $p \le p_0$ against H_1: $p > p_0$ is to reject H_0 if $S_m \ge r$ for some constant r, which at the moment need not be specified more precisely. If the sample is drawn sequentially and for some value k less than m the value of S_k already equals r, one could stop sampling immediately and reject H_0. More formally, let T denote the smallest value of k for which $S_k = r$ and put $T' = \min(T, m)$. Consider the procedure which stops sampling at the random time T' and decides that $p > p_0$ if and only if $T \le m$. If one considers these two procedures as tests of H_0 against H_1, their rejection regions, to wit $\{T \le m\}$ and $\{S_m \ge r\}$, are the same events, and hence the two tests have the same power function. Since the test which stops at the random time T' never takes more observations and may take fewer than the fixed sample test, it has a reasonable claim to be regarded as more efficient.

The preceding discussion has the appearance of delivering a positive benefit at no cost. However, the situation is not so clear if a second consideration is also to estimate p, say by means of a confidence interval. To continue the discussion with a slightly different example, suppose that $X(t)$, $t > 0$, is a Poisson process with mean value λt, and we would like to test H_0: $\lambda \le \lambda_0$ against H_1: $\lambda > \lambda_0$. This problem might be regarded as an approximation to the preceding one, for if p is small the process of failures is approximately a Poisson process. However, the Poisson formulation might also apply to a reliability analysis of items having exponentially distributed lifetimes, which (in the simplest experimental design) are put on test serially with each failed item being immediately replaced with a good one. Then λ is the reciprocal of the mean time to failure of the items. It is clear from the discussion of the preceding paragraph that instead of a fixed time test which observes $X(t)$ until $t = m$ and rejects H_0 whenever $X(m) \ge r$, one can curtail the test at the stopping time $T' = \min(T, m)$, where T denotes the first time t such that $X(t) = r$, and reject H_0 whenever $T \le m$.

Now consider the problem of giving an upper confidence bound for λ (hence a lower confidence bound for the mean lifetime of an item). The standard fixed sample $(1-\alpha) \times 100\%$ confidence bound is $\lambda_2^*[X(m)]$, where $\lambda_2^*(n)$ is defined as the unique solution of

$$P_\lambda\{X(m) \le n\} = \alpha. \tag{1.1}$$

Since

$$P_\lambda\{X(t) \le n\} = P_\lambda\{w_{n+1} > t\} \tag{1.2}$$

where w_n is the waiting time for the nth event of the Poisson process, and since λw_n has a gamma distribution with parameter n (chi-square distribution with parameter $2n$), the value of $\lambda_2^*(n)$ is easily determined. For the curtailed test having exactly the same power function as a given fixed sample test, the corresponding confidence bound is slightly different. In analogy with (1.1) (see also Problem 1.1) define $\lambda_1^*(t)$ to be the solution of

$$P_\lambda\{T > t\} = \alpha. \tag{1.3}$$

Then a $(1-\alpha) \times 100\%$ upper confidence bound for λ based on the data $(T', X(T'))$ is

$$\lambda^*[T', X(T')] = \begin{cases} \lambda_1^*(T') & \text{if } T \le m \\ \lambda_2^*[X(m)] & \text{if } T > m \end{cases} \tag{1.4}$$

(see Problem 1.2 for a proof). The relation (1.2) between $X(t)$ and w_n makes it easy to determine $\lambda_1^*(t)$.

Lower confidence bounds, $\lambda_{*2}[X(m)]$ and $\lambda_*[T', X(T')]$ may be similarly defined. Confidence intervals may be obtained by combining upper and lower confidence bounds in the usual way. It turns out that $\lambda_*[T', X(T')] \le \lambda_{*2}[X(m)]$ with equality if and only if $X(m) \le r$, so one price of curtailment is a smaller lower confidence bound for λ.

The relation between $\lambda_2^*[X(m)]$ and $\lambda^*[T', X(T')]$ is not so simple.[1] Since the Poisson distributions have monotone likelihood ratio, the confidence bound $\lambda_2^*[X(m)]$ for the fixed sample size m is optimal in the strong sense of being uniformly most accurate (see Lehmann, 1959, p. 78ff. or Cox and Hinkley, 1974, p. 213). Since the statistician who observes $X(m)$ could by sufficiency define a randomized upper confidence bound with exactly the same coverage probability as (1.4), it follows that the fixed sample upper confidence bound is uniformly more accurate than that defined by (1.4). Hence less accuracy at the upper confidence bound is also a price of curtailment. (It is easy to see that the distributions of $(T', X(T'))$ have monotone likelihood ratio and hence that the upper confidence bound (1.4) is itself uniformly most accurate in the class of procedures which depend on the sample path $X(t)$ only until time T' (cf. Problem 1.7). We shall see that the method used to define (1.4) can be

[1] The material in this paragraph plays no role in what follows. It can be omitted by anyone not already familiar with the relevant concepts.

adapted to a variety of sequential tests, but it is very rare that the resulting confidence bounds have an easily described optimal property.)

The preceding discussion illustrates qualitatively both the advantages (smaller sample size) and the disadvantages (less accurate estimation) associated with a sequential test. In Chapters III and IV these tradeoffs are studied quantitatively.

Remark 1.5. The reader interested in the foundations of statistics may find it interesting to think about various violations of the likelihood principle (Cox and Hinkley, 1974, p. 39) which occur in the sequel. One example is in the definition of confidence bounds. For a Bayesian with a prior distribution for λ which is uniform on $(0, \infty)$, an easy calculation shows that for any stopping rule τ, $\lambda_2^*[X(\tau)]$ defined above is a $1 - \alpha$ posterior probability upper bound for λ, i.e. $P\{\lambda \leq \lambda_2^*[X(\tau)] | \tau, X(\tau)\} = 1 - \alpha$. In particular, for the fixed sample experiment the confidence and posterior probability bounds agree. But for the sequential experiment, the particular stopping rule plays an important role in the determination of a confidence bound with the effect that the "confidence" of the posterior probability upper bound is strictly less than $1 - \alpha$ (see also Problem 1.5).

Although the methods described in the following chapters can be adapted to the investigation of a wide variety of sequential procedures, the primary concrete example studied in detail is the repeated significance test and some of its modifications. Let $x_1, x_2, \ldots$ be independent, normally distributed random variables with unknown mean μ and known variance σ^2, which without loss of generality can be taken equal to 1. Let $S_n = x_1 + \cdots + x_n$. The standard fixed sample .05 level significance test of H_0: $\mu = 0$ against H_1: $\mu \neq 0$ is to reject H_0 if and only if $|S_n| \geq 1.96n^{1/2}$. Here n is the arbitrary, but fixed sample size of the experiment. Suppose now that if H_1 is actually true it is desirable to discover this fact after a minimum amount of experimentation, but no similar constraint exists under H_0. Such might be the case in a clinical trial where x_i represents the difference in responses to two medical treatments in the ith pair of a paired comparison experiment. If H_0 is true, the two treatments are equally good, and from the patients' point of view the experiment could continue indefinitely. However, if H_1 is true, one or the other treatment is superior, and the trial should terminate as soon as possible so that all future patients can receive the better treatment.

An ad hoc solution to the problem of the preceding paragraph is the following. Let $b > 0$ and let m be a maximum sample size. Sample sequentially, stopping with rejection of H_0 at the first $n \leq m$, if one exists, such that $|S_n| > bn^{1/2}$. Otherwise stop sampling at m and accept (do not reject) H_0. The significance level of this procedure is

$$\alpha = \alpha(b, m) = P_0\{|S_n| > bn^{1/2} \text{ for some } n \leq m\}, \tag{1.6}$$

which means that b must be somewhat larger than 1.96 (depending on m) in order that $\alpha(b, m) = .05$.

Tests of this sort were criticized by Feller (1940), who alleged that they were used in extrasensory perception experiments without making the necessary adjustment in the value of b to account for the sequential nature of the experiment. (For these experiments, S_n might count the excess of correct over incorrect guesses by a subject who supposedly can predict the outcome of a coin toss before being informed of the result.) Feller also complained that there was no definite value of m, so that one should consider the significance level to be

$$\lim_{m \to \infty} \alpha(b, m),$$

which is known to equal 1 (for example, as a consequence of the law of the iterated logarithm). Robbins (1952) gave an upper bound for $\alpha(b, m)$ and posed the problem of giving a good approximation to α.

Such repeated significance tests were studied by Armitage *et al.* (1969) and by MacPherson and Armitage (1971), who evaluated their significance level, power, and expected sample size by lengthy numerical computations. The theoretical research from which this book has developed began with Woodroofe's (1976) and Lai and Siegmund's (1977) approximation for α (cf. (4.40)), which was followed by a series of papers approximating the power and expected sample size of repeated significance tests, extending the results to more general models, and suggesting certain modifications of the test itself (see Chapters IV and V).

As a preliminary to our study of repeated significance tests, we discuss the sequential probability ratio test in Chapter II. Although it seems unlikely that this test should be used in practice, the basic tools for studying it, to wit Wald's likelihood ratio identity (Proposition 2.24) and Wald's partial sum identity (Proposition 2.18), are fundamental for analyzing more useful procedures. So called cusum procedures for use in quality control are discussed briefly in II.6.

Chapters III–V form the core of the book. The main conceptual ideas are introduced in Chapter III in a context which minimizes the computational problems. Truncated sequential probability ratio tests and Anderson's modification of the sequential probability ratio test are also discussed. Repeated significance tests are studied in detail in Chapter IV. A number of more difficult examples are presented in Chapter V to illustrate the way one can build upon the basic theory to obtain reasonable procedures in a variety of more complicated contexts.

Chapters VI and VII deal with special topics. Chapter VI is concerned with the allocation of treatments in clinical trials, and Chapter VII briefly introduces the theory of fixed precision confidence intervals.

In order to maximize attention to statistical issues and minimize difficult probability calculations, the mathematical derivations of Chapters III and IV are essentially limited to the artificial, but simple case of a Brownian motion process. Corresponding results for processes in discrete time are given without proof and used in numerical examples. Chapters VIII–X provide the mathematical foundation for these results. Chapter XI is concerned with some miscellaneous probability calculations which are conceptually similar but technically more difficult than those which appear earlier in the book. Four appendices present some background probabilistic material.

The most obvious omission from this book is a discussion of Bayesian sequential tests. Even for the non-Bayesian, the use of prior probability distributions is a useful technical device in problems which can reasonably be treated decision-theoretically (i.e. have action spaces and loss functions). The two principal fields of application of sequential hypothesis testing are sampling inspection and clinical trials. Of these, the former seems often to admit a decision-theoretic formulation, but the latter not. (For a contrary view, see Anscombe, 1963, and for further discussion see IV.6.) Hald (1981) gives a systematic treatment of sampling inspection with ample discussion of Bayesian methods. Other general introductions to sequential Bayesian hypothesis testing without particular applications in mind are given by Ferguson (1967), Berger (1980), and especially Chernoff (1972). To avoid a substantial increase in the length of this book, the subject has been omitted here.

The formal mathematical prerequisites for reading this book have been held to a minimum—at least in Chapters II–VII. It would be helpful to have some knowledge of elementary random walk and Brownian motion theory at the level of Feller (1968), Cox and Miller (1965), or Karlin and Taylor (1975). Appendix 1 attempts to give the reader lacking this background some feeling for the essentials of Brownian motion, devoid of all details. Martingale theory makes a brief appearance in V.5. Appendix 3 presents the necessary background—again informally.

One bit of nonstandard notation that is used systematically throughout the book is $E(X; B)$ to denote $E(XI_B)$. (Here I_B denotes the indicator variable of the event B, i.e. the random variable which equals 1 if B occurs and 0 otherwise. E denotes expectation.) Some of the notation is not consistent throughout the book, but is introduced in the form most convenient for the subject under discussion. The most important example is the notation for exponential families of probability distributions, which are introduced in II.3, but parameterized slightly differently in II.6 (the origin is shifted). They reappear in the original parameterization in Chapter VIII, and in Chapter X they change again to the parameterization of II.6.

Problem sets are included at the end of each chapter. A few problems which are particularly important have been designated with $*$. Those which are somewhat more difficult or require specialized knowledge are marked †.

Problems

1.1. Suppose that the Poisson process $X(t)$ is observed until the time w_r of the rth failure. Show that $\lambda_1^*(w_r)$ is a $(1-\alpha)$ 100% upper confidence bound for λ.

1.2. Prove that for λ^* defined by (1.4)

$$P_\lambda\{\lambda^*[T', X(T')] \geq \lambda\} \geq 1 - \alpha \qquad \text{for all } \lambda.$$

Hint: Note that $\lambda_1^*(m) = \lambda_2^*(r-1)$. Consider separately the two cases $\lambda_1^*(m) \geq \lambda$ and $\lambda_1^*(m) < \lambda$.

1.3. (a) For the curtailed test of H_0: $p \leq p_0$ against H_1: $p > p_0$, show how the test can be further curtailed with *acceptance* of H_0.

(b) Discuss confidence bounds for p following a curtailed test of H_0: $p \leq p_0$ against H_1: $p > p_0$.

1.4.* Let $X(t)$, $t \geq 0$, be a Poisson process with mean λt and set $\Pi_\lambda(t, n) = P_\lambda\{X(t) \geq n\}$. For testing H_0: $\lambda = \lambda_0$ against H_1: $\lambda > \lambda_0$ based on a fixed sample of size $t = m$, the attained significance level or p-value is defined to be $\Pi_{\lambda_0}(m, X(m))$. That is, the p-value is the Type I error probability for that test with rejection region of the form $\{X(m) \geq r\}$ for some r, for which the value of $X(m)$ actually observed just barely lies in the rejection region. Small values of $\Pi_{\lambda_0}(m, X(m))$ are conventionally interpreted as providing more evidence against H_0 than large values. For a curtailed test of H_0: $\lambda \leq \lambda_0$ against H_1: $\lambda > \lambda_0$ defined by the constants m and r, suggest a definition of the attained significance of the observed value $(T', X(T'))$. For your definition, explain why data yielding a small p-value should be thought to provide strong evidence against H_0.

CHAPTER II

The Sequential Probability Ratio Test

1. Definition and Examples

We begin by recalling the Neyman–Pearson Lemma for testing a simple hypothesis against a simple alternative. Let x denote a (discrete or continuous) random variable (or vector) with probability density function f. To test H_0: $f = f_0$ against $H_1: f = f_1$, define the likelihood ratio $l(x) = f_1(x)/f_0(x)$, choose a constant $r > 0$, and

$$\begin{aligned} &\text{Reject } H_0 \quad \text{if } l(x) \geq r, \\ &\text{Accept } H_0 \quad \text{if } l(x) < r. \end{aligned}$$

This class of tests (depending on r) is optimal from both a Bayesian and a frequentist viewpoint. In particular, any test of H_0 against H_1 which is based on observing x and has significance level no larger than $\alpha = P_0\{l(x) \geq r\}$ must have power no larger than $P_1\{l(x) \geq r\}$ (cf. Cox and Hinkley, 1974, p. 91). Here P_i denotes probability under the hypothesis H_i, $i = 0, 1$.

A sequential probability ratio test of H_0 against H_1 admits a third possibility in addition to rejecting H_0 for large $l(x)$ and accepting for small $l(x)$, namely that for intermediate values of $l(x)$ one collects more data. More precisely let $x_1, x_2, \ldots$ be a sequence of random variables with joint density functions

$$P\{x_1 \in d\xi_1, \ldots, x_n \in d\xi_n\} = f_n(\xi_1, \ldots, \xi_n)\, d\xi_1, \ldots, d\xi_n \qquad (n = 1, 2, \ldots).$$

Consider testing the simple hypotheses $H_0: f_n = f_{0n}$ for all n against $H_1: f_n = f_{1n}$ for all n. Let $l_n = l_n(x_1, \ldots, x_n) = f_{1n}(x_1, \ldots, x_n)/f_{0n}(x_1, \ldots, x_n)$. Choose constants $0 < A < B < \infty$ (usually $A < 1 < B$) and sample $x_1, x_2, \ldots$ sequentially until the random time

$$(2.1) \qquad \begin{aligned} N &= \text{first } n \geq 1 \qquad \text{such that } l_n \notin (A, B) \\ &= \infty \text{ if } l_n \in (A, B) \qquad \text{for all } n \geq 1 \end{aligned}$$

Stop sampling at time N and if $N < \infty$

$$\begin{aligned} &\text{Reject } H_0 \quad \text{if } l_N \geq B, \\ &\text{Accept } H_0 \quad \text{if } l_N \leq A. \end{aligned}$$

We defer temporarily the technical issue of whether this procedure actually terminates, i.e. whether $P_i\{N < \infty\} = 1$ for $i = 0$ and 1. Assuming that it does, we have a test of size $P_0\{l_N \geq B\}$ and power $P_1\{l_N \geq B\}$. As additional properties of the test we shall consider its sample size distribution $P_i\{N = n\}$, $n = 1, 2, \ldots$ or more simply its expected sample size $E_i(N)$ for $i = 1, 2$.

It will turn out that the sequential probability ratio test has a very strong optimality property when the x_n are independent and identically distributed: it minimizes E_i (sample size) for $i = 0$ and 1 over all (sequential) tests having the same size and power. The basic idea behind this fact is very simple, although a complete proof is circuitous and difficult (see Ferguson, 1967, p. 365).

The goals of this chapter are to study the sequential probability ratio test as a test of a simple hypothesis against a simple alternative (Section 2), to extend it to certain problems involving composite hypotheses (Section 3), to give an informal discussion of its optimality property (Section 4), and to criticize it in Section 5, in preparation for the various modifications of Chapters III and IV. An introduction to the related "cusum" procedures of statistical quality control is given in Section 6.

Before proceeding we present two simple, but especially instructive examples.

Example 2.2. Let $x_1, x_2, \ldots$ be independent and normally distributed with mean μ and unit variance. For testing H_0: $\mu = \mu_0$ against H_1: $\mu = \mu_1$ (say $\mu_0 < \mu_1$), the likelihood ratio is

$$(2.3) \qquad \begin{aligned} l_n &= \prod_{k=1}^{n} \{\phi(x_k - \mu_1)/\phi(x_k - \mu_0)\} \\ &= \exp\{(\mu_1 - \mu_0)S_n - \tfrac{1}{2}n(\mu_1^2 - \mu_0^2)\}, \end{aligned}$$

where $\phi(x) = (2\pi)^{-1/2}\exp(-x^2/2)$ and $S_n = \sum_1^n x_k$. Hence the stopping rule (2.1) can be re-written

$$(2.4) \qquad \begin{aligned} N &= \text{first } n \geq 1 \text{ such that } S_n - \tfrac{1}{2}n(\mu_1 + \mu_0) \notin (a, b) \\ &= \infty \text{ if no such } n \text{ exists,} \end{aligned}$$

where $a = \log A/(\mu_1 - \mu_0)$, $b = \log B/(\mu_1 - \mu_0)$, and if $N < \infty$ the sequential probability ratio test rejects H_0 if and only if

$$S_N \geq b + \tfrac{1}{2}N(\mu_1 + \mu_0).$$

A simple special case is the symmetric one $\mu_1 = -\mu_0$, $b = -a$, for which (2.4) becomes

$$N = \text{first } n \geq 1 \text{ such that } |S_n| \geq b \tag{2.5}$$
$$= \infty \text{ if } |S_n| < b \text{ for all } n.$$

EXAMPLE 2.6. Let $x_1, x_2, \ldots$ be independent and identically distributed with $P_p\{x_k = 1\} = p$, $P_p\{x_k = -1\} = q$ $(p + q = 1)$. For testing $H_0\colon p = p_0$ against $H_1\colon p = p_1$ $(p_0 < p_1)$ the likelihood ratio is

$$l_n = (p_1 p_0^{-1})^{(n+S_n)/2}(q_1 q_0^{-1})^{(n-S_n)/2} \tag{2.7}$$
$$= (p_1 q_0 p_0^{-1} q_1^{-1})^{S_n/2}(p_1 p_0^{-1} q_1 q_0^{-1})^{n/2},$$

where $S_n = \Sigma x_k$. Again (2.1) can be expressed directly in terms of the random walk $\{S_n\}$, and a simple symmetric case occurs when $p_0 = q_1$ and $B = A^{-1}$, so (2.1) becomes

$$N = \text{first } n \geq 1 \text{ such that } |S_n| \geq b, \tag{2.8}$$
$$= \infty \text{ if } |S_n| < b \text{ for all } n,$$

where $b = (\log B)/\log(q_0 p_0^{-1})$.

2. Approximations for $P_i\{l_N \geq B\}$ and $E_i(N)$

We continue to use the notation and assumption of the preceding section. In particular we continue to assume that $P_i\{N < \infty\} = 1$ $(i = 0, 1)$, and consider the following problems

(a) Relate $\alpha = P_0\{l_N \geq B\}$ and $\beta = P_1\{l_N \leq A\}$ to A and B;

(b) Relate $E_i(N)$ to A, B for $i = 0, 1$.

We begin with a simple calculation related to (a). More general versions of this idea are one of the principal techniques developed in the following chapters. Let B_n denote the subset of n-dimensional space where $A < l_k(\xi_1, \ldots, \xi_k) < B$ for $k = 1, 2, \ldots, n-1$ and $l_n(\xi_1, \ldots, \xi_n) \geq B$. Hence $\{N = n, l_n \geq B\} = \{(x_1, \ldots, x_n) \in B_n\}$. By direct calculation

$$\alpha = P_0\{l_N \geq B\} = \sum_1^\infty P_0\{N = n, l_n \geq B\} = \sum_1^\infty \int_{B_n} f_{0n}\, d\xi_1 \cdots d\xi_n$$
$$= \sum_1^\infty \int_{B_n} \frac{f_{0n}}{f_{1n}} f_{1n}\, d\xi_1 \cdots d\xi_n = \sum_1^\infty E_1[l_n^{-1}; N = n, l_n \geq B] \tag{2.9}$$
$$= E_1[l_N^{-1}; l_N \geq B] \leq B^{-1} P_1\{l_N \geq B\} = B^{-1}(1 - \beta).$$

A similar argument with the roles of A and B interchanged leads to

$$\beta = P_1\{l_N \leq A\} \leq AP_0\{l_N \leq A\} = A(1-\alpha). \tag{2.10}$$

The inequalities (2.9) and (2.10) fail to be equalities only because l_n does not have to hit the boundaries exactly when it first leaves (A, B). However, if we agree to ignore this discrepancy and treat (2.9) and (2.10) as approximate equalities

$$\alpha \cong B^{-1}(1-\beta) \quad \text{and} \quad \beta \cong A(1-\alpha), \tag{2.11}$$

we can solve for α and β to obtain crude but extremely simple approximations:

$$\alpha \cong \frac{1-A}{B-A}; \qquad \beta \cong A\left(\frac{B-1}{B-A}\right). \tag{2.12}$$

To study the expected value of N we make the additional assumption that the observations x_n are independent and identically distributed, so $l_n = \prod_{k=1}^{n} \{f_1(x_k)/f_0(x_k)\}$, where f_i is the probability density function of x_1 under the simple hypothesis H_i $(i = 0, 1)$. In this case the log scale is convenient, and

$$\log l_n = \sum_{k=1}^{n} \log\{f_1(x_k)/f_0(x_k)\}$$

is a sum of independent, identically distributed random variables. Moreover,

$$\begin{aligned} N &= \text{first } n \geq 1 \text{ such that } \log l_n \notin (a, b), \\ &= \infty \text{ if } \log l_n \in (a, b) \text{ for all } n, \end{aligned}$$

where $a = \log A$, $b = \log B$.

The basic argument for approximating $E_i(N)$ consists of two parts. By Wald's identity given below in Proposition 2.18

$$E_i\{\log l_N\} = \mu_i E_i(N), \tag{2.13}$$

where

$$\mu_i = E_i[\log\{f_1(x_1)/f_0(x_1)\}] \qquad (i = 0, 1).$$

Moreover, the approximation we used in deriving (2.11) suggests that $\log l_N$ be regarded as a two-valued random variable taking on the values a and b, so

$$E_i\{\log l_N\} \cong aP_i\{l_N \leq A\} + bP_i\{l_N \geq B\}, \tag{2.14}$$

where the probabilities in (2.14) are given approximately in (2.12). Putting (2.12), (2.13), and (2.14) together yields the approximations

$$E_1 N \cong \mu_1^{-1}\{aA(B-1) + bB(1-A)\}/(B-A) \tag{2.15}$$

and

$$E_0 N \cong \mu_0^{-1}\{a(B-1) + b(1-A)\}/(B-A). \tag{2.16}$$

Note that $\mu_0 < 0 < \mu_1$. In fact, since $\log x \leq x - 1$ with equality if and only if $x = 1$,

$$\mu_0 = E_0[\log\{f_1(x_1)/f_0(x_1)\}] = \int \log\{f_1(\xi)/f_0(\xi)\}f_0(\xi)\,d\xi$$
$$\leq \int \{f_1(\xi)/f_0(\xi) - 1\}f_0(\xi)\,d\xi = 0$$

with equality of and only if f_0 and f_1 are the same density function. A similar argument shows $-\mu_1 < 0$. Since by the strong law of large numbers

(2.17) $$P_i\{n^{-1}\log l_n \to \mu_i\} = 1 \qquad (i = 0, 1),$$

if $A \to 0$ and $B \to \infty$ we expect to find under P_i that $\log l_n$ leaves the interval (a, b) at approximately the time and place that $n\mu_i$ does, to wit b/μ_1 under H_1 and $|a|/|\mu_0|$ under H_0. It is easy to see that this is asymptotically consistent with (2.15) and (2.16).

The following two propositions justify (2.13).

Proposition 2.18 (Wald's identity). *Let $y_1, y_2, \ldots$ be independent and identically distributed with mean value $\mu = Ey_1$. Let M be any integer valued random variable such that $\{M = n\}$ is an event determined by conditions on $y_1, \ldots, y_n$ (and is independent of $y_{n+1}, \ldots$) for all $n = 1, 2, \ldots$, and assume that $EM < \infty$. Then $E(\sum_{k=1}^{M} y_k) = \mu EM$.*

PROOF. Suppose initially that $y_k \geq 0$. We write $\sum_{k=1}^{M} y_k = \sum_{k=1}^{\infty} I_{\{M \geq k\}} y_k$, and note that $\{M \geq k\} = (\bigcup_{j=1}^{k-1} \{M = j\})^c$ is independent of $y_k, y_{k+1}, \ldots$. Hence by monotone convergence

$$E\left(\sum_1^M y_k\right) = \sum_1^\infty E(y_k; M \geq k) = \mu \sum_1^\infty P\{M \geq k\} = \mu EM.$$

For the general case we write

$$\sum_{k=1}^{M} y_k = \sum_{k=1}^{M} y_k^+ - \sum_{k=1}^{M} y_k^-,$$

where $a^+ = \max(a, 0)$, $a^- = -\min(a, 0)$, and apply the case already considered to these two terms separately. □

Proposition 2.19 (Stein's lemma). *Let $y_1, y_2, \ldots$ be independent and identically distributed with $P\{y_1 = 0\} < 1$. Let $-\infty < a < b < \infty$ and define*

$$M = \text{first } n \geq 1 \text{ such that } \sum_1^n y_k \notin (a, b)$$
$$= \infty \text{ if } \sum_1^n y_k \in (a, b) \text{ for all } n.$$

Then there exist constants $C > 0$ and $0 < \rho < 1$ such that $P\{M > n\} \leq C\rho^n$ $(n = 1, 2, \ldots)$. In particular $EM^k < \infty$ for all $k = 1, 2, \ldots$ and $Ee^{\lambda M} < \infty$ for $\lambda < \log \rho^{-1}$.

For a proof of Proposition 2.19 see Problem 2.6.

Remark 2.20. The expected sample size approximations, (2.15) and (2.16) can also be expressed in terms of the error probabilities α and β. From (2.11), (2.13), and (2.14) follow

$$E_1 N \cong \mu_1^{-1}\left\{(1-\beta)\log\left(\frac{1-\beta}{\alpha}\right) + \beta\log\left(\frac{\beta}{1-\alpha}\right)\right\} \tag{2.21}$$

and

$$E_0(N) \cong \mu_0^{-1}\left\{\alpha\log\left(\frac{1-\beta}{\alpha}\right) + (1-\alpha)\log\left(\frac{\beta}{1-\alpha}\right)\right\}. \tag{2.22}$$

The calculation (2.9) turns out to be very important in a variety of cases. It is useful to present a more abstract version for future reference.

Let $z_1, z_2, \ldots$ be an arbitrary sequence of random variables. For each n, let $\mathscr{E}_n$ denote the class of random variables determined by $z_1, \ldots, z_n$, i.e. a random variable $Y \in \mathscr{E}_n$ if and only if $Y = f(z_1, \ldots, z_n)$ for some (Borel) function f of n variables. For an event A the notation $A \in \mathscr{E}_n$ means that the indicator of A, I_A, belongs to $\mathscr{E}_n$. A random variable T with values in $\{1, 2, \ldots, +\infty\}$ is called a stopping time if $\{T = n\} \in \mathscr{E}_n$ for all n. Hence an observer who knows the values of $z_1, \ldots, z_n$ knows whether $T = n$. A random variable Y is said to be prior to a stopping time T if

$$YI_{\{T=n\}} \in \mathscr{E}_n \quad \text{for all} \quad n, \tag{2.23}$$

or equivalently $YI_{\{T \leq n\}} \in \mathscr{E}_n$ for all n. In particular T is prior to itself. Condition (2.23) has the interpretation that by the time T an observer who knows the values $z_1, z_2, \ldots, z_T$ also knows the value of Y.

Proposition 2.24 (Wald's likelihood ratio identity). *Let P_0 and P_1 denote two probabilities and assume that there exists a likelihood ratio l_n for $z_1, \ldots, z_n$ under P_1 relative to P_0 in the sense that $l_n \in \mathscr{E}_n$ and for each $Y_n \in \mathscr{E}_n$*

$$E_1(Y_n) = E_0(Y_n l_n). \tag{2.25}$$

For any stopping time T and non-negative random variable Y prior to T

$$E_1(Y; T < \infty) = E_0(Yl_T; T < \infty\}.$$

In particular, if $Y = I_A$

$$P_1(A \cap \{T < \infty\}) = E_0(l_T; A \cap \{T < \infty\}).$$

Proof. The proof basically repeats (2.9) with (2.23) and (2.25) used to justify the second equality:

$$E_1(Y; T < \infty) = \sum_{n=1}^{\infty} E_1(Y; T = n) = \sum_{n=1}^{\infty} E_0(Yl_n; T = n)$$
$$= E_0(Yl_T; T < \infty). \qquad \square$$

Remark. If $z_1, \ldots, z_n$ have joint density p_{in} under P_i $(i = 0, 1)$, then $l_n = p_{1n}/p_{0n}$. In this case for $Y_n = f_n(z_1, \ldots, z_n)$, (2.25) follows from

$$
\begin{aligned}
E_1(Y_n) &= \int f_n(\xi_1,\ldots,\xi_n)p_{1n}(\xi_1,\ldots,\xi_n)\,d\xi_1,\ldots,d\xi_n \\
&= \int f_n(\xi_1,\ldots,\xi_n)\frac{p_{1n}(\xi_1,\ldots,\xi_n)}{p_{0n}(\xi_1,\ldots,\xi_n)}p_{0n}(\xi_1,\ldots,\xi_n)\,d\xi_1,\ldots,d\xi_n \\
&= E_0\left\{Y_n\frac{p_{1n}(z_1,\ldots,z_n)}{p_{0n}(z_1,\ldots,z_n)}\right\}.
\end{aligned}
$$

Remark 2.26. We have already used Proposition 2.24 to derive the approximations (2.12). It can also be very useful in Monte Carlo studies. Consider the problem of estimating $\alpha = P_0\{l_N \geq B\}$ by simulation. The naive estimator is

$$\hat{\alpha} = n^{-1}\sum_{k=1}^{n} I\{l_{N_k} \geq B\}$$

based on n independent realizations of (N, l_N) under the probability P_0. Since α is typically small, and the standard deviation of $\hat{\alpha}$ is $[\alpha(1-\alpha)/n]^{1/2}$, it is often necessary to take large values of n to provide an accurate estimate. For example, if one wants to estimate α to within 10% of its value with probability .95, for small α n must be about $400/\alpha$.

Alternatively, by Proposition 2.24, another unbiased estimator of α is

$$\hat{\hat{\alpha}} = n^{-1}\sum_{k=1}^{n}\frac{f_{0N_k}}{f_{1N_k}} I\left\{\frac{f_{1N_k}}{f_{0N_k}} \geq B\right\},$$

where now the experiment generating the observations is conducted under P_1. By (2.11) and Proposition 2.24

$$
\begin{aligned}
n\operatorname{var}(\hat{\hat{\alpha}}) &\leq E_1\left\{\left(\frac{f_{0N}}{f_{1N}}\right)^2; \frac{f_{1N}}{f_{0N}} \geq B\right\} \\
&\leq B^{-1}E_1\left\{\frac{f_{0N}}{f_{1N}}; \frac{f_{1N}}{f_{0N}} \geq B\right\} = B^{-1}\alpha \cong \frac{\alpha^2}{1-\beta}.
\end{aligned}
$$

For this experiment only about 400 replications are required to achieve 10% relative accuracy with probability .95 no matter how small α is.

3. Tests of Composite Hypotheses

Although the sequential probability ratio test is derived as a test of a simple hypothesis against a simple alternative, it is natural to consider the consequences of using it for composite hypotheses. For example, to test $H_0\colon \theta \leq \theta^*$ against $H_1\colon \theta > \theta^*$ in a one-parameter family of distributions one might choose surrogate simple hypotheses $\theta_0 \leq \theta^*$ and $\theta_1 > \theta^*$, and use a sequential probability ratio test of θ_0 against θ_1. Then one would want to know the entire power function and expected sample size as a function of θ, in addition to their values at θ_0 and θ_1. Before discussing the general case we consider some simple examples.

Recall Example 2.6, and assume that the symmetric case with stopping rule (2.8) is to be used for the composite hypotheses H_0: $p \le \frac{1}{2}$ against H_1: $p > \frac{1}{2}$. There is no loss of generality in assuming that B is chosen so that $b = \log B/\log(q_0 p_0^{-1})$ is an integer. Since the random walk $\{S_n\}$ proceeds by steps of ± 1, $P_p\{|S_N| = b\} = 1$, so in this case the approximations of Section 2 are equalities. In particular by (2.11) and (2.15)–(2.16) (or Remark 2.20),

$$\alpha = \beta = P_{p_0}\{S_N = b\} = 1/(1 + B) = 1/[1 + (q_0/p_0)^b]$$

and

$$E_{p_0}(N) = |q_0 - p_0|^{-1} b(1 - 2\alpha).$$

[In making the required identifications it is helpful to keep in mind that $b = \log B/\log(q_0 p_0^{-1})$, for this example is proportional but not identical to $b = \log B$ in (2.15)–(2.16).]

Now suppose $p < \frac{1}{2}$, but $p \ne p_0$. Since (2.8) does not explicitly involve p_0, it is also a sequential probability ratio test of p against $1 - p$, but with a different value of B. Put another way, (2.8) can be re-written

$$\text{(2.27)} \qquad N = \text{first } n \ge 1 \text{ such that } (q/p)^{S_n} \notin \left(\left\{\left(\frac{q}{p}\right)^b\right\}^{-1}, \left(\frac{q}{p}\right)^b\right),$$

which is of the form $l_n \notin (B_1^{-1}, B_1)$, where l_n is (2.7) with $p_0 = p$, $p_1 = q$, and $B_1 = (q/p)^b$. Hence with the proper re-interpretation the approximations of Section 2 yield $P_p\{S_N = b\}$ and $E_p(N)$ for general $p \ne \frac{1}{2}$. The results are

$$\text{(2.28)} \qquad P_p\{S_N = b\} = 1/\{1 + (q/p)^b\}$$

and

$$\text{(2.29)} \qquad E_p(N) = |q - p|^{-1} b |1 - 2P_p\{S_N = b\}|,$$

for all $p \ne \frac{1}{2}$. The corresponding results for $p = \frac{1}{2}$ are easily computed by taking the limit as $p \to \frac{1}{2}$ in (2.28) and (2.29).

A similar discussion applies to Example 2.2, but now the approximations do not turn into equalities. Consider again the symmetric case (2.5) for simplicity, and note that by symmetry (2.9) becomes $\alpha < B^{-1}(1 - \alpha)$, so the approximation (2.11) is an inequality

$$\alpha = P_{\mu_0}\{S_N \ge b\} < 1/(1 + B) = 1/(1 + e^{2b|\mu_0|}).$$

In this case for arbitrary $\mu > 0$, (2.5) can be written

$$N = \text{first } n \ge 1 \text{ such that } e^{2\mu S_n} \notin (e^{-2\mu b}, e^{2\mu b}),$$

which has the form of a symmetric sequential probability ratio test of $-\mu$ against μ with $B = A^{-1} = e^{2\mu b}$ (cf. (2.3)). Hence for N defined by (2.5), for all $\mu > 0$

$$\text{(2.30)} \qquad P_{-\mu}\{S_N \ge b\} < 1/(1 + e^{2\mu b})$$

Table 2.1. Symmetric Sequential Probability Ratio Test for Normal Observations: $b = 4.91$

μ	$P_{-\mu}\{S_N \geq b\} \cong$	$E_\mu(N) \cong$
.4	.019	11.8
.3	.050	14.7
.2	.123	18.5
.1	.272	22.3
0	.500	24.1

and

$$E_\mu(N) \cong \mu^{-1} b\{(e^{2\mu b} - 1)/(e^{2\mu b} + 1)\}. \tag{2.31}$$

An approximation for $\mu = 0$ can be obtained by taking the limit as $\mu \to 0$.

Table 2.1 gives a numerical example of the approximations (2.30) and (2.31). Since (2.30) and (2.31) are only approximations, one would like to know how good they are. It may be shown (cf. Problem 2.2, also III.5, III.6, VIII.5, and X.2) that if $p(b)$ denotes the right hand side of (2.30) and $e(b)$ the right hand side of (2.31), then $p(b + .583)$ and $e(b + .583)$ give very good approximations for a wide range of values of b and μ. Thus for $\mu = .3$, in order that $P_{-.3}\{S_N \geq b\} = .05$, the right hand side of (2.30) suggests taking $b = 4.91$. However, the true value of $P_{-.3}\{S_N \geq 4.91\}$ is much closer to $p(4.91 + .583) = .036$. The approximation (2.31) is generally somewhat more satisfactory: for $\mu = .3$, $e(4.91) = 14.7$ compared to the more accurate $e(4.91 + .583) = 17.0$.

For a different example concerning the accuracy of (2.12) and (2.15)–(2.16), see Problem 2.1. A general conclusion about these approximations is the following. Usually (2.12) overestimates α and β, and the relative error is often as large as 30–70%; (2.15) and (2.16) typically underestimate the expected sample size, and except in the case of quite small samples sizes the relative error is usually on the order of 5–25%. Later we shall see that (2.12), (2.15)–(2.16), and related results can be substantially improved by approximating rather than neglecting the excess over the boundary. For example, the number .583 of the preceding paragraph is approximately the expected difference between the random variable S_N and the boundaries $\pm b$. Hence the improved approximations amount to using the no-overshoot approximations with new boundaries $\pm b'$ differing from the old ones by the average excess over the boundaries. See Chapter X for the theoretical justification.

In spite of their lack of accuracy, the remarkable generality and simplicity of the approximations of Section 2 make them quite useful.

Note that a fixed sample size test of $\mu = -.3$ against $\mu = +.3$ with significance level and Type II error probability of .036 requires 36 observations, so when $|\mu| \geq .3$ a sequential probability ratio test saves on the average more than 50% of these observations. Even when $\mu = 0$, this sequential probability ratio test has an expected sample size of only 30 observations. See Problem 2.9 for additional comparisons of fixed sample and sequential probability ratio tests.

The preceding examples show that it is sometimes possible to obtain approximations to the entire power function and expected sample size function of a sequential probability ratio test of composite hypotheses using only the theory developed for simple hypotheses. The following discussion shows that this is generally possible in the context of a one-parameter exponential family of distributions.

Consider a general sequential probability ratio test defined by (2.1) with the additional hypothesis that $x_1, x_2, \ldots$ are independent and identically distributed, so that

$$l_n = \prod_{k=1}^{n} \{f_1(x_k)/f_0(x_k)\},$$

where f_i is the density function of x_1 under H_i ($i = 0, 1$). Let f be some third probability density function, and consider the problem of evaluating $P_f\{l_N \geq B\}$. This was feasible in the preceding examples because there was yet a fourth density function f^* such that the original test of f_0 against f_1 was equivalent to a test of f against f^* with new values of A and B. Moreover, the new A and B were the old A and B raised to a power. Hence given f, assume that there exists a probability density function f^* and a $\theta_1 \neq 0$ such that

$$\frac{f^*(x)}{f(x)} = \left[\frac{f_1(x)}{f_0(x)}\right]^{\theta_1}. \tag{2.32}$$

Then for $\theta_1 > 0$ (for example)

$$\begin{aligned} N &= \text{first } n \geq 1 \text{ such that } \prod_{1}^{n} \{f_1(x_k)/f_0(x_k)\} \notin (A, B) \\ &= \text{first } n \geq 1 \text{ such that } \prod_{1}^{n} \{f^*(x_k)/f(x_k)\} \notin (A^{\theta_1}, B^{\theta_1}), \end{aligned}$$

and it follows from (2.12) that

$$P_f\{l_N \geq B\} \cong \frac{1 - A^{\theta_1}}{B^{\theta_1} - A^{\theta_1}}. \tag{2.33}$$

A similar calculation applies when $\theta_1 < 0$, and an approximation for $E_f(N)$ is also easily obtained (see Problem 2.16).

Note that (2.32) is satisfied if and only if

$$\int_{-\infty}^{\infty} \left[\frac{f_1(x)}{f_0(x)}\right]^{\theta_1} f(x)\, dx = 1. \tag{2.34}$$

Let $z(x) = \log\{f_1(x)/f_0(x)\}$, and define a function $\psi(\theta)$ by

$$e^{\psi(\theta)} = \int_{-\infty}^{\infty} e^{\theta z(x)} f(x)\, dx$$

whenever the integral converges. Then

$$f_\theta(x) = e^{\theta z(x) - \psi(\theta)} f(x)$$

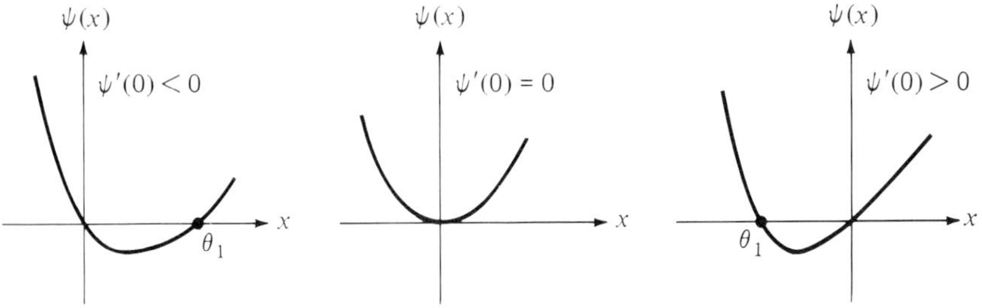

Figure 2.1.

defines an exponential family of distributions and the existence of a θ_1 satisfying (2.32) or (2.34) is equivalent to the existence of a $\theta_1 \neq 0$ such that $\psi(\theta_1) = 0$. It is easy to see by differentiation (assuming always that the integrals converge) that

$$\psi'(\theta) = \int_{-\infty}^{\infty} z(x) f_\theta(x)\, dx$$

and

$$\psi''(\theta) = \int_{-\infty}^{\infty} [z(x)]^2 f_\theta(x)\, dx - [\psi'(\theta)]^2 \geq 0,$$

so ψ is convex. Hence under some modest convergence assumptions ψ has the appearance of one of the three functions in Figure 2.1; and the desired $\theta_1 \neq 0$ satisfying (2.34) exists if $\psi'(0) = \int_{-\infty}^{\infty} z(x) f(x)\, dx = E_f z(x_1) \neq 0$, but not if $\psi'(0) = 0$. Note that the case $\psi'(0) = 0$ corresponds to cases where we obtained approximations to $P\{l_N \geq B\}$ and $E(N)$ by continuity in the two examples at the beginning of this section. That technique works quite generally, but for a different approach see Problems 2.10 and 2.11.

To see how this general discussion relates to the examples, observe that if $f_i(x) = \phi(x - (-1)^i \mu_0)$ for some $\mu_0 \neq 0$, and if f is a normal density with mean $\mu \neq 0$, then f^* is normal with mean $-\mu$. Similarly in the symmetric (about $\frac{1}{2}$) Bernoulli example, if f is Bernoulli with success probability $p \neq \frac{1}{2}$, then f^* is Bernoulli with success probability $1 - p$.

Example. Morgan *et al.* (1951) made a detailed study of the sequential probability ratio test as a method of grading raw milk prior to pasteurization. The classification process involved counting bacterial clumps in each of several fields of a film of milk in order to determine their average density. After deciding that subject to certain qualifications the number of bacterial clumps per field was reasonably approximated by a Poisson distribution, the authors proposed a sequential probability ratio test to determine the acceptability of the milk as (for example) Grade A milk in the state of Connecticut.

Let x_i denote the number of clumps of bacteria in the ith field, and assume that $x_1, x_2, \ldots$ are independent Poisson random variables with mean λ. The standard fixed sample plan for accepting milk as Grade A was equivalent to taking a sample of size $m = 30$ and testing H_0: $\lambda = .1808$ against H_1: $\lambda = .5249$ with Type I and Type II error probabilities of .05. Let $S_n = x_1 + \cdots + x_n$, $n = 1, 2, \ldots$. A simple calculation shows that a sequential probability ratio test of H_0: $\lambda = \lambda^{(0)}$ against H_1: $\lambda = \lambda^{(1)}$ ($\lambda^{(0)} < \lambda^{(1)}$) is defined by a stopping rule of the form $N = \text{first } n \geq 1$ such that $S_n - \lambda^* n \notin (a, b)$, where $\lambda^* = (\lambda^{(1)} - \lambda^{(0)})/\log(\lambda^{(1)}/\lambda^{(0)})$, $b = \log B/\log(\lambda^{(1)}/\lambda^{(0)})$, and $a = \log A/\log(\lambda^{(1)}/\lambda^{(0)})$. For $\lambda^{(0)} = .1808$ and $\lambda^{(1)} = .5249$, $\lambda^* = .3229$. In order to achieve the desired .05 error probabilities the approximations (2.11) suggest the values $A = B = 19$, so $-a = b = 2.7629$.

Moreover, the arguments of this section give approximations to the power function and expected sample size, as follows. For each $\lambda_0 < \lambda^*$ there exists a $\lambda_1 > \lambda^*$ such that $(\lambda_1 - \lambda_0)/\log(\lambda_1/\lambda_0) = \lambda^*$, and conversely. One such pair is, of course, $\lambda_0 = \lambda^{(0)}$ and $\lambda_1 = \lambda^{(1)}$. The given test is also a sequential probability ratio test of λ_0 against λ_1 with error probabilities

$$p_{\lambda_0}(a, b) = P_{\lambda_0}\{S_N - \lambda^* N \geq b\} \cong \frac{1 - (\lambda_1/\lambda_0)^a}{(\lambda_1/\lambda_0)^b - (\lambda_1/\lambda_0)^a},$$

$$1 - p_{\lambda_1}(a, b) = P_{\lambda_1}\{S_N - \lambda^* N \leq a\} \cong \frac{1 - (\lambda_0/\lambda_1)^b}{(\lambda_0/\lambda_1)^a - (\lambda_0/\lambda_1)^b},$$

and expected sample size

$$E_{\lambda_i}(N) \cong (\lambda_i - \lambda^*)^{-1}[(b - a)p_{\lambda_i}(a, b) + a].$$

In particular, for $\lambda_0 = .1808$ and $\lambda_1 = .5249$, $E_{\lambda_0}(N) \cong 17.5$ and $E_{\lambda_1}(N) \cong 12.3$, both of which are considerably less than the fixed sample size of 30. Even for $\lambda = \lambda^*$, the expected sample size is only about 23.6.

In order to evaluate these approximations, Morgan *et al.* tried out the sequential probability ratio test (truncated at a maximum of 60 observations—cf. III.6, especially Table 3.6) on the data they had previously gathered to confirm the adequacy of the Poisson model. The "true" value of λ was determined by counting the clumps per field for 100 fields, and then several sequential tests were run on that group of 100 fields by starting at different points in the sequence. The authors' general conclusions were that the error probabilities were generally slightly smaller and the expected sample sizes slightly larger than the approximations suggest, and that the sequential test did indeed result in an overall savings in sampling cost.

4. Optimality of the Sequential Probability Ratio Test

For testing a simple hypothesis against a simple alternative with independent, identically distributed observations, a sequential probability ratio test is optimal in the sense of minimizing the expected sample size both under H_0 and under H_1 among all tests having no larger error probabilities. To understand this result it is helpful to consider first a degenerate version of the problem in which only one error probability and expected sample size appear.

Suppose $x_1, x_2, \ldots$ are independent observations from f, which may be either f_0 or f_1. Assume also that if f_0 is the true density, sampling costs nothing, and our preferred action is to observe $x_1, x_2, \ldots$ *ad infinitum*. On the other hand, if f_1 is the true density each observation costs a fixed amount, so in this case we want to stop sampling as soon as possible and reject the hypothesis $H_0: f = f_0$.

It may help to imagine that a new drug is being marketed under the hypothesis that its side effects are insignificant. However, physicians prescribing the drug must record and report the side effects. As long as the hypothesis of insignificant side effects ($f = f_0$) remains tenable, no action is required. If it even appears that the level of side effects is unacceptably high ($f = f_1$), this must be announced and the drug withdrawn from use.

A "test" of $H_0: f = f_0$ in the sense described above is a stopping time T. If $T < \infty$, H_0 is rejected. We seek a stopping time for which $P_0\{T < \infty\}$ is acceptably small, say no larger than some prescribed α, and for which $E_1(T)$ is a minimum. A candidate is a "one-sided" sequential probability ratio test, i.e. the stopping time

$$N = \text{first } n \geq 1 \text{ such that } l_n \geq B \qquad (2.35)$$
$$= \infty \text{ if } l_n < B \text{ for all } n,$$

where as usual $l_n = \prod_{k=1}^{n} \{f_1(x_k)/f_0(x_k)\}$. By letting $A \to 0$ in (2.9) and (2.15), or by a direct argument along the lines of Section 2,

$$P_0\{N < \infty\} \leq B^{-1} \qquad (2.36)$$

and

$$E_1(N) \cong \log B / E_1[\log\{f_1(x_1)/f_0(x_1)\}]. \qquad (2.37)$$

The following proposition gives a lower bound for $E_1(T)$ for any stopping time for which $P_0\{T < \infty\} < 1$. It implies that if (2.36) and (2.37) were actual equalities, the stopping time N of (2.35) would achieve the lower bound and hence would be optimal.

Proposition 2.38. *For any stopping time T with $P_0\{T < \infty\} < 1$,*

$$E_1(T) \geq -\log P_0\{T < \infty\} / E_1[\log\{f_1(x_1)/f_0(x_1)\}].$$

Proof. We may assume $E_1(T) < \infty$; otherwise the result is trivially true. Note that for any random variable y with mean μ, since $e^{\xi} \geq 1 + \xi$, $Ee^{(y-\mu)} \geq 1 + E(y - \mu) = 1$ and hence $Ee^{y} > e^{Ey}$. The following computation uses Propositions 2.24 and 2.18 (Wald's identities):

$$\begin{aligned}
P_0\{T < \infty\} &= E_1 \exp\left(-\sum_1^T \log\{f_1(x_k)/f_0(x_k)\}\right) \\
&\geq \exp\left(-E_1\left[\sum_1^T \log\{f_1(x_k)/f_0(x_k)\}\right]\right) \\
&= \exp(-E_1 T\, E_1 \log\{f_1(x_1)/f_0(x_1)\}),
\end{aligned}$$

which immediately implies the proposition, since $E_1[\log\{f_1(x_1)/f_0(x_1)\}] > 0$. □

Now consider a conventional (sequential) test of $H_0: f = f_0$ against $H_1: f = f_1$ with error probabilities $\alpha = P_0\{\text{Reject } H_0\}$ and $\beta = P_1\{\text{Accept } H_0\}$. For a sequential probability ratio test we have the approximate relations (2.21) and (2.22) between the expected sample sizes and the error probabilities. Theorem 2.39 generalizes Proposition 2.38 in asserting that these expected sample sizes are approximately minimal.

Theorem 2.39. *Let T be the stopping-time of any test of $H_0: f = f_0$ against $H_1: f = f_1$ with error probabilities α, β $(0 < \alpha < 1, 0 < \beta < 1)$. Assume $E_i(T) < \infty$ $(i = 0, 1)$. Then*

$$E_1(T) \geq \mu_1^{-1}\left\{(1-\beta)\log\left(\frac{1-\beta}{\alpha}\right) + \beta \log\left(\frac{\beta}{1-\alpha}\right)\right\}$$

and

$$E_0(T) \geq \mu_0^{-1}\left\{\alpha \log\left(\frac{1-\beta}{\alpha}\right) + (1-\alpha)\log\left(\frac{\beta}{1-\alpha}\right)\right\},$$

where

$$\mu_i = E_i[\log\{f_1(x_1)/f_0(x_1)\}] \qquad (i = 0, 1).$$

Proof. Let $R = \{\text{Reject } H_0\}$, $R^c = \{\text{Accept } H_0\}$. As in the proof of Proposition 2.38, by Wald's likelihood ratio identity (2.24)

$$\begin{aligned}
\alpha = P_0(R) &= E_1\left\{\prod_1^T \frac{f_0(x_k)}{f_1(x_k)}; R\right\} \\
&= E_1\{e^{-\log l_T} | R\} P_1(R) \geq \exp[-E_1(\log l_T | R)](1-\beta) \\
&= \exp[-E_1(\log l_T; R)/(1-\beta)](1-\beta).
\end{aligned}$$

Taking logarithms yields

$$(1-\beta)\log\left(\frac{\alpha}{1-\beta}\right) \geq -E_1(\log l_T; R).$$

A similar calculation gives

$$\beta \log\left(\frac{1-\alpha}{\beta}\right) \geq -E_1(\log l_T; R^c),$$

so by addition and Wald's identity

$$(1-\beta)\log\left(\frac{\alpha}{1-\beta}\right) + \beta \log\left(\frac{1-\alpha}{\beta}\right) \geq -E_1(\log l_T) = -\mu_1 E_1(T),$$

which is equivalent to the first assertion of the proposition, since $\mu_1 > 0$. The second assertion is proved similarly. □

Since in general no precise meaning is attached to the approximate equalities (2.21) and (2.22), we have not really proved the optimality of the sequential probability ratio test. Indeed a complete proof is quite difficult and involves the introduction of several auxiliary concepts (see, for example, Ferguson (1967), p. 365). For the very special symmetric Bernoulli case of Example 2.6, the approximations (2.21) and (2.22) are actual equalities, so the preceding discussion contains a completely rigorous proof.

5. Criticism of the Sequential Probability Ratio Test and the Anscombe–Doeblin Theorem

Although the optimality of the sequential probability ratio test is a remarkably strong property in some respects, it applies only to simple hypotheses. For applications involving composite hypotheses the test has noteworthy deficiencies. One is the open continuation region, which leads occasionally to very large sample sizes, especially when $E\{\log[f_1(x_1)/f_0(x_1)]\} \cong 0$. Another is the difficulty associated with estimating a parameter when the data are obtained from a sequential probability ratio test.

The first of these problems can be treated in principle by truncating the stopping rule to take no more than some maximum number of observations. The analysis of such a test is more difficult, but no new statistical concepts are involved (see Chapter III).

The problem of estimation exists to some extent with all sequential tests. If one wants to stop sampling as soon as it is possible to tell in which of two subsets of the parameter space a parameter lies, there presumably are cases where the amount of data necessary to make this rather coarse distinction is inadequate for estimation. A possible solution is to enforce artificially a larger sample size when estimation is of interest. Investigation of the issue is complicated by the fact that even in very simple cases, when one would ordinarily consider unbiased estimators, sequentially stopped versions of the estimators are biased, and their sampling distributions can be quite complicated.

The problems of estimation following sequential tests are discussed in detail in Chapters III and IV. Here we prove a crude but useful result, which shows that randomly stopped averages are asymptotically normal under quite general conditions.

Theorem 2.40 (Anscombe, 1952, Doeblin, 1938). *Let $x_1, x_2, \ldots$ be independent and identically distributed with mean μ and variance $\sigma^2 \in (0, \infty)$. Let $S_n = \sum_{k=1}^n x_k$. Suppose M_c, $c \geq 0$, are positive integer valued random variables such that for some constants $m_c \to \infty$, $M_c/m_c \xrightarrow{P} 1$. Then as $c \to \infty$*

$$P\{M_c^{-1/2}(S_{M_c} - \mu M_c) \leq x\} \to \Phi(x/\sigma),$$

where Φ denotes the standard normal distribution function.

Remark 2.41. Consider the symmetric sequential probability ratio test of (2.5) for the mean of a normal distribution. Assume $\mu > 0$. As $b \to \infty$, it follows from (2.30) that with probability close to one

$$S_{N-1} < b \leq S_N.$$

Divide by N and note that by the strong law of large numbers $N^{-1}S_{\tilde{N}} \to \mu$ *with probability one, where* $\tilde{N}$ is either N or $N - 1$. Hence as $b \to \infty$

$$\mu b^{-1} N \xrightarrow{P} 1. \tag{2.42}$$

It follows from Theorem 2.40 that $N^{-1/2}(S_N - N\mu)$ is approximately normally distributed, and $N^{-1}S_N \pm 1.645N^{-1/2}$ is an approximate 90% confidence interval for μ for large b. Unfortunately this approximation is very poor for moderate values of b (see III.4, 6).

Proof of Theorem 2.40. Without loss of generality assume that $\mu = 0$ and $\sigma^2 = 1$. Also recall that if Z_n converges in law to Z, $\xi_n \xrightarrow{P} 1$, and $\eta_n \xrightarrow{P} 0$, then $\xi_n Z_n + \eta_n$ converges in law to Z (cf. Cramér, 1946, p. 254). Hence, since

$$M^{-1/2}S_M = (m/M)^{1/2} m^{-1/2} S_m + (m/M)^{1/2} m^{-1/2}(S_M - S_m),$$

and $m^{-1/2}S_m$ is asymptotically normally distributed by the central limit theorem, it suffices to show

$$m^{-1/2}(S_M - S_m) \xrightarrow{P} 0. \tag{2.43}$$

Let $\varepsilon, \delta \in (0, 1)$. Let $m_1 = m(1 - \delta)$, $m_2 = m(1 + \delta)$. Then

$$\{m^{-1/2}|S_M - S_m| > \varepsilon\} \subset \{|m^{-1}M - 1| > \delta\} \cup \left\{\max_{m_1 \leq n \leq m_2} |S_n - S_{m_1}| > m^{1/2}\varepsilon/2\right\}.$$

By hypothesis $P\{|m^{-1}M - 1| > \delta\} \to 0$ for all $\delta > 0$. Also by Kolmogorov's inequality

$$P\left\{\max_{m_1 \leq n \leq m_2} |S_n - S_{m_1}| > m^{1/2}\varepsilon/2\right\} \leq 4(m_2 - m_1)/m\varepsilon^2 \leq 8\delta/\varepsilon^2.$$

Hence

$$\limsup_{c \to 0} P\{m^{-1/2}|S_M - S_m| > \varepsilon\} \leq 8\delta/\varepsilon^2.$$

Since δ can be made arbitrarily small, this proves (2.43) and hence the Theorem. □

Remark. The preceding proof is deceptively simple. It seems remarkably difficult by comparison to obtain any more accurate approximation, e.g. an Edgeworth type asymptotic expansion, even for fairly simple stopping rules.

6. Cusum Procedures

Imagine a process which produces a potentially infinite sequence of observations $x_1, x_2, \ldots$. Initially the process is "in control" in the sense that an observer is satisfied to record the x's without taking any action. At some unknown time ν the process changes and becomes "out of control." The observer would like to infer from the x's that this change has taken place and take appropriate action "as soon as possible" after time ν.

To give this problem a simple, precise formulation, assume that the x_i are independent random variables and that for some $\nu \geq 1$, $x_1, x_2, \ldots, x_{\nu-1}$ have the probability density function f_0, whereas $x_\nu, x_{\nu+1}, \ldots$ have the probability density function f_1.

Let P_ν denote probability when the change from f_0 to f_1 occurs at the νth observation, $\nu = 1, 2, \ldots$; and let P_0 denote probability when there is no change, i.e. $\nu = \infty$, so $x_1, x_2, \ldots$ are independent and indentically distributed with probability density function f_0. We seek a stopping rule τ which makes the P_ν distributions of $(\tau - \nu)^+$ stochastically small, $\nu \geq 1$, subject to the constraint that the P_0 distribution of τ be stochastically large. A simple formal requirement is to minimize

$$\sup_{\nu \geq 1} E_\nu(\tau - \nu + 1 | \tau \geq \nu) \tag{2.44}$$

subject to

$$E_0 \tau \geq B \tag{2.45}$$

for some given (large) constant B.

An *ad hoc* proposal to solve this problem approximately is the following. Suppose that $x_1, \ldots, x_n$ have been observed. Consider for $1 \leq \nu \leq n$ the hypotheses H_ν that $x_1, \ldots, x_{\nu-1}$ are distributed according to f_0 and $x_\nu, \ldots, x_n$ according to f_1, and H_0 the hypothesis of no change. The log likelihood ratio for testing H_ν against H_0 is $\sum_{k=\nu}^n \log\{f_1(x_k)/f_0(x_k)]$. For testing the composite hypothesis that at least one of the H_ν hold $(1 \leq \nu \leq n)$ against H_0, the log likelihood ratio statistic is

$$\max_{0 \leq k \leq n} (\tilde{S}_n - \tilde{S}_k) = \tilde{S}_n - \min_{0 \leq k \leq n} \tilde{S}_k, \tag{2.46}$$

where $\tilde{S}_n = \sum_{j=1}^n \log[f_1(x_j)/f_0(x_j)]$. An intuitively appealing stopping rule based on (2.46) is

$$\tau = \inf\left\{n: \tilde{S}_n - \min_{0 \le j \le n} \tilde{S}_j \ge b\right\}. \tag{2.47}$$

Note that $\tilde{S}_n - \min_{0 \le k \le n} \tilde{S}_k$ measures the current height of the random walk $\tilde{S}_j$, $j = 0, 1, \ldots$ above its minimum value. Whenever the random walk establishes a new minimum, i.e. $\tilde{S}_n = \min_{0 \le k \le n} \tilde{S}_k$, the process forgets its past and starts over in the sense that for all $j \ge 0$, $\tilde{S}_{n+j} - \min_{0 \le k \le n+j} \tilde{S}_k = \tilde{S}_{n+j} - \tilde{S}_n - \min_{0 \le k \le j}(\tilde{S}_{n+k} - \tilde{S}_n)$.

This renewal property has several important consequences. First, it implies that for τ defined by (2.47), $\sup_{\nu \ge 1} E_\nu(\tau - \nu + 1 | \tau \ge \nu) = E_1 \tau$, because at all times after $\nu - 1$ the process (2.46) must be at least as large as if there had been a renewal at $\nu - 1$. Hence to evaluate (2.44) and (2.45) one must calculate $E_\nu(\tau)$ for $\nu = 0, 1$, and for each of these "extreme" cases $x_1, x_2, \ldots$ are identically distributed. Second, it means that τ can be defined in terms of a sequence of sequential probability ratio tests as follows. Let

$$N_1 = \inf\{n: \tilde{S}_n \notin (0, b)\}. \tag{2.48}$$

If $\tilde{S}_{N_1} \ge b$, then $\tau = N_1$. Otherwise $\tilde{S}_{N_1} = \min_{0 \le k \le N_1} \tilde{S}_k$ and we define

$$N_2 = \inf\{n: n \ge 1, \tilde{S}_{N_1 + n} - \tilde{S}_{N_1} \notin (0, b)\}.$$

If $\tilde{S}_{N_1 + N_2} - \tilde{S}_{N_1} \ge b$, then $\tau = N_1 + N_2$. Otherwise $\tilde{S}_{N_1 + N_2} \le \tilde{S}_{N_1}$, and $\tilde{S}_{N_1 + N_2} = \min_{0 \le k \le N_1 + N_2} \tilde{S}_k$. In general let

$$N_k = \inf\{n: n \ge 1, \tilde{S}_{N_1 + \cdots + N_{k-1} + n} - \tilde{S}_{N_1 + \cdots + N_{k-1}} \notin (0, b)\}. \tag{2.49}$$

It is easy to see that

$$\tau = N_1 + \cdots + N_M, \tag{2.50}$$

where

$$M = \inf\{k: \tilde{S}_{N_1 + \cdots + N_k} - \tilde{S}_{N_1 + \cdots + N_{k-1}} \ge b\}. \tag{2.51}$$

(See Figure 2.2.) Hence under any probability P which makes $x_1, x_2, \ldots$ independent and identically distributed, in particular for $P = P_1$ or P_0, Wald's identity (2.18) and (2.50) yield

$$E\tau = EN_1 EM.$$

Moreover, by (2.51), M is geometrically distributed, with $E(M) = 1/P\{\tilde{S}_{N_1} \ge b\}$. As a consequence we obtain the basic identity

$$E(\tau) = E(N_1)/P(\tilde{S}_{N_1} \ge b), \tag{2.52}$$

which expresses $E(\tau)$ in terms of the expected sample size and error probability of a single sequential probability ratio test having lower (log) boundary $a = 0$.

For (2.52) the Wald approximations of Section 2 yield the expression 0/0 when $a = 0$, but it is possible to give an approximation by evaluating the ratio

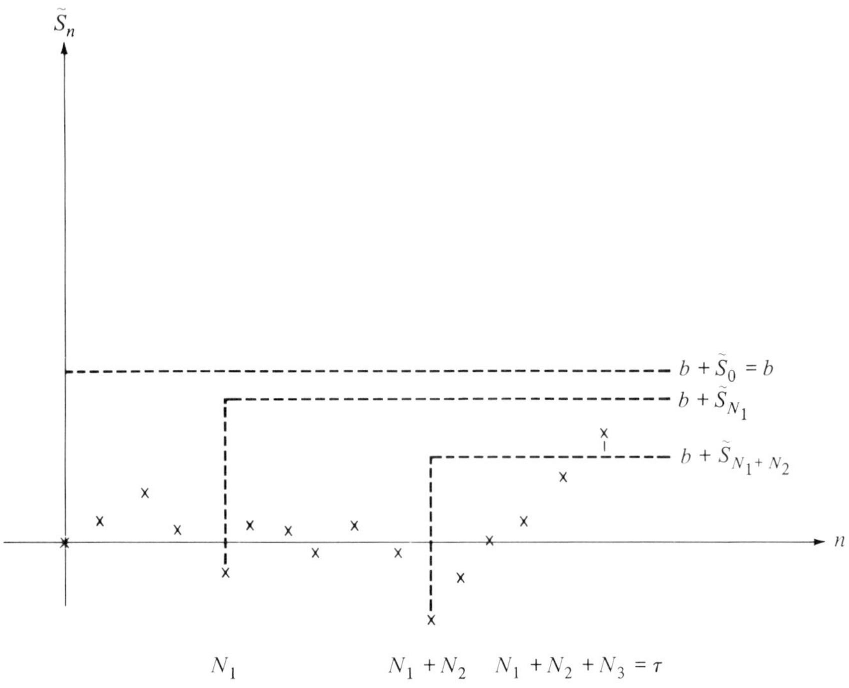

Figure 2.2.

on the right hand side for arbitrary negative a and letting $a \to 0$. The results of these calculations, using (2.12), (2.15), and (2.16), are

$$(2.53)\quad E_0(\tau) = |e^b - b - 1|/|\mu_0| \qquad \text{and} \qquad E_1(\tau) = (e^{-b} + b - 1)/\mu_1,$$

where

$$\mu_i = \int \{\log\{f_1(x)/f_0(x)]\} f_i(x)\, dx \qquad (i = 0, 1).$$

If f_0 and f_1 belong to a one-parameter exponential family of distributions, as in Section 3, it is possible to obtain an analogous approximation for the expectation of τ when $x_1, x_2, \ldots$ are independent and identically distributed according to a fixed distribution in that family. Specifically, let $z(x) = \log f_1(x)/f_0(x)$ and assume that the P_θ distribution of $z_n = z(x_n)$ is of the form

$$(2.54)\qquad P_\theta\{z_n \in dx\} = dF_\theta(x) = \exp\{\theta z(x) - \psi(\theta)\}\, dF(x)$$

relative to some fixed distribution function $F(x)$. It is convenient to assume that F is standardized to satisfy $\int z(x)\, dF(x) = 0$, so $\psi'(0) = 0$. This can be accomplished by a change of origin on the θ axis. Let θ_0 and θ_1 denote *conjugate* pairs of θ values defined by $\theta_0 < 0 < \theta_1$ and $\psi(\theta_0) = \psi(\theta_1)$. See the middle graph in Figure 2.1.

Now suppose $\theta \neq 0$. If $\theta < 0$, it is denoted by θ_0 and its conjugate by θ_1.

Otherwise it is denoted by θ_1 and its conjugate by θ_0. The stopping rule (2.47) can be written

$$\tau = \inf\left\{n: \Delta\tilde{S}_n - \min_{0\le k\le n} \Delta\tilde{S}_k \ge b'\right\}, \tag{2.55}$$

where $\Delta = \theta_1 - \theta_0$ and $b' = \Delta b$. Since by (2.54)

$$\Delta\tilde{S}_n = \sum_1^n \log\{dF_{\theta_1}(x_k)/dF_{\theta_0}(x_k)\},$$

(2.55) is again of the form (2.47), but with b' in place of b. It follows immediately from (2.53) (in different, but self-evident notation) that

$$E_{\theta_i}(\tau) \cong |\exp[(-1)^i\Delta b] - (-1)^i\Delta b - 1|/\Delta|\psi'(\theta_i)|. \tag{2.56}$$

Taking a limit as $\theta_i \to 0$ yields

$$E_0(\tau) \cong b^2\psi''(0).$$

Consider the special case $f_i(x) = \phi(x + (-1)^i/2)$ and $dF(x) = \phi(x)\,dx$, so $z(x) = x$ and the family of distributions (2.54) is just the normal family with mean θ and variance 1. In this case (2.56) specializes to

$$E_\theta(\tau) \cong |\exp(-2\theta b) + 2\theta b - 1|/(2\theta^2). \tag{2.57}$$

Unfortunately these approximations are not especially accurate. For example for $b = 6$, (2.57) gives $E_{-.4}(\tau) \cong 362$ and $E_{.4}(\tau) \cong 11.9$. Van Dobben de Bruyn (1968) has calculated these quantities numerically and has obtained 940 and 14.9, respectively.

Often it is possible to improve upon (2.56), and for the normal case the improvement is very similar to that suggested in Section 3. If the right hand side of (2.57) is denoted by $e(b)$, the improved approximation for $E_\theta(\tau)$ is $e(b + 2\cdot 0.583)$. For $b = 6$ as above, this gives $E_{-.4}(\tau) \cong 944$ and $E_{.4}(\tau) \cong 14.8$. See X.3 for justification and generalization of this approximation.

The preceding discussion is concerned with a change in distribution from F_0 to F_1. Suppose now that $F_i = F_{\theta_i}$ for $\{F_\theta\}$ the exponential family (2.54). (Note that θ_0 and θ_1 no longer have the meaning of the preceding paragraphs.) If $\theta_1 > \theta_0$ ($\theta_1 < \theta_0$), the stopping rule (2.47) can be used to signal a change from θ_0 in the direction $\theta > \theta_0$ ($\theta < \theta_0$). In order to signal a change in either direction one can splice together a pair of one-sided stopping rules as follows.

Suppose that the process is in control if $\theta = \theta_0$ and that (2.54) has been normalized so that $\theta_0 = 0 = \psi(\theta_0) = \psi'(\theta_0)$. (This is always possible by a linear transformation.) To detect a change in θ in either direction, let $\theta_1 < 0 < \theta_2$ and define

$$\tau_i = \inf\left\{n: \theta_i S_n - n\psi(\theta_i) - \min_{0\le k\le n} [\theta_i S_k - k\psi(\theta_i)] \ge b_i\right\}, \tag{2.58}$$

where $S_n = x_1 + \cdots + x_n$. The stopping rule τ_i is of the form (2.47) for detecting a change from $\theta_0 = 0$ in the direction of θ_i, $i = 1, 2$. To detect a change in either direction let $\tau = \min(\tau_1, \tau_2)$.

In a number of special cases there is a simple relation among $E\tau$, $E\tau_1$, and $E\tau_2$. Here E denotes expectation under any probability P which makes $x_1, x_2, \ldots$ independent and identically distributed. It is shown in the somewhat technical Lemma 2.64 below that if

$$|\theta_1|^{-1}\psi(\theta_1) + \theta_2^{-1}\psi(\theta_2) \geq \big||\theta_1^{-1}|b_1 - \theta_2^{-1}b_2\big| \tag{2.59}$$

then

$$\tau = \tau_i \Rightarrow \theta_j S_\tau - \tau\psi(\theta_j) = \min_{0 \leq k \leq \tau} [\theta_j S_k - k\psi(\theta_j)] \qquad (j \neq i) \tag{2.60}$$

(see Problem 2.20). Obviously for $j \neq i$

$$\tau_i = \tau + I_{\{\tau=\tau_j\}}(\tau_i - \tau).$$

By (2.60) and the renewal property of $\tilde{S}_n - \min_{0\leq k\leq n} \tilde{S}_k$ described above, the conditional distribution of $\tau_i - \tau$ given $\tau = \tau_j$ is the same as the unconditional distribution of τ_i $(i \neq j)$. Hence for $i \neq j$,

$$E\tau_i = E\tau + E(\tau_i - \tau; \tau = \tau_j) = E\tau + P\{\tau = \tau_j\}E\tau_i.$$

Since $P\{\tau = \tau_1\} + P\{\tau = \tau_2\} = 1$, one can solve these two equations for $E\tau$ to obtain

$$(E\tau)^{-1} = (E\tau_1)^{-1} + (E\tau_2)^{-1}. \tag{2.61}$$

Summarizing this argument yields the following result.

Theorem 2.62. *Let $x_1, x_2, \ldots$ be independent and identically distributed. Let τ_i $(i = 1, 2)$ be defined by* (2.58) *and* $\tau = \min(\tau_1, \tau_2)$. *If* (2.59) *holds then $E\tau$, $E\tau_1$, and $E\tau_2$ satisfy the relation* (2.61).

EXAMPLE 2.63. Wilson *et al.* (1979) have used cusum techniques to minitor the quality of radioimmunoassays. A laboratory makes repeated assays in the form of an average of 2 or 3 independent measurements of a concentration in a plasma. In order to maintain quality, a plasma of known concentration is occasionally submitted to be assayed. From these assays of known concentration one obtains observations $x_1, x_2, \ldots$ which are assumed to be independent and normally distributed with mean 0 and (known) variance σ_0^2— as long as the procedure remains in control. A two-sided cusum procedure can be used to detect a change in the mean in either direction from its target value of 0.

It is also important to detect a change in the variance σ^2 from σ_0^2 to some larger value. However, the target value σ_0^2 is a much fuzzier concept than the target value of 0 for the mean error. Presumably σ_0^2 is determined from prior experimentation, but it may be revised as experience accumulates. Given σ_0^2, one has associated with each x_n an independent y_n; and when the process is under control, $y_1/\sigma_0^2, y_2/\sigma_0^2, \ldots$ are independent and have a χ^2 distribution with one or two degrees of freedom, according as the original assays involve 2 or 3 measurements.

For an approximation to the average run length of a one-sided cusum procedure to detect an increase in σ^2, see Problem 2.17.

Lemma 2.64. *If* (2.59) *holds, then* (2.60) *follows.*

PROOF. Suppose $\tau = \tau_2 = n$ and that l denotes the largest k, $0 \le k \le n$ such that

$$\theta_2 S_k - k\psi(\theta_2) = \min_{0 \le j \le n} [\theta_2 S_j - j\psi(\theta_2)].$$

Then

$$S_n - n\psi(\theta_2)/\theta_2 - [S_l - l\psi(\theta_2)/\theta_2] \ge b_2/\theta_2 \tag{2.65}$$

but

$$S_i - i\psi(\theta_2)/\theta_2 - \min_{0 \le j \le i} [S_j - j\psi(\theta_2)/\theta_2] < b_2/\theta_2 \tag{2.66}$$

for all $i < n$. Since $\theta_1 < 0$, (2.60) is equivalent to

$$S_n - n\psi(\theta_1)/\theta_1 = \max_{0 \le j \le n} [S_j - j\psi(\theta_1)/\theta_1]. \tag{2.67}$$

For any $l \le j \le n$

$$\begin{aligned} S_n - n\psi(\theta_1)/\theta_1 - [S_j - j\psi(\theta_1)/\theta_1] &= S_n - n\psi(\theta_2)/\theta_2 - (S_j - j\psi(\theta_2)/\theta_2) \\ &\quad + (n-j)[\psi(\theta_2)/\theta_2 - \psi(\theta_1)/\theta_1] > 0, \end{aligned}$$

because by the definition of l, (2.65), and (2.66)

$$S_n - n\psi(\theta_2)/\theta_2 - S_j + j\psi(\theta_2)/\theta_2 \ge 0,$$

and by the normalization $\theta_0 = 0 = \psi(\theta_0) = \psi'(\theta_0)$ and the convexity of ψ, $\psi(\theta_i) \ge 0$ for $i = 1, 2$. Hence if (2.67) is not satisfied there exists $0 \le j < l$ such that

$$S_n - n\psi(\theta_1)/\theta_1 < S_j - j\psi(\theta_1)/\theta_1. \tag{2.68}$$

By (2.68) and (2.65)

$$\begin{aligned} &S_l - l\psi(\theta_1)/\theta_1 - [S_j - j\psi(\theta_1)/\theta_1] \\ &\quad < S_l - l\psi(\theta_1)/\theta_1 - [S_n - n\psi(\theta_1)/\theta_1] \\ &\quad = S_l - l\psi(\theta_2)/\theta_2 - [S_n - n\psi(\theta_2)/\theta_2] \\ &\qquad - (n-l)[\psi(\theta_2)/\theta_2 - \psi(\theta_1)/\theta_1] \\ &\quad \le -b_2/\theta_2 - [\psi(\theta_2)/\theta_2 + \psi(\theta_1)/|\theta_1|] \le b_1/\theta_1, \end{aligned}$$

where the last inequality follows from (2.59). But then $\tau \le \tau_1 \le l$, contradicting the hypothesis that $\tau = n > l$. □

Remarks. Condition (2.59) can be interpreted as a measure of symmetry. It is trivially satisfied if $\theta_1 = -\theta_2$ and $b_1 = b_2$. Lemma 2.64 was first proved at this level of generality by van Dobben de Bruyn (1968), who also pointed out that (2.59) is necessary for (2.60).

Problems

2.1.* Let $x_1, x_2, \ldots$ be independent with probability density function $f_\theta(x) = \theta e^{-\theta x}$ for $x \geq 0$ and $= 0$ otherwise. Consider a "one-sided" sequential probability ratio test of H_0: $\theta = \theta_0$ against H_1: $\theta = \theta_1$, where $\theta_1 < \theta_0$ (cf. (2.35)). Argue that by the lack of memory property of the exponential distribution, the P_{θ_1} distribution of $\log l_N - \log B$ is exponential. Hence $P_{\theta_0}\{N < \infty\} = \theta_1/B\theta_0$ and $E_{\theta_1}(N) = [\log B + \theta_0\theta_1^{-1} - 1]/[\theta_0\theta_1^{-1} - 1 - \log \theta_0\theta_1^{-1}]$. Note that for $\theta_0 \cong 1.5\theta_1$, say, there is a considerable discrepancy between this result and the Wald approximation, $P_{\theta_0}\{N < \infty\} \cong B^{-1}$.

2.2. Investigate the numerical accuracy of the modifications of (2.30) and (2.31) suggested in Section 3. One possibility is to compare results of the modified approximations with numerically computed values given by Barraclough and Page (1959) or Kemp (1958). Another is to make comparisons with simulated values using the technique of Remark 2.26 to increase the accuracy of the simulations.

2.3. For $i = 1, 2$, let $x_{i1}, x_{i2}, \ldots$ be independent Bernoulli variables with $P\{x_{ij} = 1\} = p_i$, $P\{x_{ij} = 0\} = q_i = 1 - p_i$ for $j = 1, 2, \ldots$. Assume that the x_{1j} and x_{2j} are also independent and that observations are made in pairs (x_{11}, x_{21}), (x_{12}, x_{22}), ... Suppose $p \neq \frac{1}{2}$ and $p + q = 1$. Find a sequential probability ratio test of H_0: $p_1 = p$, $p_2 = q$ against H_1: $p_1 = q$, $p_2 = p$. Calculate the error probability and expected sample size for arbitrary p_1, p_2 (a) by using the theory developed in the text and (b) by reducing this problem to the one considered in Example 2.6.

2.4. (a) Let x_1, x_2, ... be independent random variables with exponential distribution, $P_\lambda\{x_k \in dx\} = \lambda e^{-\lambda x}\,dx$ for $x > 0$. Let $\lambda^{(0)} < \lambda^{(1)}$. Show that a sequential probability ratio test of H_0: $\lambda = \lambda^{(0)}$ against H_1: $\lambda = \lambda^{(1)}$ is defined by a stopping rule of the form

$$N = \text{first } n \geq 1 \qquad \text{such that } n - \lambda^* \sum_1^n x_k \notin [a, b],$$

where $\lambda^* = (\lambda^{(1)} - \lambda^{(0)})/\log(\lambda^{(1)}/\lambda^{(0)})$. Find approximations to the power function and expected sample size. Denote them by $p_\lambda(a, b)$ and $e_\lambda(a, b)$. It is shown in Chapter X that improved approximations similar to those suggested in Section 3 for normal variables are given by $p_\lambda(a - 1, b + \frac{1}{3})$ and $e_\lambda(a - 1, b + \frac{1}{3})$. Compare numerically the approximations in the case $b = \infty$ with the exact results of Problem 1. (Note the difference in notation between the two problems.)

(b) Now suppose that observations are made continuously on a Poisson process with intensity λ. Find a sequential probability ratio test of $\lambda^{(0)}$ against $\lambda^{(1)}$. What is the relation between the "no excess" approximations in this case and in part (a)? Now because the process does not jump over a, improved approximations are given by $p_\lambda(a, b + \frac{1}{3})$ and $\lambda e_\lambda(a, b + \frac{1}{3})$. See X.3 or Hald (1981), p. 266, for a numerical comparison.

2.5. Assume that m items with exponentially distributed lifetimes are simultaneously put on test. Let $x_1, x_2, \ldots, x_m$ denote their failure times, so one observes sequentially $y_1 = \min(x_1, \ldots, x_m)$, y_2 = second smallest of the x_i, etc. Show how the theory of the preceding problem can be used to set up a sequential probability ratio test for this experimental situation, where now, however, there is a maximum number of m observations. (The effect of truncation on sequential probability ratio tests is discussed in III.6.) What is the relation between the expected number of failures and the expected length of the test measured in real time? Discuss also the situation where an item which fails is replaced immediately by a good item, so until the test is terminated there are always m items on test.
Hint: Show that $(m - i + 1)(y_i - y_{i-1})$, $i = 1, 2, \ldots, m$ $(y_0 = 0)$ are independent and exponentially distributed and that the likelihood function after observing $y_1, \ldots, y_k$ can be expressed in terms of $\Sigma(m - i + 1)(y_i - y_{i-1})$, $i = 1, 2, \ldots, k$.

2.6. Prove Proposition 2.19. *Hint:* Suppose $\delta > 0$ is such that $P\{y_1 \geq \delta\} \geq \delta$ and let $r > (b - a)/\delta$. Note that $P\{y_1 + \cdots + y_r > b - a\} \geq \delta^r$. Let $a \leq 0 \leq b$ and compare M with the "geometric" random variable

$$\tilde{M} = \text{first value } mr \ (m \geq 1) \qquad \text{such that } \sum_{i=(m-1)r+1}^{mr} y_i > b - a.$$

2.7. Suppose that l_n is the likelihood ratio of $z_1, \ldots, z_n$ under P_1 relative to P_0. Show that if $P_0\{l_n > 0\} = 1$, then l_n^{-1} is the likelihood ratio of $z_1, \ldots, z_n$ under P_0 relative to P_1 (cf. (2.23)). Note that it is unnecessary to assume $P_1\{l_n > 0\} = 1$. Why?

2.8. Use Proposition 2.24 and a symmetry argument to show that the right hand side of (2.31) is actually a lower bound for $E_\mu(N)$.

2.9. Show that the Wald approximation for $E_0(N)$ for the stopping rule (2.5) in the symmetric normal case is $E_0(N) \cong b^2$. Use this result but with b replaced by $b + .583$ (cf. Section 3) to show that even for error probabilities as small as .01, $E_0(N)$ is smaller than the sample size of a competing fixed sample test.

2.10.* Use Wald's identity (2.18) to derive an approximation to $P_f\{l_N \geq B\}$ when $E_f\{\log l_1\} = 0$.

2.11.† Prove Wald's identity for the second moment: Let $y_1, y_2, \ldots$ be independent and identically distributed with $Ey_1 = 0$ and $\sigma^2 = Ey_1^2 < \infty$. If T is any stopping time with $ET < \infty$, then $E[(\sum_1^T y_i)^2] = \sigma^2 ET$.
Hint: For a *bounded* stopping time T this can be proved using the representation $\sum_1^T y_i = \sum_1^\infty I_{\{T \geq i\}} y_i$ as in the proof of (2.18). To extend this to a general stopping time with $ET < \infty$, show that

$$\lim_{m<n,\, m\to\infty,\, n\to\infty} E\left[\left(\sum_{i=T\wedge m+1}^{T\wedge n} y_i\right)^2\right] = 0,$$

and use Fatou's lemma.

2.12. Use Problem 11 to obtain an approximation for $E_f(N)$ when $E_f\{\log l_1\} = 0$. Check that this answer is consistent with the special case in Problem 9.

2.13. Let $x_1, x_2, \ldots$ be independent with probability density function of the form

$$f_\theta(x) = \exp[\theta x - \psi(\theta)]f(x),$$

where f is some given density function. Assume that the parameter space is an open interval $(\underline{\theta}, \bar{\theta})$ and that $\psi(\theta) \to \infty$ as $\theta \to \bar{\theta}$ or $\underline{\theta}$. Consider a sequential probability ratio test of H_0: $\theta = \theta^{(0)}$ against H_1: $\theta = \theta^{(1)}$. Find approximations for the power function and the expected sample size function of the test. Specialize the results to the Bernoulli and Normal examples considered in the text.

Remark. Although one can apply the general theory developed in the text to solve this problem, it is probably more instructive to proceed from first principles with the ideas given in the chapter as guidelines. An important and particularly simple special case occurs when $\psi(\theta^{(0)}) = \psi(\theta^{(1)})$.

2.14.† For detecting a change of distribution, the following alternative to (2.47) has been suggested by Shiryayev (1963) and Roberts (1966):

$$T = \inf\left\{n: \sum_{k=0}^{n-1} \prod_{i=k+1}^{n} \frac{f_1(x_i)}{f_0(x_i)} \geq B\right\}.$$

Show that

$$E_\infty T = E_\infty\left\{\sum_{0}^{T-1} \prod_{k+1}^{T} \frac{f_1(x_i)}{f_0(x_i)}\right\},$$

so by neglecting the excess over the boundary $E_\infty T \cong B$. Obtaining a reasonable approximation to $E_1 T$ is difficult, although it is at least intuitively apparent that a crude approximation is given by

$$E_1 T \sim \log B / E_1[\log\{f_1(x_1)/f_0(x_1)\}] \qquad \text{as } B \to \infty.$$

2.15. Prove the following generalization of Theorem 2.40 (Anscombe, 1952). Suppose that $Y_n \xrightarrow{\mathscr{L}} Y$ and that the sequence $\{Y_n\}$ is slowly changing in the sense that for each $\varepsilon > 0$ there exists $\delta > 0$ such that for all sufficiently large m

$$P\left\{\max_{m \leq n < m(1+\delta)} |Y_n - Y_m| \geq \varepsilon\right\} < \varepsilon. \tag{2.69}$$

Suppose $\{M(c), c \geq 0\}$ is a family of positive integer valued random variables such that for some constants $\rho(c) \to \infty$, $M(c)/\rho(c) \xrightarrow{P} 1$. Then $Y_{M(c)} \xrightarrow{\mathscr{L}} Y$. (See IX.2 for a condition similar to (2.69) in a different context.)

2.16.* Show that under the condition (2.32)

$$E_f(N) \cong [\log(BA^{-1})P_f\{l_N \geq B\} + \log A]/E_f\{\log[f_1(x_1)/f_0(x_1)]\}$$

(where $P_f\{l_N \geq B\}$ is given approximately by (2.33) in the case $\theta_1 > 0$ and by $(A^{\theta_1} - 1)/(A^{\theta_1} - B^{\theta_1})$ in the case $\theta_1 < 0$).

2.17. Let $f_\lambda(x) = \lambda e^{-\lambda x}$ for $x \geq 0$ and 0 otherwise. Consider the problem of detecting a change in distribution from $\lambda = 1$ to $\lambda = \lambda^{(1)} < 1$. (This is equivalent to detecting an increase in σ^2 in Example 2.63 when two degrees of freedom are available from each assay for estimating σ^2.) Let $\lambda^* = (1 - \lambda^{(1)})/\log(1/\lambda^{(1)})$, $\theta = 1 - \lambda/\lambda^*$, and $\psi(\theta) = -(\theta + \log(1 - \theta))$. Show that the stopping rule N_1 of (2.48) equals $\inf\{n: \sum_1^n (\lambda^* x_k - 1) \notin (0, b')\}$, where $b' = b/\log(1/\lambda^{(1)})$, and that the distribution of $z_n' = \lambda^* x_n - 1$ in the form (2.54) is given by

$$\exp[\theta x - \psi(\theta)] \exp[-(x+1)]\,dx$$

for $x \geq -1$. Evaluate the approximation (2.56) for arbitrary λ. (If this approximation is denoted by $e(b')$, an improved approximation taking excess over the boundary into account is $e(b' + \frac{4}{3})$ (see X.2).)

2.18. Discuss Problem 2.17 if one seeks to detect a change from $\lambda = 1$ to $\lambda = \lambda^{(1)} > 1$. Now $\lambda^* = (\lambda^{(1)} - 1)/\log \lambda^{(1)}$, $\theta = \lambda/\lambda^* - 1$, $\psi(\theta) = \theta - \log(1 + \theta)$, $b' = b/\log \lambda^{(1)}$, and $z'_n = 1 - \lambda^* x_n$. (The correction for excess over the boundary is again of the form $e(b' + \frac{4}{3})$. For a more complete discussion of this problem which contains a different approximation to the expected run length and some numerical examples, see Lorden and Eisenberger, 1973.)

2.19.* Consider the general model of Section 1 and an arbitrary sequential test of the simple hypotheses $H_0: f_n = f_{0n}, n = 1, 2, \ldots$ against $H_1: f_n = f_{1n}, n = 1, 2, \ldots$. Let T denote the stopping time, A the acceptance region, and R the rejection region of the test. Let P denote probability and E expectation when f_n $(n = 1, 2, \ldots)$ is the true sequence of joint densities, not necessarily either f_{0n} or f_{1n}. Suppose that $P\{T < \infty\} = 1$. Show that the total error probability $\alpha + \beta$ satisfies $\alpha + \beta \geq E\{\min(f_{0T}/f_T), (f_{1T}/f_T)\}$, and that there is equality for a sequential probability ratio test provided $A < 1 < B$. (For an application of this inequality see III.7.)

2.20. For the normal distribution (i.e. (2.54) with $dF(x) = \phi(x)\,dx$, $\psi(\theta) = \theta^2/2$, $z(x) = x$), demonstrate (2.60) geometrically by an appropriate picture in the symmetric case $|\theta_1| = \theta_2$, $b_1 = b_2$.

CHAPTER III

Brownian Approximations and Truncated Tests

1. Introduction

The central role of the normal distribution in statistics arises because of its simplicity and usefulness as an approximation to other probability distributions. In sequential analysis considerable additional simplification results from approximating sums of independent random variables $x_1 + \cdots + x_n$ in discrete time by a Brownian motion process $\{W(t), 0 \leq t < \infty\}$ in continuous time. Although this approximation is rarely quantitatively adequate (cf. Section 5), its comparative simplicity leads to appreciable qualitative insight; and quantitatively it does provide a first, crude approximation which can often be used as a basis for subsequent refinement. This chapter is concerned primarily with sequential tests for the mean of a Brownian motion process. Various truncated modifications of the sequential probability ratio test will be introduced, and problems of estimates and attained significance levels relative to sequential tests will be discussed.

A Brownian motion with drift μ (and unit variance) is a family of random variables $\{W(t), 0 \leq t < \infty\}$ with the following properties:

(i) $W(0) = 0$;
(ii) $P_\mu\{W(t) - W(s) \leq x\} = \Phi[(x - \mu(t - s))/(t - s)^{1/2}], 0 \leq s < t < \infty$;
(iii) for all $0 \leq s_1 < t_1 \leq s_2 < t_2 \leq \cdots \leq s_n < t_n < \infty, n = 2, 3, \ldots$ the random variables $W(t_i) - W(s_i)$, $i = 1, 2, \ldots, n$ are stochastically independent;
(iv) $W(t)$, $0 \leq t < \infty$ is a continuous function of t.

In (ii) Φ denotes the standard normal distribution function. Occasionally it will be convenient to consider Brownian motion with drift μ and variance σ^2 per unit time. For this process (ii) is replaced by (ii′) $P_{\mu,\sigma}\{W(t) - W(s) \leq x\} = \Phi[(x - \mu(t - s))/\sigma(t - s)^{1/2}]$, $0 \leq s < t < \infty$. It is easy to see that (ii) and

(iii) are equivalent to the conditions that the joint distribution of $W(t_1), \ldots, W(t_n)$ is Gaussian, $E_\mu[W(t)] = \mu t$, and $\mathrm{Cov}(W(t_i), W(t_j)) = \min(t_i, t_j)$ for all n and $t_1, \ldots, t_n$.

If $x_1, x_2, \ldots$ are independent and normally distributed with mean μ and variance 1, and if $\{W(t), 0 \le t < \infty\}$ is a Brownian motion with drift μ, then $S_n = x_1 + \cdots + x_n$, $n = 0, 1, \ldots$ and $W(n)$, $n = 0, 1, \ldots$ have the same joint distributions. The reader not already familiar with Brownian motion should think in terms of this analogy: Brownian motion is an interpolation of the discrete time random walk S_n, $n = 0, 1, \ldots$ which preserves the Gaussian distribution and which makes the paths of the process continuous in time. To the extent that many random walks are approximately normally distributed for large n, the Brownian motion process may be used as an asymptotic approximation to a large class of random walks and hence of log likelihood ratios. For a more complete discussion of this point, see Appendix 1.

Although the intuitive meaning is usually clear, there are technical mathematical problems associated with the continuous time analogues of the basic concepts of Chapter II: the class $\mathscr{E}_t$ of random variables defined by $W(s)$ for $s \le t$, the likelihood ratio of $\{W(s), s \le t\}$ under P_{μ_1} relative to P_{μ_2}, the class of random variables prior to a stopping rule T, etc. For a systematic discussion of the relevant concepts, see Freedman (1971). Here we shall proceed by analogy with Chapter II and take the following results as our starting point (cf. (2.3), Proposition 2.18, Proposition 2.24, and Problem 2.11).

Let $\{W(t), 0 \le t < \infty\}$ be a Brownian motion with drift μ.

Proposition 3.1. *For any* $-\infty < \mu_i < \infty$ $(i = 1, 2)$, *and* $t > 0$, *the likelihood ratio of* $\{W(s), s \le t\}$ *under* P_{μ_1} *relative to* P_{μ_2} *is*

$$l(t, W(t); \mu_1, \mu_2) = \exp\left[(\mu_1 - \mu_2)W(t) - \frac{t}{2}(\mu_1^2 - \mu_2^2)\right].$$

Proposition 3.2 (Likelihood ratio identity). *For any* $-\infty < \mu_i < \infty$ $(i = 1, 2)$, *stopping rule* T *and* $Y \in \mathscr{E}_T$ *(the class of random variables prior to* T*),*

$$E_{\mu_1}(Y; T < \infty) = E_{\mu_2}\{Y l(T, W(T); \mu_1, \mu_2); T < \infty\};$$

in particular if $Y = I_A \in \mathscr{E}_T$

$$P_{\mu_1}(A\{T < \infty\}) = E_{\mu_2}[l(T, W(T); \mu_1, \mu_2); A\{T < \infty\}],$$

where l *is given in Proposition* 3.1.

Proposition 3.3 (Wald's identities). *For any stopping rule* T *with* $E_\mu(T) < \infty$,

$$E_\mu W(T) = \mu E_\mu T$$

and

$$E_\mu[(W(T) - \mu T)^2] = E_\mu T.$$

Remark. For a stopping rule T taking values in a finite set $0 \le t_1 < t_2 < \cdots < t_n \le \infty$, Propositions 3.2 and 3.3 are special cases of the corresponding discrete time results proved in Chapter II. The proofs in continuous time require a rather technical approximation argument involving a sequence of finite subsets of $[0, \infty)$ which become dense in the limit. An identical argument proves Proposition 3.12 given below. See Freedman (1971) for details.

2. Sequential Probability Ratio Test for the Drift of Brownian Motion

Let $\{W(t), 0 \le t < \infty\}$ be Brownian motion with drift μ. A sequential probability ratio test of H_0: $\mu = \mu_0$ against H_1: $\mu = \mu_1$ is defined by constants $A < 1 < B$, the stopping rule

$$T = \inf\{t\colon l(t, W(t); \mu_1, \mu_0) \notin (A, B)\}, \tag{3.4}$$

and the decision rule to reject H_0 if and only if $l(T, W(T); \mu_1, \mu_0) \ge B$. Here l is given by (3.1). Obviously T does not exceed the analogous stopping rule N in which observation of $W(t)$ is restricted to discrete instants $t = 1, 2, \ldots$, and this N is the stopping rule of a discrete time sequential probability ratio test for the mean of a normal distribution with unit variance. In particular $P_\mu\{T < \infty\} = 1$ for all μ. By the continuity of $W(t)$, $0 \le t < \infty$, and (3.4) $l(T, W(T); \mu_1, \mu_0)$ equals A or B with probability one, and hence the approximations of Chapter II become equalities here.

To summarize the important results it will be convenient to assume that $\mu_0 < \mu_1$ and introduct the notation $b = (\mu_1 - \mu_0)^{-1} \log B$, $a = (\mu_1 - \mu_0)^{-1} \log A$, and $\theta = \mu - \frac{1}{2}(\mu_1 + \mu_0)$. Then (3.4) may be re-written

$$T = \inf\{t\colon W(t) - \tfrac{1}{2}(\mu_1 + \mu_0)t \notin (a, b)\} \qquad (a < 0 < b), \tag{3.5}$$

and under P_μ, $W(t) - \frac{1}{2}(\mu_1 + \mu_0)t$ is a Brownian motion with drift θ. Propositions 3.2 and 3.3 allow us to repeat the arguments of Chapter II, but now with inequalities and approximations replaced by equalities to obtain

Theorem 3.6. *Let T be defined by* (3.5) *and set* $\theta = \mu - \frac{1}{2}(\mu_1 + \mu_0)$. *Then*

$$P_\mu\{W(T) - \frac{1}{2}(\mu_1 + \mu_0)T = b\} = \frac{1 - e^{-2a\theta}}{e^{-2b\theta} - e^{-2a\theta}} \qquad (\theta \ne 0)$$

$$= \frac{|a|}{|a| + b} \qquad (\theta = 0)$$

and

$$E_\mu(T) = [b(1 - e^{-2a\theta}) + a(e^{-2b\theta} - 1)]/\theta(e^{-2b\theta} - e^{-2a\theta}) \qquad (\theta \ne 0)$$

$$= |a|b \qquad (\theta = 0).$$

For the special case $a = -b$ these become

$$P_\mu\{W(T) - \tfrac{1}{2}(\mu_1 + \mu_0)T = b\} = \frac{1}{1 + e^{-2b\theta}} \qquad (\theta \neq 0)$$

$$= 1/2 \qquad (\theta = 0)$$

and

$$E_\mu(T) = \frac{b}{\theta}\left(\frac{1 - e^{-2b\theta}}{1 + e^{-2b\theta}}\right) \qquad (\theta \neq 0)$$

$$= b^2 \qquad (\theta = 0).$$

The heuristic considerations of optimality in II.4 become complete proofs for Brownian motion. For the stopping rule $\tilde{T}$ of any sequential test of H_0: $\mu = \mu_0$ against H_1: $\mu = \mu_1$, the lower bounds for $E_{\mu_i}(\tilde{T})$ $(i = 0, 1)$ in terms of the error probabilities given in Theorem 2.39 go through unchanged in the present context. Since the relations of Theorem 3.6 are exact, the argument of II.4 applied here shows that a sequential probability ratio test achieves the lower bound on both $E_{\mu_0}(\tilde{T})$ and $E_{\mu_1}(\tilde{T})$ for the given error probabilities.

3. Truncated Sequential Tests

The absence of a definite upper bound on the stopping rule of a sequential probability ratio test presents difficulties in application, and hence it is natural to consider the effect of truncation on the properties of the test. Since the probability calculations associated with truncated sequential probability ratio tests are complicated in certain details, we shall begin by introducing a class of one-sided stopping rules which allow us to introduce a number of important concepts in the simplest possible context.

Let $\{W(t), 0 \leq t < \infty\}$ be Brownian motion with drift μ and consider the problem of testing H_0: $\mu \leq \mu_0$ against H_1: $\mu > \mu_0$. Suppose also that for μ considerably larger than μ_0, say $\mu \geq \mu_1 > \mu_0$, the collection of data is expensive and a small sample size is desirable, but if $\mu \leq \mu_0$ data are relatively inexpensive, so a larger (perhaps fixed) sample would be appropriate.

It is easy to imagine a variety of experimental situations for which these considerations may be relevant. In quality control, if testing is destructive it may be desirable to minimize the amount of inspection of a lot which will ultimately be accepted, but considerably less important to have a small sample from a rejected lot. In demonstrating reliability before beginning mass production, rejection of H_0 may indicate a return to research and development, which should begin as soon as possible. However, if H_0 is accepted and mass production begun, one may want a larger amount of data to estimate μ accurately, so that predictions of future reliability or comparisons of field reliability of mass produced items with experimental reliability of prototypes may be more adequately assessed. From a different viewpoint Butler and Lieberman (1981) have argued that an "early accept" option in problems of acceptance sampling may be used as an incentive to producers to maintain high quality or reliability.

Suppose that in a clinical trial $W(t)$ measures the cumulative difference in response between patients receiving a new treatment and those receiving a standard treatment (or placebo), H_0: $\mu = 0$ denotes the null hypothesis of no difference between treatments, and H_1: $\mu > 0$ indicates the superiority of the new treatment. If H_1 is true, a small sample is desirable, so that all future patients can receive the new treatment, but from the patients' viewpoint the trial can continue forever if H_0 is ture. (In clinical trials one is often interested in the two-sided alternative H_1: $\mu \neq 0$, which allows for the possibility that either treatment may be superior. It will turn out that a reasonable solution to the one-sided problem can be adapted to the two-sided version. See Remark 3.18 and Chapter IV.)

A different possibility in clinical trials is that patient response is very complicated, and a large trial seems advisable to answer a variety of different questions. Suppose, however, that a new treatment may have severe side effects. Let μ denote the mean difference in side effects between the new treatment and the standard. Under H_0: $\mu = 0$ the fear of these side effects is unfounded, and the trial can proceed to its conclusion. However, large values of μ may be sufficiently serious that the trial should be terminated early.

The reverse situation arises in drug screening experiments on animals and to some extent in phase II clinical trials. Here one is interested in estimating a treatment effect (or perhaps a toxicity level) if a substantial one exists, and for this a large sample size is desirable. However, if the null hypothesis of no treatment effect appears to be true, one wants to minimize the cost of experimentation by stopping the experiment as soon as possible.

One possible approach to these problems is a test designed to perform like a sequential probability ratio test under H_1 and a fixed sample test under H_0.

Let $\mu_1 > \mu_0$ and for $B > 1$ define the stopping rule

$$\tau = \inf\{t\colon l(t, W(t); \mu_1, \mu_0) \geq B\}.$$

Here l is the likelihood ratio given in Proposition 3.1. With the notation $b = (\log B)/(\mu_1 - \mu_0)$ and $\eta = \frac{1}{2}(\mu_0 + \mu_1)$ the definition of τ can be re-written

$$\tau = \inf\{t\colon W(t) \geq b + \eta t\}. \tag{3.7}$$

Given $m > 0$ and $c < b + \eta m$ we shall consider the test of H_0: $\mu \leq \mu_0$ against H_1: $\mu > \mu_0$ defined by the following: stop sampling at $\tau \wedge m = \min(\tau, m)$; reject H_0 if either $\tau \leq m$ or $\tau > m$ and $W(m) > c$, and otherwise do not reject H_0. The most obvious choice of c is $b + \eta m$, which means that one rejects H_0 if and only if $\tau \leq m$. However, it is no more difficult and will prove useful to consider the more general class of procedures defined above.

The power of this test is

$$P_\mu\{\tau \le m\} + P_\mu\{\tau > m, W(m) > c\} \\ = P_\mu\{W(m) > c\} + P_\mu\{\tau < m, W(m) \le c\}. \tag{3.8}$$

The essential ingredient in computing (3.8) is the conditional probability

$$P_\mu\{\tau < m | W(m) = \xi\} \qquad (\xi < b + \eta m), \tag{3.9}$$

from which (3.8) may be obtained by unconditioning. Since

$$P_\mu\{W(m) \in d\xi\} = \phi[(\xi - m\mu)/m^{1/2}]\, d\xi/m^{1/2}$$

we have

$$P_\mu\{\tau < m, W(m) \le c\} = \int_{-\infty}^{c} P_\mu\{\tau < m | W(m) = \xi\}\phi[(\xi - m\mu)/m^{1/2}]\, d\xi/m^{1/2}. \tag{3.10}$$

The usual argument for computing first passage distributions for Brownian motion uses the "reflection principle" (e.g. Karlin and Taylor, 1975, p. 345 ff.). To illustrate the basic similarity with untruncated sequential probability ratio tests, we shall use a likelihood ratio argument.

First note that since $W(m)$ is sufficient for μ, the conditional probability in (3.9) does not depend on μ. We consider a family of probabilities parameterized by ξ in (3.9). A more suitable notation is to write

$$P_\xi^{(m)}(A) = P_0\{A | W(m) = \xi\} \qquad (A \in \mathscr{E}_m).$$

Direct computations show that for all $\xi_0 \ne \xi_1$, $t < m$, and $0 < t_1 < t_2 < \cdots < t_k = t$, the likelihood ratio of $\{W(t_i), i \le k\}$ under $P_{\xi_0}^{(m)}$ relative to $P_{\xi_1}^{(m)}$ is (cf. Appendix 1)

$$l^{(m)}(t, W(t); \xi_0, \xi_1) = \exp\left\{\left[(\xi_0 - \xi_1)W(t) - \frac{t}{2m}(\xi_0^2 - \xi_1^2)\right]\Big/(m - t)\right\}. \tag{3.11}$$

Standard arguments now show that (3.11) gives the likelihood ratio of the entire path $\{W(s), 0 \le s < t\}$. For these conditional Brownian motions the basic likelihood ratio identity analogous to Proposition 2.24 and Proposition 3.2 is given by

Proposition 3.12. *For any* $-\infty < \xi_0 \ne \xi_1 < \infty$, $m > 0$, *stopping rule* T *and event* $A \in \mathscr{E}_T$,

$$P_{\xi_0}^{(m)}(A \cap \{T < m\}) = E_{\xi_1}^{(m)}[l^{(m)}(T, W(T); \xi_0, \xi_1); A \cap \{T < m\}],$$

where $l^{(m)}$ *is given explicitly in* (3.11).

Consider τ defined by (3.7). For $\xi > b + \eta m$, $P_\xi^{(m)}\{\tau < m\} = 1$. For $\xi = \xi_0 < b + \eta m$ let $\xi_1 = 2(b + \eta m) - \xi_0$ denote ξ_0 reflected about $b + \eta m$, so $\xi_0 - \xi_1 = -2(b + \eta m - \xi)$, $\frac{1}{2}(\xi_0 + \xi_1) = b + \eta m$, and by (3.11)

$$l^{(m)}(t, W(t); \xi, \xi_1) = \exp\{-2(b + \eta m - \xi)[W(t) - (b + \eta m)t/m]/(m - t)\}.$$

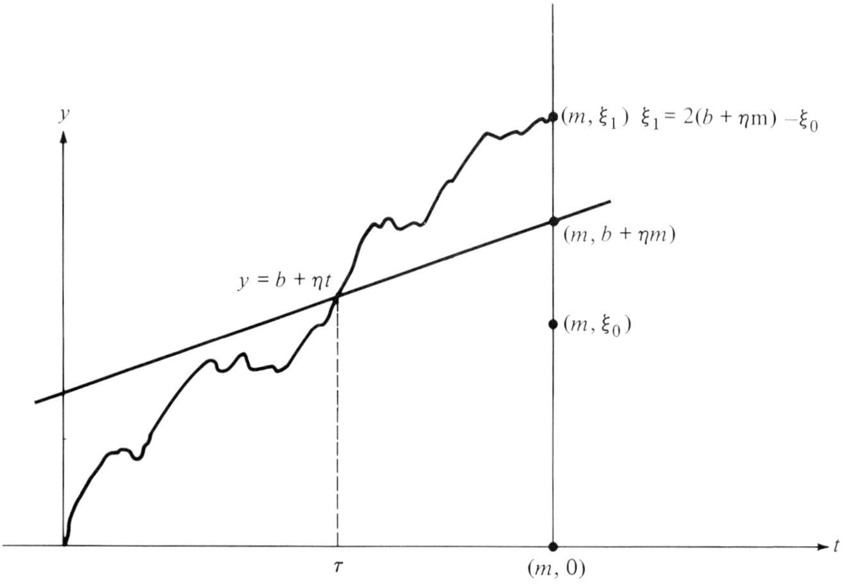

Figure 3.1.

Since $W(\tau) = b + \eta\tau$ if $\tau < m$, Proposition 3.12 yields

$$P_{\xi}^{(m)}\{\tau < m\} = \exp\{-2b(b + \eta m - \xi)/m\} P_{\xi_1}^{(m)}\{\tau < m\} = \exp\{-2b(b + \eta m - \xi)/m\}. \tag{3.13}$$

(See Figure 3.1.) Substituting (3.13) into (3.10) gives

$$P_\mu\{\tau < m, W(m) < c\} = e^{2b(\mu-\eta)}\Phi[(c - 2b)/m^{1/2} - \mu m^{1/2}]. \tag{3.14}$$

For the special case $c = b + \eta m$ (3.14) together with (3.8) leads to

$$P_\mu\{\tau \le m\} = 1 - \Phi(bm^{-1/2} - (\mu - \eta)m^{1/2}) + e^{2b(\mu-\eta)}\Phi(-bm^{-1/2} - (\mu - \eta)m^{1/2}). \tag{3.15}$$

Differentiation of (3.15) gives the probability density function

$$P_\mu\{\tau \in dt\} = bt^{-3/2}\phi(bt^{-1/2} - (\mu - \eta)t^{1/2})\, dt. \tag{3.16}$$

Given (3.16), the calculation of $E_\mu(\tau \wedge m)$ is straightforward in principle, albeit somewhat tedious in detail (see Problem 3.1). The following argument is more complicated than the present case requires, but it can be adapted to problems in discrete time and to nonlinear boundaries. (see Theorem 4.27).

Consider first the special case $\eta = 0$, $\mu > 0$. Obviously

$$E_\mu(\tau \wedge m) = E_\mu(\tau) - E_\mu(\tau - m; \tau > m).$$

By conditioning on $\tau > m$ and $W(m)$ we have

Table 3.1. Linear Boundary: $b = 6.05$; $\eta = .3$; $m = 49$; $\alpha = .025$ (One-Sided Test)

μ	Power (3.15)	$E_\mu(\tau \wedge m)$ (3.17)	Power of Fixed Sample Test ($m = 49$)	$m(\mu)$
.7	.99	15	1.00	38
.6	.95	19	.99	36
.4	.63	32	.80	33
.3	.39	39	.56	31
.2	.19	44	.29	29

$$E_\mu(\tau \wedge m) = E_\mu(\tau) - \int_{(-\infty,b)} E_\mu(\tau - m|\tau > m, W(m) = \xi)P_\mu\{\tau > m, W(m) \in d\xi\}.$$

By Wald's identity $E_\mu(\tau) = b/\mu$ and $E_\mu(\tau - m|\tau > m, W(m) = \xi) = (b - \xi)/\mu$. By (3.14)

$$P_\mu\{\tau > m, W(m) \in d\xi\} = \{\phi(\xi m^{-1/2} - \mu m^{1/2}) - e^{2b\mu}\phi[(\xi - 2b)m^{-1/2} - \mu m^{1/2}]\}\, d\xi/m^{1/2}.$$

Hence direct calculation yields

$$E_\mu(\tau \wedge m) = b\mu^{-1}[P_\mu\{W(m) \geq b\} - P_\mu\{\tau < m, W(m) < b\}] + mP_\mu\{\tau > m\}.$$

The case $\mu < 0$ is somewhat different. The appropriate decomposition is

$$E_\mu(\tau \wedge m) = E_\mu(\tau; \tau < \infty) - E_\mu(\tau - m; m < \tau < \infty) + mP_\mu\{\tau = \infty\}.$$

By Propositions 3.2 and 3.3

$$E_\mu(\tau; \tau < \infty) = E_{|\mu|}(\tau e^{-2|\mu| W(\tau)}; \tau < \infty) = e^{-2|\mu|b}E_{|\mu|}\tau = b|\mu|^{-1}e^{-2|\mu|b}.$$

Similar arguments to evaluate $E_\mu(\tau - m; m < \tau < \infty)$ and $P_\mu\{\tau < \infty\}$ (or see (3.15)) yield $E_\mu(\tau \wedge m)$ in the case $\mu < 0$. The case $\mu = 0$ follows from the others by continuity. The results for all μ can be be summarized by

$$\begin{aligned} E_\mu(\tau \wedge m) &= (b/|\mu|)e^{b(\mu - |\mu|)}P_{|\mu|}\{\tau \leq m\} + mP_\mu\{\tau < m\} \\ &\quad - (2b/|\mu|)e^{b(\mu + |\mu|)}\Phi(-bm^{-1/2} - |\mu|m^{1/2}) \qquad (\mu \neq 0), \\ &= mP_0\{\tau > m\} + 2b[m^{1/2}\phi(bm^{-1/2}) - b\Phi(-bm^{-1/2})] \qquad (\mu = 0), \end{aligned} \tag{3.17}$$

where $P_\mu\{\tau \leq m\}$ is evaluated in (3.15). For $\eta \neq 0$, $E_\mu(\tau \wedge m)$ is given by (3.17) with μ replaced by $\mu - \eta$.

For a numerical example let $b = 6.05$, $\eta = .3$, $m = 49$, and $c = b + \eta m = 20.75$. These parameters give a significance level $\alpha = .025$ for $\mu_0 = 0$ and power .95 at $\mu_1 = 2\eta = .6$. Table 3.1 gives the power function and expected sample size of this test. Also included are two comparisons with fixed sample size tests. The first gives the power of a .025 level fixed sample test with sample

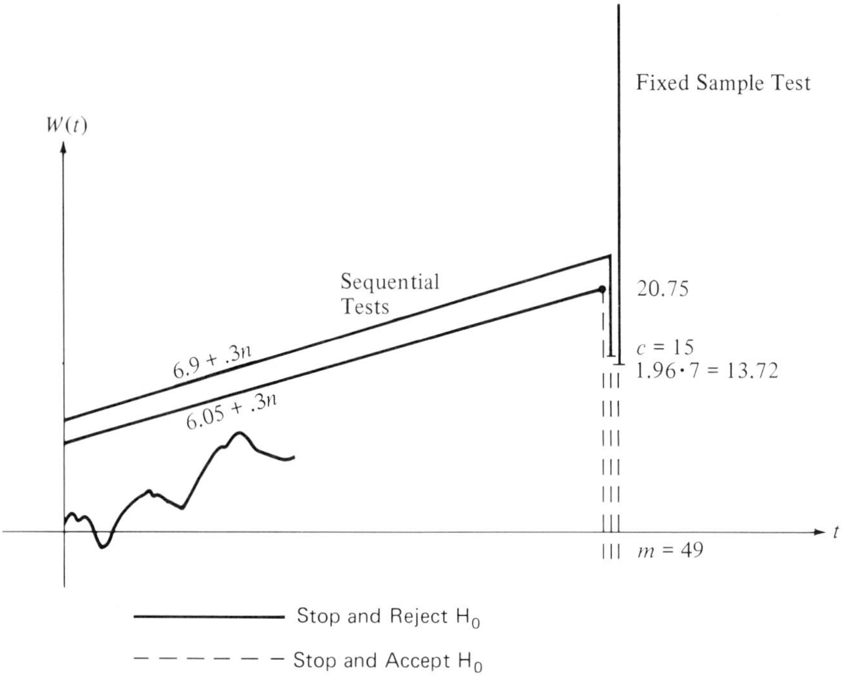

Figure 3.2.

size $m = 49$. The second gives that fixed sample size $m(\mu)$ which would yield the same power at μ as our sequential test (for $\alpha = .025$). The first comparison shows the sequential test trading a loss of power for a decrease in expected sample size. The second shows that the sequential test has an expected sample size which is smaller for large μ and larger for small μ than any of the several corresponding fixed sample tests.

Either a loss of power or an increase in the maximum sample size of a sequential test with respect to a fixed sample test is a serious disadvantage in some applications. For large values of μ the primary contribution to (3.8) is $P_\mu\{W(m) > c\}$, which increases as c decreases. This suggests the possibility of choosing a somewhat larger value of b and a suitable $c < b + \eta m$ in an attempt to obtain a maximum sample size and power function more like a given fixed sample test while maintaining an expected sample size function resembling that of the sequential test. See Figure 3.2 and Table 3.2 for an example. Since for the composite alternative H_1: $\mu > 0$, the value of μ_1, hence of $\eta = \frac{1}{2}\mu_1$ is somewhat arbitrary, an alternative possibility is to use a smaller value of η. An intuitively appealing choice is $\eta = 0$ (see Problem 3.17). A more thorough discussion of this issue is contained in IV.4.

Remark 3.18. In sequential clinical trials one frequently wants to consider the two treatments symmetrically and hence to have a two-sided test of H_0: $\mu = 0$

Table 3.2. Linear Boundary: $b = 6.9$; $c = 15$; $\eta = .3$; $m = 49$; $\alpha = .025$ (One-Sided)

μ	Power (3.8) and (3.14)	$E_\mu(\tau \wedge m)$ (3.17)
.7	1.00	17
.6	.98	22
.4	.76	35
.3	.51	41
.2	.26	46

against $H_1: \mu \neq 0$. If we are satisfied with the test discussed above for a one-sided problem, it is easy to splice together two such tests for the corresponding two-sided problem. For example, define

$$T = \inf\{t: |W(t)| \geq b + \eta t\},$$

stop sampling at $\min(T, m)$ and reject H_0 if any only if $T \leq m$. If we denote the significance level of the corresponding one-sided test by α, the significance level of the two-sided test is less than 2α and usually will be very close to 2α. The difference will involve sample paths which cross both the upper and the lower stopping boundary before time m, and at least whenever $\eta \geq 0$ this probability is less than α^3, which will be small compared to α. Similarly one can usually approximate the power and expected sample size when μ is not too close to 0 by appropriate calculations for the one-sided case (see Sections 6 and 7).

Sections 4 and 6 of Chapter IV attempt a critical discussion of the adequacy of these and the related repeated significance tests for sequential clinical trials.

Remark 3.19. The parameter values of the preceding numerical example were selected to be in an intermediate range between small and larger sample sizes. However, one may use the (easily proved) fact that for each $\gamma > 0$ $\{\gamma^{-1/2} W(\gamma t), 0 \leq t < \infty\}$ is a new Brownian motion with drift $\gamma^{1/2}\mu$ to construct with essentially no additional computations a family of related examples. For example, putting $\gamma^{1/2} = \frac{2}{3}$ and replacing b by $b/\gamma^{1/2}$, η by $\eta\gamma^{1/2}$, and m by m/γ gives a new test defined by $b' = (6.05)(1.5) = 9.08$, $\eta' = .2$, and $m' = 110$, which has significance level $\alpha = .025$, power $= .95$ at $\mu' = (.6)(\frac{2}{3}) = .4$, expected sample size $(19)(3/2)^2 \cong 43$ at $\mu' = .4$, etc.

4. Attained Significance Level and Confidence Intervals

Comparisons of the power functions and expected sample size functions of (sequential) tests are basically considerations of experimental design. They provide guidance in planning an experiment. In some cases (for example, acceptance sampling for quality control), there may be no other considerations. But if the purpose of performing an experiment is to make inferences about a probability model for the data, a simple accept-reject dichotomy may not seem sufficiently informative. In this section we consider attained significance levels and confidence intervals as means of making more informative inferential statements.

If in a fixed sample test a null hypothesis H_0 is rejected for large values of a statistic S, and if $\bar{F}_0(s)$ denotes the maximum probability under H_0 that $S \geq s$, then the attained significance level or p-value of the observation S is defined to be $\bar{F}_0(S)$. This is the smallest significance level at which H_0 is rejected by the datum S. The ingredients of the p-value are the ordering of points of the sample space indicated by taking larger values of S as more evidence against H_0 than small values and the probability calculation involved in evaluating $\bar{F}_0$.

Consider now a Brownian motion $\{W(t), 0 \leq t < \infty\}$ with drift μ and a test of the one-sided hypotheses H_0: $\mu \leq \mu_0$ against H_1: $\mu > \mu_0$ defined by the stopping rule T (e.g. $T = \tau \wedge m$, where τ is defined by (3.7)). A natural ordering of the outcome $(T, W(T))$ is in terms of the value of $W(T)/T$: large values of $W(T)/T$ are considered to be more evidence against H_0 than small values. If T is defined to be the time at which the Brownian process exits from a region R in the half-plane $\{(t, w)\colon t \geq 0, -\infty < w < \infty\}$, this is often tantamount to ordering the points on the boundary of R to be increasing as one moves around the boundary in a counterclockwise direction. Then the probability calculation involves the event that W exits from R through a "larger" point on the boundary than the observed point $(T, W(T))$.

For example, consider the test of Section 3 defined by the stopping rule $\tau \wedge m$, where τ is given by (3.7). Ordering the observation $(\tau \wedge m, W(\tau \wedge m))$ according to $W(\tau \wedge m)/\tau \wedge m$ (with large values giving more evidence against H_0) is equivalent to the following prescription. If $\tau = t < m$ and hence $W(\tau \wedge m) = b + \eta\tau$, the attained significance level is

$$\sup_{\mu \leq \mu_0} P_\mu\{\tau \leq t\} = P_{\mu_0}\{\tau \leq t\},$$

which is evaluated in (3.15). If $\tau > m$ and $W(\tau \wedge m) = c < b + \eta m$, the attained significance level is

$$\sup_{\mu \leq \mu_0} [P_\mu\{\tau \leq m\} + P_\mu\{\tau > m, W(m) \geq c\}]$$
$$= P_{\mu_0}\{\tau \leq m\} + P_{\mu_0}\{\tau > m, W(m) \geq c\},$$

which can be evaluated by using (3.8) and (3.14).

For example, for the test of Table 3.1, if $\tau = 16$ the attained significance level is $P_0\{\tau \leq 16\} = .013$. Since the overall significance level of the test is $P_0\{\tau \leq 49\} = .025$, we see that termination of the test at approximately $\frac{1}{3}$ the maximum sample size yields a p-value more than half as large as the significance level of the test. The evidence against H_0 as measured by the p-value is perhaps not as persuasive as early termination of the test ought to indicate. To some extent this is in the nature of a sequential test which is designed to stop sampling as soon as the evidence against H_0 is strong (but before it becomes overwhelming). We return to this point in IV.4.

For the sequential probability ratio test defined by (3.5) and used to test $H_0\colon \mu \leq \mu_0$ against $H_1\colon \mu > \mu_0$, a similar ordering of the points on the stopping boundary seems appropriate, but the probability calculation is more complicated and will be considered in Section 6.

We now turn to the related but more complicated problem of estimation. It will be apparent that the main ideas are more generally applicable, but we begin with the very specific stopping rule τ defined by (3.7).

Suppose that quite apart from any consideration of hypothesis testing, we would like to estimate μ based on observation of $(\tau \wedge m, W(\tau \wedge m))$. In a sequential clinical trial where positive values of μ indicate a beneficial effect of the experimental treatment, estimation may be of greatest interest when $\mu > 0$. On the other hand, the stopping rule $\tau \wedge m$ may have seemed attractive in the first place because we were particularly interested in estimating μ if μ is small; and the larger sample size for small μ should permit more accurate estimation.

For the most part we shall consider estimation by confidence intervals, but we begin with some remarks about point estimation. For general problems of point estimation one must consider simultaneously the bias and the variance of a proposed estimator. For most simple problems attention naturally focuses on estimators which are unbiased, or almost unbiased, and then one can concentrate on the variance as a measure of goodness of an estimator. For the sequential problems we discuss in most of this book, however, problems of bias will be very important.

For any stopping rule T for Brownian motion $\{W(t), 0 \leq t < \infty\}$ the likelihood function for μ may be taken to be

$$l(T, W(T); \mu, 0) = \exp\{\mu W(T) - \mu^2 T/2\},$$

and hence the maximum likelihood estimator of μ is $\hat{\mu}(T) = W(T)/T$. For any fixed sample size $T \equiv t$, $\hat{\mu}(t)$ is unbiased. However, Starr and Woodroofe (1968) have shown for essentially any curve $c(t)$ with $c(0) > 0$ and stopping rule T defined by $T = \inf\{t\colon W(t) = c(t)\}$, if $P_\mu\{T < \infty\} = 1$, then $E_\mu[W(T)/T] > \mu$.

For τ defined by (3.7) one can use (3.14)–(3.16) to calculate the bias of $W(\tau \wedge m)/\tau \wedge m$ exactly. However, the result is rather complicated, and to see that the bias can be substantial, we shall consider only the extreme case $m = \infty$. Then for $\mu \geq \eta$

$$E_\mu[W(\tau)/\tau] = bE_\mu(1/\tau) + \eta = b\int_0^\infty E_\mu e^{-\lambda\tau}\, d\lambda + \eta,$$

and some calculation (cf. Problem 3.1) shows that

$$E_\mu[W(\tau)/\tau] = \mu + b^{-1} \qquad (\mu \geq \eta). \tag{3.20}$$

For μ large enough that $P_\mu\{\tau > m\}$ is approximately 0 the bias of $W(\tau \wedge m)/\tau \wedge m$ is about b^{-1}, which in practice can be significant. For the numerical example of Table 3.1 with $b = 6.05$, the bias is about .16; i.e., for $\mu = .7$ it is about $.25\mu$. Since the expected sample size in this case is 15, a crude assessment of the standard error of an estimator of μ would be $1/\sqrt{15} \cong .26$. Hence the bias appears to be more than 50% as large as the standard error of estimation. Of course the bias would be smaller for larger values of b and vice versa.

When μ is small, so that $P_\mu\{\tau \leq m\}$ is approximately 0, $W(\tau \wedge m)/\tau \wedge m$ is essentially just $W(m)/m$, and hence is almost unbiased with variance m^{-1}. An unbiased estimator of μ based on the observation of $(\tau \wedge m, W(\tau \wedge m))$ has been given by Ferebee (1980), who discusses several other stopping rules as well.

Consider now the problem of finding a confidence interval for μ based on $(\tau \wedge m, W(\tau \wedge m))$. According to a theorem of Anscombe (1952)—see II.5—$(\tau \wedge m)^{1/2}[\hat{\mu}(\tau \wedge m) - \mu]$ has approximately a standard normal distribution when m and b are both large. This permits one to give an approximate confidence interval for μ, which in effect disregards completely the sequential nature of the sampling scheme. The previous discussion of bias indicates that one should be somewhat skeptical of this procedure for moderate sample sizes. Consequently we shall consider a method for obtaining an exact interval.

We begin by recalling the standard method for obtaining a one-sided confidence bound for the parameter of a one-parameter exponential family (cf. Cox and Hinkley, 1974, p. 212 or Lehmann, 1959, p. 79). For a normal mean it seems unnecessarily complicated, but it also works in the sequential case, where appropriate exact pivotal quantities seem not to exist. Let $0 < \gamma < \frac{1}{2}$, and define

$$\mu_*(x) = \inf\{\mu\colon P_\mu\{W(m) \geq x\} \geq \gamma\}.$$

Then $\mu_*(W(m))$ is a $(1 - \gamma)100\%$ lower confidence bound, i.e.

$$P_\mu\{\mu_*[W(m)] \leq \mu\} = 1 - \gamma$$

for all $-\infty < \mu < \infty$. Similarly $\mu^*(W(m))$ is a $(1 - \gamma)100\%$ upper confidence bound, where

$$\mu^*(x) = \sup\{\mu\colon P_\mu\{W(m) \leq x\} \geq \gamma\},$$

and $[\mu_*(W(m)), \mu^*(W(m)]$ is a $(1 - 2\gamma)100\%$ confidence interval for μ. This method in effect orders points in the sample space of $W(m)$ as being more or less consistent with a given μ and excludes from a lower one sided interval (upper one sided interval) those μ so small (large) that the probability under μ of obtaining a larger (smaller) value of $W(m)$ than the observed one is less than the required γ.

This method can be applied directly to data obtained by a sequential stopping rule once we agree on the ordering of points in the sample space. A reasonable ordering is the one defined by $\hat{\mu}$, which we have already considered in our discussion of attained significance levels. For the specific stopping rule

Table 3.3. 90% Confidence Interval: $b = 6.05$; $\eta = .3$; $m = 49$; $\gamma = .05$

Data	$\hat{\mu}(\tau \wedge m) \pm 1.645/(\tau \wedge m)^{1/2}$	Exact Interval
$\tau = 19$	(.24, 1.00)	(.13, .93)
$\tau > 49$, $W(49) = 16.7$	(.11, .58)	(.06, .56)

$\tau \wedge m$ with τ given by (3.7), this leads to the following definitions. For $0 < t \leq m$ let

$$\mu_{*1}(t) = \inf\{\mu: P_\mu\{\tau \leq t\} \geq \gamma\} \tag{3.21}$$

and

$$\mu_1^*(t) = \sup\{\mu: P_\mu\{\tau \geq t\} \geq \gamma\}.$$

For $-\infty < x < b + \eta m$ let

$$\mu_{*2}(x) = \inf\{\mu: P_\mu\{\tau \leq m\} + P_\mu\{\tau > m, W(m) \geq x\} \geq \gamma\} \tag{3.22}$$

and

$$\mu_2^*(x) = \sup\{\mu: P_\mu\{\tau > m, W(m) \leq x\} \geq \gamma\}. \tag{3.23}$$

Then define

$$\begin{aligned} \mu_*(\tau \wedge m, W(\tau \wedge m)) &= \mu_{*1}(\tau) && \text{if } \tau \leq m \\ &= \mu_{*2}(W_m) && \text{if } \tau > m \end{aligned} \tag{3.24}$$

and

$$\begin{aligned} \mu^*(\tau \wedge m, W(\tau \wedge m)) &= \mu_1^*(\tau) && \text{if } \tau \leq m \\ &= \mu_2^*(W_m) && \text{if } \tau > m. \end{aligned} \tag{3.25}$$

It is easy to see that $(\mu_*, +\infty)$ gives a $(1-\gamma)100\%$ lower confidence limit for μ, $(-\infty, \mu^*)$ a $(1-\gamma)100\%$ upper limit, and (μ_*, μ^*) a $(1-2\gamma)100\%$ confidence interval.

To verify, for example, that μ_* gives a lower confidence bound, note that μ_{*1} is continuous and strictly decreasing, and $P_{\mu_{*1}(m)}\{\tau \leq m\} = \gamma$. Also μ_{*2} is continuous, strictly increasing, and $\mu_{*2}(b + \eta m) = \mu_{*1}(m)$. Hence for $\mu \geq \mu_{*1}(m)$, $\mu_* > \mu$ if and only if $\mu_{*1}(\tau) > \mu$, so by (3.21)

$$P_\mu\{\mu_* > \mu\} = P_\mu\{\tau \leq \mu_{*1}^{-1}(\mu)\} = \gamma.$$

For $\mu < \mu_{*1}(m)$, $\mu_* > \mu$ if either $\tau \leq m$ or $\tau > m$ and $\mu_{*2}(W_m) > \mu$, so by (3.22)

$$P_\mu\{\mu_* > \mu\} = P_\mu\{\tau \leq m\} + P_\mu\{\tau > m, W_m \geq \mu_{*2}^{-1}(\mu)\} = \gamma.$$

Table 3.3 gives a numerical example which compares the naive asymptotic interval $\hat{\mu}(\tau \wedge m) \pm z_\gamma/(\tau \wedge m)^{1/2}$ with the interval defined by (3.24) and (3.25). Notice that the exact interval has both endpoints to the left of those of the naive interval, which reflects the positive bias of $\hat{\mu}$. The left endpoint is shifted

Table 3.4. Probability of Non-Coverage for $W(\tau \wedge m)/\tau \wedge m \pm z_\gamma/(\tau \wedge m)^{1/2}$: $b = 6.05$; $\eta = .3$; $m = 49$; $\gamma = .05$

μ	Lower Endpoint	Upper Endpoint	Total
.7	.077	.023	.10
0	.062	.050	.11

more than the right endpoint, so the exact intervals are slightly longer than the naive intervals. Finally, these adjustments are greater when $\tau \leq m$ than when $\tau > m$.

A second illuminating comparison comes from computing the exact coverage probability of the naive interval. Let z_γ denote the $1 - \gamma$ percentile of the standard normal distribution ($z_{.05} = 1.645$). The probability that the lower endpoint of the naive interval exceeds μ is

$$P_\mu\{W(\tau \wedge m)/\tau \wedge m - z_\gamma/(\tau \wedge m)^{1/2} > \mu\}$$
$$= P_\mu\{\tau \leq m, b\tau^{-1} - z_\gamma \tau^{-1/2} > (\mu - \eta)\} + P_\mu\{\tau > m, W(m) - m\mu > z_\gamma m^{1/2}\}.$$
(3.26)

For $\mu \geq \eta + (bm^{-1} - z_\gamma m^{-1/2})^+$ the second probability on the right hand side of (3.26) is zero and the first can be re-written as $P_\mu\{\tau \leq m_1 \wedge m\}$, where $m_1^{1/2} = \{-z_\gamma + [z_\gamma^2 + 4b(\mu - \eta)]^{1/2}\}/2(\mu - \eta)$. For other values of μ one can obtain related but more complicated expressions, which can be evaluated using the results of Section 3. A similar calculation gives the probability of non-coverage at the upper endpoint of the interval. Table 3.4 gives some numerical results.

Qualitatively these figures are not surprising. Because of the positive bias of $\hat{\mu}$ the non-coverage probability at the lower endpoint is greater than the nominal .05, and at the upper endpoint it is less. For large μ the discrepancies are greater than for small μ because the stopping rule τ plays a relatively greater role and truncation at m a relatively lesser role.

There are also two interesting peculiarities of the stopping rule τ defined by (3.7), which are not true in general. (i) For $\mu \geq \eta + bm^{-1} + z_\gamma m^{-1/2}$ the total non-coverage probability of the naive interval is exactly the nominal 2γ because the discrepancies at the two endpoints cancel (see Problem 3.16). (ii) For an interval of values of μ which are less than η it can happen that the set of values $(\tau \wedge m, W(\tau \wedge m))$ for which the lower endpoint of the naive interval does not cover the true value of μ has the following unexpected form: $\tau < m_1$, or $m_2 < \tau \leq m$, or $\tau > m$ and $W(m) > c$ for some $m_1 < m_2 < m$ and $c < b + \eta m$. This possibility occurs when small values of τ lead to non-coverage because $\hat{\mu}$ is too large, intermediate values lead to coverage, but larger values again lead to non-coverage because the assessment of accuracy, $z_\gamma/(\tau \wedge m)^{1/2}$, is too small.

Other examples of interval estimation relative to sequential tests are given in III.6 and IV.5.

5. Group Sequential Tests and the Accuracy of Brownian Approximations

Brownian motion makes it possible to discuss a large number of important concepts without getting lost in difficult probability calculations. However, Brownian approximations are frequently not especially accurate. The most obvious reason is that the test statistics under consideration may not be approximately normally distributed, but even for exactly normal data, the discreteness of the time scale can have a substantial effect. (In some cases these two factors cancel and make the Brownian approximation look very good, but this should be regarded as a fortunate accident and not something to be expected.) The effect of discrete time is especially important whenever data are grouped and the nth "observation" x_n is actually a sufficient statistic summarizing the data in the nth group.

Suppose that $x_1, x_2, \ldots$ are independent and normally distributed with mean μ and variance 1. Let $S_n = x_1 + \cdots + x_n$, and in analogy with (3.7) put

$$\tilde{\tau} = \inf\{n: S_n \geq b + \eta m\} \tag{3.27}$$

Consider the test of $H_0: \mu \leq 0$ against $H_1: \mu > 0$ which stops sampling at $\min(\tilde{\tau}, m)$ and rejects H_0 if and only if $\tilde{\tau} \leq m$. The power function of this test is $P_\mu\{\tilde{\tau} \leq m\}$, for which $P_\mu\{\tau \leq m\}$ given by (3.15) is a convenient approximation (see Appendix 1).

In order to check the accuracy of this and other approximations, one may compute $P_\mu\{\tilde{\tau} \leq m\}$ numerically. Let $f_n(x) = dP_\mu\{\tilde{\tau} > n, S_n \leq x\}/dx$ for $x < b + \eta n$. Then $f_1(x) = \phi(x - \mu)$ $(x < b + \eta)$ and for $n > 1$

$$f_n(x) = \int_{-\infty}^{b+\eta(n-1)} f_{n-1}(y)\phi(x - y - \mu)\,dy.$$

Also

$$P_\mu\{\tilde{\tau} = n\} = \int_{-\infty}^{b+\eta(n-1)} f_{n-1}(y)[1 - \Phi(b + \eta n - y - \mu)]\,dy$$

$$\text{for all } n = 1, 2, \ldots.$$

This algorithm may be easily adapted to other stopping rules. It has been applied by several authors in order to study various sequential tests numerically (e.g. Armitage *et al.*, 1969; Pocock, 1977; Aroian and Robison, 1969; Samuel-Cahn, 1974).

Table 3.5 compares the Brownian approximation to $P_\mu\{\tilde{\tau} \leq m\}$ and $E_\mu(\tilde{\tau} \wedge m)$ given by (3.15) and (3.17) with the exact values computed numerically by Samuel-Cahn (1974). To allow for the possibility that the data are examined periodically rather than after each new observation, one should think of m as the maximum number of (equally spaced) times a sequence of observations may be examined. Even for small m the actual sample size may be large if the data are grouped and the "observation" x_n denotes the sum of the

Table 3.5. Approximations to $P_\mu\{\tilde{\tau} \le m\}$ and $E_\mu(\tilde{\tau} \wedge m)$ ($\eta = 0$ in all cases)

			Brownian Approx.		Exact Value[a]		Modified Approx.	
b	m	μ	P	E	P	E	P	E
3.68	5	0	.10	4.8	.063	4.8	.065	4.8
3.68	5	.5	.41	4.3	.33	4.6	.33	4.6
3.68	5	1	.80	3.3	.74	3.9	.75	3.7
5.20	10	0	.10	9.7	.071	9.8	.072	9.8
5.20	10	.5	.59	7.9	.52	8.5	.53	8.4
5.20	10	1	.96	5.1	.94	5.9	.95	5.7

[a] Computed numerically by Samuel-Cahn (1974).

observations in the nth group. Table 3.5 indicates that the Brownian approximations can be rather poor, with the approximation to $E_\mu(\tilde{\tau} \wedge m)$ being somewhat better than the approximation to $P_\mu\{\tilde{\tau} \le m\}$.

Chapter X proposes modifications of the Brownian approximations to give better numerical accuracy in discrete time. The suggested approximation to $P_\mu\{\tilde{\tau} < m, S_m \le c\}$ is (cf. 3.14 and 10.45)

$$P_\mu\{\tilde{\tau} < m, S_m \le c\} \cong e^{2(b+\rho)(\mu-\eta)}\Phi\{[c - 2(b + \rho)]/m^{1/2} - \mu m^{1/2}\} \quad (c \le b + \eta m), \tag{3.28}$$

where

$$\rho = -\pi^{-1}\int_0^\infty t^{-2}\log\{2(1 - e^{-t^2/2})/t^2\}\,dt \cong .583. \tag{3.29}$$

Putting $c = b + \eta m$ yields an improved approximation to

$$P_\mu\{\tilde{\tau} \le m\} = P_\mu\{S_m \ge b + \eta m\} + P_\mu\{\tilde{\tau} < m, S_m < b + \eta m\}. \tag{3.30}$$

It is interesting to note that (3.28) is just (3.14) with b replaced by $b + \rho$. Although (3.30) does not directly lend itself to this suggestive interpretation, substitution of (3.28) into (3.30) and an expansion show that up to terms which are $o(m^{-1/2})$ the latter approximation can also be written

$$\begin{aligned} P_\mu\{\tilde{\tau} \le m\} &\cong 1 - \Phi[(b + \rho)m^{-1/2} - (\mu - \eta)m^{1/2}] \\ &\quad + e^{2(b+\rho)(\mu-\eta)}\Phi[-(b + \rho)m^{-1/2} - (\mu - \eta)m^{1/2}]. \end{aligned} \tag{3.30*}$$

Presumably (3.30)* is slightly less accurate than using (3.28) in (3.30), but it is in the form of (3.15) with b replaced by $b + \rho$. These considerations suggest approximating $E_\mu(\tau \wedge m)$ by (3.17) with b replaced by $(b + \rho)$.

In Chapter X it is shown that ρ is approximately $E_\mu[S_{\tilde{\tau}} - (b + \eta\tilde{\tau})]$. Hence these corrected approximations are tantamount to using the exact formulae for Brownian motion applied to a boundary $b' + \eta n$, where $b' - b$ is the expected excess of the discrete process over the boundary.

The last two columns of Table 3.5 contain these modified approximations ((3.28) and (3.30) are used for $P_\mu\{\tilde{\tau} \le m\}$). They seem to be considerably better than the Brownian pproximations (see also Table 3.7).

The modified approximations make it easy to study the effect of examining the data periodically. Instead of (potentially) observing $x_1, x_2, \ldots, x_m$, suppose that for some integer $k > 1$ we observe

$$x_i^* = k^{-1/2} \sum_{j=(k-1)i+1}^{ki} x_j \quad \left(i = 1, 2, \ldots, m^* = \frac{m}{k}\right).$$

The stopping rule (3.27) is now defined with $S_n^* = \sum_{j=1}^n x_j^*$ in place of S_n and $\eta k^{1/2}$ in place of η. Truncation takes place after m^* observations. The scaling property of Brownian motion discussed in Remark 3.19 together with the observation that the modified approximation (3.30)* is just the Brownian approximation with $b + \rho$ in place of b suggests that the effect of grouping the observations will prove to be insignificant, at least when $c = b + \eta m$.

For a numerical example, both for the ungrouped test with $m = 49$, $b = 5.48$, $k = 1$, and $\eta = .3$, and for the grouped test with $m^* = k = 7$, $b = 1.73$, and $\eta k^{1/2} = .3 \cdot 7^{1/2} = .79$, the approximation (3.28) with $\mu = 0$ gives significance levels of .025. These two tests are designed to be comparable to that in Table 3.1, and the modified approximations indicate that all three have essentially the same power and expected sample size over the range of parameter values of Table 3.1. Hence the effect of examining the data in 7 groups of 7 observations instead of continuously appears to be negligible.

Obviously this conclusion fails in extreme cases. For example, for $m^* = 2$ and $k = 25$, the expected sample size when $\mu = .7$ must be at least 25 and hence is substantially larger than the 15 of Table 3.1.

6. Truncated Sequential Probability Ratio Test

In this section we study the effect of truncation on the properties of a sequential probability ratio test. For the most part only Brownian motion is considered in detail, but the suggestions of Section 5 may be easily adapted to correct for discrete time. (Corrections for non-normality are discussed in X.3.)

Let $\{W(t), 0 \le t < \infty\}$ be Brownian motion with drift μ and for given $\delta > 0$ consider a sequential probability ratio test of H_0: $\mu = -\delta$ against H_1: $\mu = \delta$ defined by the stopping rule

$$T = \inf\{t \colon l(t, W(t); \delta, -\delta) \notin (A, B)\}.$$

By (3.1) $l(t, W(t); \delta, -\delta) = \exp(2\delta W(t))$, and hence with the notation $a = \log A/2\delta$, $b = \log B/2\delta$ this stopping rule may be expressed as

$$T = \inf\{t \colon W(t) \notin (a, b)\}. \tag{3.31}$$

(Although the hypotheses $-\delta$ and δ seem special, (3.5) shows that a sequential probability ratio test of μ_0 against μ_1 ($\mu_0 < \mu_1$) can always be put into this special form with $\delta = \frac{1}{2}(\mu_1 - \mu_0)$.)

The absence of a definite upper bound on T is a practical disadvantage of the sequential probability ratio test, and hence it is natural to consider the effect of truncation on the properties of the test. Let $m > 0$ and define a new stopping rule $T \wedge m = \min(T, m)$, i.e. we stop sampling at time T defined by (3.31) if $T \le m$

and otherwise at time m. If $T \leq m$ we reject H_0 if $W(T) = b$ and accept H_0 if $W(T) = a$. If $T > m$, so our final observation is $W(m) \in (a, b)$, we must decide which points of (a, b) should be part of the rejection region. In some cases the simple hypotheses $-\delta$ and δ will be surrogates for a symmetric problem of deciding whether $\mu \leq 0$ or $\mu > 0$. Then it may be natural to take $a = -b$, and when sampling is truncated at time m to reject H_0 if and only if $W(m) > 0$. A complete description of this test is to stop samping at $T \wedge m$ and reject H_0 if and only if $W(T \wedge m) > 0$. On the other hand, if we are comparing an experimental treatment with a control, then H_0 may be a simple hypothesis of no treatment effect and H_1 may be surrogate for the composite hypothesis that the experimental treatment has a positive effect. Symmetry conditions are not obviously relevant, and we may, for example, decide that when $T > m$ we will not reject H_0. The test would be to stop at $T \wedge m$ and reject H_0 if and only if $T \leq m$ and $X(T) = b$, with power $P_\mu\{T \leq m, X(T) = b\}$.

In general, given $a < 0 < b$, $m > 0$, and $a \leq c \leq b$, define T by (3.31) and consider the test which stops sampling at $\min(T, m)$ and rejects H_0 if either $T \leq m$ and $W(T) = b$ or if $T > m$ and $W(m) > c$. Its power function is

$$P_\mu\{T \leq m, W(T) = b\} + P_\mu\{T > m, W(m) > c\}. \tag{3.32}$$

Exact evaluation of (3.32) and the expected sample size $E_\mu(T \wedge m)$ is complicated, but it is possible to obtain very useful approximations. One is especially good when m is small, another when m is large compared to a and b. For a broad range of values of a, b, and m, both are quite good. The small m approximation relates T to the one-sided stopping rule of Section 3 by neglecting paths which first hit one boundary and then cross the interval (a, b) to hit the other boundary before time m. We begin with this argument and then proceed to the exact evaluation of (3.32), which leads to an approximation for large m.

For $x \neq 0$, let $\tau_x = \inf\{t: W(t) = x\}$. Then

$$P_\mu\{T \leq m, W(T) = b\} = P_\mu\{\tau_b \leq m\} - P_\mu\{W(T) = a, \tau_b \leq m\}. \tag{3.33}$$

Since the second probability on the right hand side of (3.33) involves sample paths which first hit a, then cross the interval (a, b) to hit b before time m, it is often negligible—especially for small m. A simple upper bound is given by

$$\begin{aligned} P_\mu\{W(T) = a, \tau_b \leq m\} &= \int_{[0,m]} P_\mu\{T \in dt, W(T) = a\} P_\mu\{\tau_{b+|a|} \leq m - t\} \\ &\leq \int_{[0,m]} P_\mu\{\tau_a \in dt\} P_\mu\{\tau_{b+|a|} \leq m - t\} \\ &= e^{2\mu a} \int_{[0,m]} P_{-\mu}\{\tau_a \in dt\} P_\mu\{\tau_{b+|a|} \leq m - t\} \\ &= e^{2\mu a} \int_{[0,m]} P_\mu\{\tau_{|a|} \in dt\} P_\mu\{\tau_{b+|a|} \leq m - t\} \\ &= e^{2\mu a} P_\mu\{\tau_{b+2|a|} \leq m\}. \end{aligned} \tag{3.34}$$

More generally, (3.32) equals

$$
\begin{aligned}
&P_\mu\{W(T \wedge m) \geq c\} \\
&\quad = P_\mu\{W(\tau_b \wedge m) \geq c\} - P_\mu\{W(\tau_b \wedge m) \geq c, W(T \wedge m) < c\} \\
&\quad = P_\mu\{W(m) \geq c\} + P_\mu\{\tau_b < m, W(m) < c\} \\
&\qquad - P_\mu\{W(\tau_b \wedge m) \geq c, W(T \wedge m) < c\}.
\end{aligned} \tag{3.35}
$$

The last probability in (3.35) equals

$$
\begin{aligned}
&P_\mu\{\tau_a < m, \tau_b > m, W(m) \geq c\} + P_\mu\{\tau_a < \tau_b < m\} \\
&\quad = P_\mu\{\tau_a < m, W(m) \geq c\} + P_\mu\{\tau_a < \tau_b < m, W(m) < c\} \\
&\qquad - P_\mu\{\tau_b < \tau_a < m, W(m) \geq c\},
\end{aligned}
$$

of which the last two terms are often negligible. Ignoring them leads to approximating (3.35), hence (3.32) by

$$
P_\mu\{W(m) \geq c\} + P_\mu\{\tau_b < m, W(m) < c\} - P_\mu\{\tau_a < m, W(m) \geq c\}. \tag{3.36}
$$

A similar argument applies to

$$
\begin{aligned}
E_\mu(T \wedge m) = {} & E_\mu(\tau_b; \tau_b \leq m, W(T) = b) + E_\mu(\tau_a; \tau_a \leq m, W(T) = a) \\
& + m[1 - P_\mu\{T \leq m, W(T) = b\} - P_\mu\{T \leq m, W(T) = a\}].
\end{aligned}
$$

Since $E_\mu(\tau_b; \tau_b \leq m, W(T) = b) = E_\mu(\tau_b; \tau_b \leq m) - E_\mu(\tau_b; W(T) = a, \tau_b \leq m)$, and $P_\mu\{T \leq m, W(T) = b\}$ is given by (3.33), simple algebra yields

$$
\begin{aligned}
E_\mu(T \wedge m) = {} & E_\mu(\tau_b \wedge m) + E_\mu(\tau_a \wedge m) - m \\
& + E_\mu(m - \tau_b; W(T) = a, \tau_b \leq m) \\
& + E_\mu(m - \tau_a; W(T) = b, \tau_a \leq m).
\end{aligned} \tag{3.37}
$$

Since the last two expectations in (3.37) involve the event of hitting one boundary, then crossing the interval (a, b) to hit the other boundary before time m, it seems plausible that at least for small m most of the allotted time will be exhausted and the resulting expectations will be negligible. Hence $E(T \wedge m)$ can often be approximated by the first three terms on the right hand side of (3.37), which are evaluated in (3.17).

Tables 3.6 and 3.7 indicate that the preceding approximations are usually adequate, but for completeness we consider the problem of computing $P_\mu\{T \leq m, W(T) = b\}$ exactly. As in Section 3 it suffices to compute the conditional probability $P_\xi^{(m)}\{T \leq m, W(T) = b\}$ for all ξ and then obtain the unconditional probability by integrating with respect to the marginal distribution of $W(m)$. It may be assumed without loss of generality that $\xi < b$, because if $\xi \geq b$, the argument given below suffices to compute the complementary probability $P_\xi^{(m)}\{T \leq m, W(T) = a\}$; the desired probability is then obtained by subtraction.

Put $\xi_0 = \xi$, and let $\xi_1 = 2b - \xi > b$.

Proposition 3.12 yields (cf. 3.13)

$$
\begin{aligned}
&P_{\xi_0}^{(m)}\{T < m, W(T) = b\} \\
(3.38)\qquad &= \exp\{-2b(b - \xi_0)/m\} P_{\xi_1}^{(m)}\{T < m, W(T) = b\} \\
&= \exp\{-2b(b - \xi_0)/m\}(1 - P_{\xi_1}^{(m)}\{T < m, W(T) = a\}),
\end{aligned}
$$

where the second equality follows from the fact that $P_{\xi_1}^{(m)}\{T < m\} = 1$. Hence the probability we seek is given in terms of another probability of the same kind with the roles of a and b interchanged. By putting $\xi_2 = 2a - \xi_1 = 2(a - b) + \xi < a$, we obtain similarly

$$
\begin{aligned}
&P_{\xi_1}^{(m)}\{T < m, W(T) = a\} \\
(3.39)\qquad &= \exp\{-2a(a - \xi_1)/m\}(1 - P_{\xi_2}^{(m)}\{T < m, W(T) = b\}).
\end{aligned}
$$

For the ordinary sequential probability ratio test we obtained a similar pair of (approximate) equations in Chapter II—(2.9) and (2.10). There, however, we had the good fortune that the probability corresponding to $P_{\xi_2}^{(m)}$ in (3.39) was identical with the probability corresponding to $P_{\xi_0}^{(m)}$ in (3.38). Hence there were two (approximate) equations for two unknown probabilities, which could be solved simultaneously. Now $\xi_2 \neq \xi_0$ and we can only iterate the preceding argument to obtain an infinite number of recursive equations, which yields an infinite series for $P_{\xi}^{(m)}\{T < m, W(T) = b\}$. In general $\xi_{2i} = -2i(b - a) + \xi$, $\xi_{2i+1} = 2[(i + 1)b - ia] - \xi$ $(i = 0, 1, \ldots)$; and for each i

$$
\begin{aligned}
&P_{\xi_{2i}}^{(m)}\{T < m, W(T) = b\} \\
(3.40)\qquad &= \exp\{-2b(b - \xi_{2i})/m\}(1 - P_{\xi_{2i+1}}^{(m)}\{T < m, W(T) = a\}),
\end{aligned}
$$

$$
\begin{aligned}
&P_{\xi_{2i+1}}^{(m)}\{T < m, W(T) = a\} \\
(3.41)\qquad &= \exp\{-2a(a - \xi_{2i+1})/m\}(1 - P_{\xi_{2i+2}}^{(m)}\{T < m, W(T) = b\}).
\end{aligned}
$$

It is easy to see that

$$
\lim_{i\to\infty} P_{\xi_{2i}}^{(m)}\{T < m, W(T) = b\} = 0 = \lim_{i\to\infty} P_{\xi_{2i+1}}^{(m)}\{T < m, W(T) = a\},
$$

so (3.40) and (3.41) can be solved recursively. The result of considerable algebra is recorded as

Theorem 3.42. *Let* $p(m, \xi; a, b) = P_{\xi}^{(m)}\{T < m, W(T) = b\}$. *For* $\xi < b$

$$
\begin{aligned}
p(m, \xi, a, b) = e^{\xi^2/2m} \sum_{i=1}^{\infty} \Bigg(&\exp\left\{-\frac{1}{2m}[\xi - 2a - 2i(b - a)]^2\right\} \\
&- \exp\left\{-\frac{1}{2m}[\xi - 2i(b - a)]^2\right\}\Bigg).
\end{aligned}
$$

For $\xi > b$ *(and hence* $-\xi < -b < -a$*)*

$$
p(m, \xi, a, b) = 1 - p(m, -\xi, -b, -a).
$$

Since $P_\xi^{(m)}\{T < m, W(T) = a\} = p(m, -\xi, -b, -a)$, Theorem 3.42 yields an expression for

$$P_\xi^{(m)}\{T < m\} = P_\xi^{(m)}\{T < m, W(T) = b\} + P_\xi^{(m)}\{T < m, W(T) = a\},$$

from which follows

Corollary 3.43. *For* $a < \xi < b$

$$\begin{aligned} P_\mu\{T > m, W(m) \in d\xi\} &= P_\mu\{W(m) \in d\xi\}(1 - P_\xi^{(m)}\{T \le m\}) \\ &= \exp(\mu\xi - \mu^2 m/2) m^{-1/2} \sum_{i=-\infty}^{\infty} (\phi\{[\xi - 2i(b-a)]/m^{1/2}\} \\ &\quad - \phi\{[\xi - 2a - 2i(b-a)]/m^{1/2}\})\, d\xi. \end{aligned}$$

The series in Corollary 3.43 converges slowly for large values of m, but it can be expressed in an alternative form which is often very useful for numerical computation. Either from its probabilistic meaning or by direct inspection, one can easily see that the series vanishes for $\xi = a$ and for $\xi = b$. Moreover, the series is a continuous periodic function of ξ and hence equals its Fourier sine series. The result of some computation is

Corollary 3.44 (alternative form of 3.43). *For* $a < \xi < b$

$$P_\mu\{T > m, W(m) \in d\xi\} = \exp(\mu\xi - \mu^2 m/2) P_0\{T > m, W(m) \in d\xi\},$$

where

$$P_0\{T > m, W(m) \in d\xi\} = 2 \sum_{k=1}^{\infty} \exp[-k^2\pi^2 m/2(b-a)^2] \times \sin\left[\frac{k\pi|a|}{b-a}\right] \sin\left[\frac{k\pi(\xi - a)}{b-a}\right] d\xi/(b-a).$$

For large m all terms of this series are negligible in comparison with the first, which hence can be used to give simple and accurate approximations. The most useful results are recorded in the following sequence of corollaries. Explicit estimates of the error made in these approximations are easily obtained by analyzing more terms of the series in Corollary 3.44.

Corollary 3.45. *For* $a \le \xi \le b$, as $m \to \infty$,

$$\begin{aligned} P_\mu\{T > m, W(m) \le \xi\} &\sim \exp\{\mu a - \tfrac{1}{2} m[\mu^2 + \pi^2/(b-a)^2]\} \\ &\cdot \left\{ e^{\mu(\xi - a)} \sin\left[\frac{\pi(\xi - a)}{b-a}\right] \Big/ \mu(b-a) \right. \\ &\left. + \frac{\pi}{\mu^2(b-a)^2}\left[1 - e^{\mu(\xi-a)} \cos\left[\frac{\pi(\xi - a)}{b-a}\right]\right]\right\} \\ &\cdot 2 \sin\left[\frac{\pi|a|}{b-a}\right] \Big/ [1 + \pi^2/\mu^2(b-a)^2], \end{aligned}$$

which for $\xi = b$ specializes to

$$P_\mu\{T > m\} \sim \exp\left\{\mu a - \frac{1}{2}m[\mu^2 + \pi^2/(b-a)^2]\right\}[e^{\mu(b-a)} + 1]$$
$$\times \sin\left[\frac{\pi|a|}{(b-a)}\right]\cdot\frac{2\pi}{\pi^2 + \mu^2(b-a)^2}.$$

Since

$$E_\mu(T \wedge m) = E_\mu(T) - E_\mu(T - m; T > m) = E_\mu(T) - \int_m^\infty P_\mu\{T > m'\}\,dm',$$

integration of Corollary 3.45 over (m, ∞) yields

Corollary 3.46. $E_\mu(T \wedge m) = E_\mu(T) - E_\mu(T - m; T > m)$, *where* $E_\mu(T)$ *is given in Theorem* 3.6 *and as* $m \to \infty$

$$E_\mu(T - m; T > m) \sim \exp\{\mu a - \tfrac{1}{2}m[\mu^2 + \pi^2/(b-a)^2]\}[e^{\mu(b-a)} + 1]$$
$$\times \sin\left(\frac{\pi|a|}{b-a}\right)\cdot\frac{4\pi(b-a)^2}{[\pi^2 + \mu^2(b-a)^2]^2}.$$

The decomposition

$$P_\mu\{T \le m, W(T) = b\} = P_\mu\{W(T) = b\} - \int_a^b P_\mu\{T > m,$$
$$W(m) \in d\xi\}P_\mu\{W(T) = b \mid T > m, W(m) = \xi\}$$

and a simple calculation give

Corollary 3.47. $P_\mu\{T \le m, W(T) = b\} = P_\mu\{W(T) = b\} - P_\mu\{T > m, W(T) = b\}$, *where* $P_\mu\{W(T) = b\}$ *is given by Theorem* 3.6 *and as* $m \to \infty$

$$P_\mu\{T > m, W(T) = b\} \sim \exp\{-\tfrac{1}{2}m[\mu^2 + \pi^2/(b-a)^2] + \mu b\}$$
$$\times \sin\left(\frac{\pi|a|}{b-a}\right)\cdot\frac{2\pi}{\pi^2 + \mu^2(b-a)^2}.$$

Table 3.6 contains a numerical example with $-a = b = 5$, $c = 0$, and $m = 49$. This test has a power function and expected sample size very similar to an untruncated sequential probability ratio test with $-a = b = 4.9$ (cf. Table 2.1). The truncation value m is about twice the expected sample size of the untruncated test when $\mu = 0$. It is easy to see from Corollary 3.45 that this has the effect of truncating about the upper 10% of the P_0 distribution of T, while having little effect on the expected sample size.

The first entry in each cell of Table 3.6 is an approximation given by Corollaries 3.45–3.47. The second is that given by (3.36) or the first three terms

Table 3.6. Truncated Sequential Probability Ratio Test for Brownian Motion: $b = -a = 5$; $c = 0$; $m = 49$

μ	$P_\mu\{W(T \wedge m) \le 0\}$	$E_\mu(T \wedge m)$
.4	.019, .019	12.0, 11.8
.3	.051, .051	14.9, 14.5
.2	.129, .130	18.4, 17.8
.1	.282, .282	21.1, 20.7
0	.500, .500	22.7, 22.0

Table 3.7. Truncated Test in Discrete Time: $b = -a = 2.944$; $m = 11$

μ	$P_\mu\{S_{T\wedge m} \le 0\}$	$E_\mu(T \wedge m)$	$P_\mu\{T > m\}$
.75	.009, .009, .009	4.6, 4.6, 4.8	.04, .03, .03
.50	.055, .055, .056	6.0, 6.0, 6.2	.14, .13, .14
.25	.214, .211, .211	7.5, 7.5, 7.6	.33, .31, .32
0	.5, .5, .5	8.1, 8.1, 8.3	.43, .41, .42

Table 3.8. 90% Confidence Interval Following Truncated Sequential Probability Ratio Test: $b = -a = 5$; $m = 49$; $P_{.3}\{W(T \wedge m) \le 0\} \cong .05$

Data	90% Confidence Interval
$T = 10$, $W(T) = 5$	$(-.15, .94)$
$T = 15$, $W(T) = 5$	$(-.23, .68)$
$T = 20$, $W(T) = 5$	$(-.26, .55)$
$T = 49$, $W(T) = 5$	$(-.29, .32)$

on the right hand side of (3.37). For this combination of test parameters, which has m fairly large compared to b, the first entry is usually more accurate, although the differences are quite small.

The results of Section 5 suggest ways to modify these approximations for processes in discrete time, to wit: use the Brownian motion approximations developed above and apply them to tests with boundaries displaced to $b + \rho$ and $a - \rho$, where $\rho \cong .583$ (see also X.3). Table 3.7 compares such approximations with results computed numerically by Aroian and Robison (1969) for the problem of testing a normal mean with unit variance. The three entries in each cell are, respectively, the large m approximations given by Corollaries 3.45–3.47, the small m approximations given by (3.36) or (3.37), and the numerical computation of Aroian and Robison.

It is easy to adapt the arguments of Section 4 to obtain a confidence interval for μ based on observing $(T \wedge m, W(T \wedge m))$. A numerical example for the test of Table 3.6 is given in Table 3.8. If one considers this test to be essentially a test of $\mu = -.3$ against $\mu = +.3$, where the error probabilities are about .05, the results of Table 3.8 indicate that very little interesting information is obtained from a 90% confidence interval, which for most values of the data covers most of the interval $(-.3, +.3)$. The situation is somewhat better if one makes the error probabilities of the test smaller than the conventional .05 to force a larger sample size (or makes the confidence coefficient smaller). We return to this point in a slightly different context in IV.5.

7. Anderson's Modification of the Sequential Probability Ratio Test

In order to reduce the maximum expected sample size of the sequential probability ratio test, Anderson (1960) proposed a modification in which the stopping boundaries are no longer restricted to be parallel. Let $a < 0 < b$, and $-\infty < \eta_0, \eta_1 < \infty$. If $\eta_0 > \eta_1$ define m^* by $a + \eta_0 m^* = b + \eta_1 m^*$, i.e. m^* is the point t at which the lines $a + \eta_0 t$ and $b + \eta_1 t$ meet. Otherwise let $m^* = +\infty$. Let $0 < m \leq m^*$ and let $c \in [a + \eta_0 m, b + \eta_1 m]$. Define the stopping rule

$$T = \inf\{t\colon W(t) \notin (a + \eta_0 t, b + \eta_1 t)\} \tag{3.48}$$

and consider the sequential test which stops sampling at $\min(T, m)$ and rejects the one sided hypothesis $H_0\colon \mu \leq \mu_0$ if either $T \leq m$ and $W(T) = b + \eta_1 T$ or if $T > m$ and $W(m) > c$.

The sequential probability ratio test is a special case of Anderson's test with $\eta_0 = \eta_1$ and $m = \infty$. The truncated test studied earlier in this section is the special case $\eta_0 = \eta_1 = 0$ and $m < \infty$. Even the one-sided test of Section 3 can be considered a special case if we put $a = -\infty$.

Again the essential ingredient in calculating the power function and expected sample size of the test is

$$P_{\xi_0}^{(m)}\{T < m, W(T) = b + \eta_1 T\} = P_0\{T < m, W(T) = b + \eta_1 T \mid W(m) = \xi_0\} \tag{3.49}$$

for $-\infty < \xi_0 < \infty$.

Evaluating (3.49) in general requires only trivial modification of the argument leading to Theorem 3.42 (and some lengthy algebraic simplification). Again there is no loss of generality in assuming that $\xi_0 < b + \eta_1 m$. (The result for $\xi_0 = b + \eta_1 m = a + \eta_0 m$ must be obtained by continuity.) Let $\xi_1 = 2(b + \eta_1 m) - \xi_0 > b + \eta_1 m$. Then by Proposition 3.12 (cf. (3.13) and (3.38))

$$\begin{aligned} &P_{\xi_0}^{(m)}\{T < m, W(T) = b + \eta_1 T\} \\ &\qquad = \exp\{-2b(b + \eta_1 m - \xi_0)/m\}(1 - P_{\xi_1}^{(m)}\{T < m, W(T) = a + \eta_0 T)\}. \end{aligned} \tag{3.50}$$

Putting $\xi_2 = 2(a + \eta_0 m) - \xi_1 = 2[a - b + (\eta_0 - \eta_1)m] + \xi_0$ yields

$$P_{\xi_1}^{(m)}\{T < m, W(T) = a + \eta_0 T\}$$
$$= \exp\{-2a(a + \eta_0 m - \xi_1)/m\}(1 - P_{\xi_2}^{(m)}\{T < m, W(T) = b + \eta_1 T\}). \tag{3.51}$$

Inductively one obtains an infinite sequence of recursive relations similar to (3.40) and (3.41), which can be solved by straighforward but tedious algebra to give the probability (3.49) in the form of an infinite series, which resembles but is rather more complicated than Theorem 3.42. (See Anderson, 1960, for details.)

Simpler results can be obtained in the triangular case $a + \eta_0 m = b + \eta_1 m$ and especially the symmetric triangular case, for which $b = -a$, $\eta_0 = -\eta_1 = \frac{1}{2}\eta > 0$, and $m = 2\eta/b$. Given $\mu_0 < 0 < \mu_1$, the likelihood ratios for testing H_0: $\mu = \mu_i$ against H_1: $\mu = 0$ are $l(t, W(t); 0, \mu_i) = \exp[-\mu_i W(t) + \frac{1}{2}\mu_i^2 t]$, so the stopping rule $T = \min(\tau_0, \tau_1)$, where

$$\tau_i = \inf\{t: l(t, W(t); 0, \mu_i) \geq e^{|\mu_i| b_i}\}$$
$$= \inf\{t: (-1)^i W(t) \geq b_i - \tfrac{1}{2}|\mu_i| t\}$$

is of the form (3.48) with $a = -b_1$, $b = b_0$, and $\eta_i = -\frac{1}{2}\mu_i$ $(i = 0, 1)$. Since τ_i is optimal in the sense of II.4 for testing $\mu = \mu_i$ against $\mu = 0$ (cf. (2.36)–(2.38)), it seems plausible that $T = \tau_0 \wedge \tau_1$ should have approximately minimum sample size when $\mu = 0$ subject to constraints on the errors probabilities at μ_0 and μ_1, at least in moderately symmetric situations.

Remark. In the symmetric triangular case it is possible to show that the error probabilities at the values $\mu = \pm\eta$ are given by

$$P_\eta\{W(T) = -b + \tfrac{1}{2}\eta T\} = \tfrac{1}{2}e^{-b\eta} \tag{3.52}$$

(cf. Problem 3.5). A correction for discrete time is obtained by replacing b by $b + .583$ on the right hand side of (3.52). It is remarkable how close this approximation is to that obtained by Lorden (1976) by numerical and curve fitting methods.

In the triangular case $a + \eta_0 m = b + \eta_1 m$, for $\xi_0 \neq a + \eta_0 m$ the value ξ_2 in (3.51) equals ξ_0 in (3.50), so that, like the ordinary sequential probability ratio test, there are only two equations to be solved simultaneously. Putting $A = \{W(T) = a + \eta_0 T\}$ and $B = \{W(T) = b + \eta_1 T\}$, we obtain for $a + \eta_0 m = b + \eta_1 m \neq \xi_0$.

$$P_{\xi_0}^{(m)}(B) = \frac{\exp\{-2m^{-1}a(b + \eta_1 m - \xi_0)\} - 1}{\exp\{2m^{-1}(b - a)(b + \eta_1 m - \xi_0)\} - 1} = 1 - P_{\xi_0}^{(m)}(A). \tag{3.53}$$

As always

$$P_\mu(B) = \int_{-\infty}^{\infty} P_\xi^{(m)}(B)\phi[(\xi - m\mu)/m^{1/2}]\, d\xi/m^{1/2}, \tag{3.54}$$

so an easy numerical integration yields the power function of the test. The following analysis yields a similar expression for $E_\mu(T)$.

By proposition 3.2 and (3.48), for $\tilde{\mu} \neq \mu$

$$P_{\tilde{\mu}}(B) = E_\mu\{\exp[(\tilde{\mu} - \mu)W(T) - \tfrac{1}{2}(\tilde{\mu}^2 - \mu^2)T]; B\}$$
$$= \exp[(\tilde{\mu} - \mu)b]E_\mu\{\exp[\{(\tilde{\mu} - \mu)\eta_1 - \tfrac{1}{2}(\tilde{\mu}^2 - \mu^2)\}T]; B\}.$$

Differentiating with respect to $\tilde{\mu}$ and putting $\tilde{\mu} = \mu$ leads to

$$\left.\frac{d}{d\tilde{\mu}} P_{\tilde{\mu}}(B)\right|_{\tilde{\mu}=\mu} = bP_\mu(B) + (\eta_1 - \mu)E_\mu(T; B).$$

Hence by differentiating (3.54) one obtains for all $\mu \neq \eta_1$

$$E_\mu(T; B) = (\eta_1 - \mu)^{-1} \int_{-\infty}^{\infty} (\xi - m\mu - b)P_\xi^{(m)}(B)\phi[(\xi - m\mu)/m^{1/2}]\, d\xi/m^{1/2}, \tag{3.55}$$

where $P_\xi^{(m)}(B)$ is given by (3.53). Similarly for $\mu \neq \eta_0$

$$E_\mu(T; A) = (\eta_0 - \mu)^{-1} \int_{-\infty}^{\infty} (\xi - m\mu - a)P_\xi^{(m)}(A)\phi[(\xi - m\mu)/m^{1/2}]\, d\xi/m^{1/2}. \tag{3.56}$$

These results are summarized as

Theorem 3.57. *For the triangular case* $a + \eta_0 m = b + \eta_1 m$ *of* (3.48), *let* $B = \{W(T) = b + \eta_2 T\}$ *and* $A = \{W(T) = a + \eta_0 T\}$. *The power function* $P_\mu(B)$ *and expected sample size* $E_\mu(T)$ *are given by* (3.54) *and* (3.55)–(3.56), *where* $P_\xi^{(m)}(B)$ *and* $P_\xi^{(m)}(A)$ *are evaluated in* (3.53).

In the symmetric triangular case $b = -a$, $\eta_0 = -\eta_1 = \frac{1}{2}\eta > 0$, and $m = 2b/\eta$, *the error probabilities at* $\pm\eta$ *are given by* (3.52). *Moreover,*

$$E_0(T) = m - 4\eta^{-1}I$$

and

$$E_\eta(T) = m/3 - 4e^{-b\eta}I/(3\eta),$$

where $I = \int_{-\infty}^{\infty} z\phi(z)[1 + \exp(-cz)]^{-1}\, dz$, *with* $c = (2b\eta)^{1/2}$.

Tables 3.9 and 3.10 compare Anderson's test in the symmetric triangular case with symmetric, truncated sequential probability ratio tests. (In Table 3.9 the truncated sequential probability ratio test is precisely that of Table 3.6.) Roughly speaking, Anderson's test has an expected sample size which is slightly smaller for $\mu = 0$ and slightly larger for large $|\mu|$. The triangular tests have slightly larger maximum sample sizes. However, the probability that the triangular test requires almost its maximum sample size to terminate is so small that with minor adjustments one can truncate the triangular region without changing the power and expected sample size significantly. To study

Table 3.9. Truncated Sequential Probability Ratio Test: $b = -a = 5$; $c = 0$; $m = 49$; and Triangular Test: $b = 7.68$; $\eta = .3$; $m = 51.2$

μ	$P_\mu\{W(T \wedge m) \le 0\}$	$E_\mu(T \wedge m)$	$P_\mu\{W(T) \le 0\}$	$E_\mu(T)$
0.3	.05	14.9	.05	16.1
0.0	.50	22.7	.50	21.5

Table 3.10. Truncated Sequential Probability Ratio Test: $b = -a = 8$; $c = 0$; $m = 79$; and Triangular Test: $b = 13.04$; $\eta = .3$; $m = 86.9$

μ	$P_\mu\{W(T \wedge m) \le 0\}$	$E_\mu(T \wedge m)$	$P_\mu\{W(T) \le 0\}$	$E_\mu(T)$
0.3	.01	26.0	.01	28.7
0.0	.50	49.6	.50	44.9

these truncated tests, simple approximations along the lines of (3.36)–(3.37) usually are adequate.

The preceding discussion permits one to test H_0: $\mu \le 0$ against H_1: $\mu > 0$ (or $\mu \le \mu^*$ against $\mu > \mu^*$) with a test having specified error probabilities at $\mu = \mu_0 < 0$ and $\mu = \mu_1 > 0$. At least in symmetric problems the test has a slightly smaller expected sample size at $\mu = 0$ than a truncated sequential probability ratio test. The following special case of an inequality of Hoeffding (1960) (see also Problem 3.14) places a bound on further reductions in the expected sample size.

Theorem 3.58. *Let $\tilde{T}$ denote the stopping time of a sequential test of H_0: $\mu \le 0$ against H_1: $\mu > 0$ with error probabilities α and β at $\mu_0 < 0 < \mu_1$, respectively. Assume that $\alpha + \beta < 1$. Then*

$$E_0(\tilde{T}) \ge \left[\frac{-\frac{1}{2}|\mu_0 - \mu_1| + \{\frac{1}{4}|\mu_0 - \mu_1|^2 - 2\max(\mu_0^2, \mu_1^2)\log(\alpha + \beta)\}^{1/2}}{\max(\mu_0^2, \mu_1^2)}\right]^2.$$

Proof of Theorem 3.58. By Problem 2.19

$$\begin{aligned}\alpha + \beta &\ge E_0\{\exp[\min(\mu_0 W(\tilde{T}) - \tfrac{1}{2}\mu_0^2 \tilde{T}, \mu_1 W(\tilde{T}) - \tfrac{1}{2}\mu_1^2 \tilde{T})]\} \\ &\ge E_0\{\exp[\min(\mu_0 W(\tilde{T}), \mu_1 W(\tilde{T})) - \tfrac{1}{2}\max(\mu_0^2, \mu_1^2)\tilde{T}]\}.\end{aligned}$$

There is no loss of generality in assuming that $E_0(\tilde{T}) < \infty$. Since $\min(u, v) = \frac{1}{2}(u + v - |u - v|)$ and for any random variable y, $E(e^y) \ge e^{E(y)}$, it follows (cf. the proof of Proposition 2.38) from Wald's identity that

$$\alpha + \beta \ge \exp\{-\tfrac{1}{2}|\mu_0 - \mu_1| E_0|W(\tilde{T})| - \tfrac{1}{2}\max(\mu_0^2, \mu_1^2)E_0(\tilde{T})\}.$$

By the Schwarz inequality and Wald's identity for the second moment

$$E_0|W(\tilde{T})| \le \{E_0[W^2(\tilde{T})]\}^{1/2} = [E_0(\tilde{T})]^{1/2},$$

so

$$\alpha + \beta \geq \exp\{-\tfrac{1}{2}|\mu_0 - \mu_1|[E_0(\tilde{T})]^{1/2} - \tfrac{1}{2}\max(\mu_0^2, \mu_1^2)E_0(\tilde{T})\}.$$

Taking logarithms and solving the quadratic inequality in $[E_0(\tilde{T})]^{1/2}$ complete the proof. □

Remark. For the examples of Tables 3.9 and 3.10 the lower bounds for $E_0(T)$ provided by Theorem 3.58 are 20.8 and 43.1, respectively. Hence at least for these cases the symmetric triangular test comes very close to minimizing the expected sample size at $\mu = 0$ subject to symmetric constraints on the error probabilities. An examination of the preceding argument shows that in the symmetric triangular case only the Schwarz inequality fails to be an equality. Since the Schwarz inequality is in general rather crude, it is surprising that the lower bound is so close to the correct value.

8. Early Stopping to Accept H_0

In Section 2 we studied a test which was designed to behave like a sequential probability ratio test under H_1: $\mu > 0$ and a fixed sample test under H_0: $\mu = 0$. In the specific example of clinical trials, if H_1 indicates the superiority of an experimental treatment compared with, say, placebo, ethical considerations suggest termination of the test when H_1 appears to be true. Under H_0 there is no ethical mandate, and a large sample size may actually be desirable. More specifically, μ may measure some overall treatment effect of considerable importance, for which a linear model is used to combine data from different substrata of the population. In the absence of evidence that μ differs from zero it may yet be interesting to examine endpoints of secondary importance or consider different substrata separately, and for this a large sample size is advisable.

However, in other cases there is incentive to stop testing if H_0 appears to be true. For example, if a treatment is already acknowledged to be successful and one is comparing two different doses to find an optimal level, in the absence of evidence for the superiority of one of these doses, one may wish to terminate the trial to permit a new trial with a new dose to begin as soon as possible. Another example occurs in drug screening, where a principal consideration is the cost of time and laboratory animals. If it appears that an experiment is going to lead to acceptance of H_0, that a drug does not have some minimal level of activity, there is little point in prolonging the experiment to examine endpoints of secondary importance or individual strata of the population.

For a one-sided formulation of this problem where one wants to test the hypothesis of no treatment effect ($\mu = 0$) against the hypothesis of positive effect ($\mu > 0$), a natural candidate is a truncated sequential probability ratio test of H_0: $\mu = 0$ against H_1: $\mu = \mu_1$ for suitable $\mu_1 > 0$. Let $A < 1 < B$ and define

Table 3.11. Truncated Sequential Probability Ratio Test: $b = -a = 6.7$; $c = 15$; $\eta = .3$; $m = 49$; $\alpha = .025$

μ	Power	$E_\mu(T \wedge m)$
.7	1.00	16
.6	.97	21
.4	.74	31
.3	.49	35
.2	.25	31
0	.025	21

$$T = \inf\{t: \exp[\mu_1 W(t) - \mu_1^2 t/2] \notin (A, B)\}$$
$$= \inf\{t: W(t) \notin (a + \eta t, b + \eta t)\},$$

where $a = (\log A)/\mu_1 < 0 < b = (\log B)/\mu_1$ and $\eta = \mu_1/2$. Let $m > 0$ and $a + \eta m \le c \le b + \eta m$. Consider the test which stops sampling at $\min(T, m)$ and rejects H_0 if and only if $T < m$ and $W(T) = b + \eta T$ or $T > m$ and $W(m) \ge c$.

The basic theory for this test was developed in Section 6. There, however, without a specific scientific problem in mind it seemed natural to let the symmetry of the mathematical problem lead to an emphasis on symmetric tests, i.e. those with $a = -b$ and $c = \eta m$. The present problem is clearly asymmetric, so the choice of a, b, η, m, and c becomes a matter for some thought.

Suppose we begin with the test of Table 3.2, which has a power function comparable to a fixed sample test with $m = 49$ and a much smaller sample size for large μ. Consider inserting a lower stopping boundary $a + \eta t$. Since for large μ the power of the test is essentially $P_\mu\{W(m) \ge c\}$ it seems reasonable to use the same value of c as before. For illustrative purposes suppose that $a = -b$, and then $E_{\mu_1}(T \wedge m) = E_0(T \wedge m)$. The value $b = 6.7$ yields the same significance level $\alpha = .025$ as in Table 3.2. Table 3.11 shows that the loss of power compared with Table 3.2 is slight but the decrease in expected sample size for μ close to 0 is quite substantial. By using a slightly smaller c and larger b the loss of power can probably be made completely negligible.

9. Brownian Approximation with Nuisance Parameters

Since Brownian motion is the natural limiting process associated with sums of independent random variables, it can be used in a fairly straightforward way to obtain simple, albeit crude approximations to the operating characteristics of sequential tests based on log likelihhood ratios—at least in the case of a one-parameter family of distributions. Here we discuss the problem of testing hypotheses about a parameter θ in the presence of a nuisance parameter λ. Although the following approximation should not be expected to be especially precise, its simplicity and potential generality for dealing with quite complex problems make it appealing.

Suppose $x_1, x_2, \ldots, x_n, \ldots$ are independent with probability density function $f(x; \theta, \lambda)$, where θ is a real parameter. For simplicity assume that λ is also real; the extension to the case of a vector nuisance parameter is straightforward. Let $l(n; \theta, \lambda) = \sum_{k=1}^{n} \log f(x_k; \theta, \lambda)$ denote the log likelihood function; let

$$\dot{l}(n; \theta, \lambda) = (\dot{l}_\theta(n; \theta, \lambda), \dot{l}_\lambda(n; \theta, \lambda))'$$

be the gradient of l, and

$$i(n; \theta, \lambda) = \begin{pmatrix} i_{\theta\theta}(n; \theta, \lambda), & i_{\theta\lambda}(n; \theta, \lambda) \\ i_{\lambda\theta}(n; \theta, \lambda), & i_{\lambda\lambda}(n; \theta, \lambda) \end{pmatrix}$$

the matrix of second partial derivatives of $-l(n; \theta, \lambda)$.

Consider the problem of sequentially testing H_0: $\theta = \theta_0$ against $H_1: \theta > \theta_0$. Let $\theta_1 > \theta_0$. If λ were known, one could base a test on

$$l(n; \theta_1, \lambda) - l(n; \theta_0, \lambda).$$

For unknown λ one possibility is to use $l(n; \theta_1, \hat{\lambda}_n) - l(n; \theta_0, \hat{\lambda}_n)$, where $(\hat{\theta}_n, \hat{\lambda}_n)$ is a maximum likelihood estimator for (θ, λ) (see Cox, 1963). Here we discuss a related method suggested by Whitehead (1978) to base a test on the score statistic

$$\dot{l}_\theta(n, \theta_0, \hat{\lambda}_{0n}),$$

where $\hat{\lambda}_{0n}$ is a maximum likelihood estimator of λ under the hypothesis that $\theta = \theta_0$.

Under regularity conditions which allow one to develop a large sample theory of maximum likelihood estimation, for θ in a neighborhood of θ_0 of size $0(n^{-1/2})$, a Taylor series expansion of $\dot{l}_\theta(n, \theta_0, \hat{\lambda}_{0n})$ about (θ, λ) yields

$$\dot{l}_\theta(n, \theta_0, \hat{\lambda}_{0n}) = \dot{l}_\theta(n; \theta, \lambda) + (\theta - \theta_0) i_{\theta\theta}(n; \theta, \lambda) - (\hat{\lambda}_{0n} - \lambda) i_{\theta\lambda}(n; \theta, \lambda) + o_p(n^{1/2}).$$

Since $\dot{l}_\theta(n; \theta_0, \lambda) = (\hat{\lambda}_{0n} - \lambda) i(n; \theta_0, \lambda) + o_p(n^{1/2})$, and $\dot{l}_\lambda(n; \theta_0, \lambda) = \dot{l}_\lambda(n; \theta, \lambda) + (\theta - \theta_0) i_{\theta\lambda}(n; \theta, \lambda) + o_p(n^{1/2})$, some algebra gives

$$\begin{aligned}\dot{l}_\theta(n; \theta_0, \hat{\lambda}_{0n}) = &\, \dot{l}_\theta(n; \theta, \lambda) - \dot{l}_\lambda(n; \theta, \lambda) i_{\theta\lambda}(n; \theta_0, \lambda)/i_{\lambda\lambda}(n; \theta_0, \lambda) \\ &+ (\theta - \theta_0)\{i_{\theta\theta}(n; \theta_0, \lambda) - i_{\theta\lambda}^2(n; \theta_0, \lambda)/i_{\lambda\lambda}(n; \theta_0, \lambda)\} + o_p(n^{1/2}).\end{aligned} \tag{3.59}$$

Since each element of the matrix $i(n; \theta_0, \lambda)$ is a sum of independent, identically distributed random variables it follows from the law of large numbers that

$$n^{-1} i(n; \theta_0, \lambda) \to I = \begin{pmatrix} I_{\theta_0\theta_0} & I_{\theta_0\lambda} \\ I_{\lambda\theta_0} & I_{\lambda\lambda} \end{pmatrix} \tag{3.60}$$

in $P_{\theta,\lambda}$-probability, where

$$I_{\theta\theta} = -E_{\theta,\lambda}\left\{\frac{\partial^2}{\partial\theta^2}\log f(x_1;\theta,\lambda)\right\} = E_{\theta,\lambda}\left[\frac{\partial}{\partial\theta}\log f(x_1;\theta,\lambda)\right]^2,$$

etc. Let $i^*(n;\theta,\lambda) = i_{\theta\theta}(n;\theta,\lambda) - i_{\theta\lambda}^2(n;\theta,\lambda)/i_{\lambda\lambda}(n;\theta,\lambda)$ and $I^*(\theta,\lambda) = I_{\theta\theta} - I_{\theta\lambda}^2/I_{\lambda\lambda}$. Then from (3.60) we see that the right hand side of (3.59) behaves asymptotically like a random walk with mean value $n(\theta - \theta_0)I^*(\theta_0,\lambda)$ and variance $nI^*(\theta_0,\lambda)$, and thus like a Brownian motion process. A different way of expressing this result is that if $\dot{l}_\theta(n;\theta_0,\hat{\lambda}_{0n})$ is plotted as a function of $nI^*(\theta_0,\lambda)$, then for large n the process behaves like $W(nI^*(\theta_0,\lambda))$, where W is Brownian motion with drift $\theta - \theta_0$.

This result is not immediately applicable because $I^*(\theta_0,\lambda)$ depends on the unknown nuisance parameter λ. Of course λ can be estimated by $\hat{\lambda}_{0n}$, although then the "time" scale of the Brownian motion process becomes a random variable. In many special cases it seems more reasonable to use the observed Fisher information $i^*(n;\theta,\lambda)$ in preference to $nI^*(\theta_0,\lambda)$ as a measure of variability (cf. Efron and Hinkley, 1978). Then $\dot{l}_\theta(n;\theta_0,\hat{\lambda}_{0,n})$ is approximated by $W(i^*(n;\theta_0,\hat{\lambda}_{0n}))$, where W is a Brownian motion with drift $\theta - \theta_0$.

A precise description of such a procedure is complicated by the fact that the "time scale" may run backwards, i.e. $i^*(n;\theta_0,\hat{\lambda}_{0n})$ may decrease due to variation of $\hat{\lambda}_{0n}$. For example, consider a group sequential test of the kind discussed in Section 5. Now the size of each group is determined by the rate of increase in $i^*(n;\theta_0,\hat{\lambda}_{0n})$. Specifically, if at most k "equally spaced" observations are to be made, let

$$m_j = \inf\{n\colon i^*(n;\theta_0,\hat{\lambda}_{0n}) \geq jm/k\},$$

$j = 1, 2, \ldots, k, \ldots$, and define

$$\tau = \inf\{m_j\colon \dot{l}_\theta(m_j;\theta_0,\hat{\lambda}_{0m_j}) \geq b + \eta i^*(m_j;\theta_0,\hat{\lambda}_{0m_j})\}.$$

Stop sampling at $\min(\tau, m_k)$ and reject H_0 if and only if $\tau \leq m_k$. The power function of this test is given approximately by (3.28) and (3.30). The expected sample size approximation, (3.17) with b replaced by $b + \rho$ to correct for excess over the boundary, yields an approximation to $E_{\theta,\lambda}\{i^*(\tau \wedge m_k;\theta_0,\lambda)\}$, which by (3.60) is approximately $I^*(\theta_0,\lambda)E_{\theta,\lambda}(\tau \wedge m_k)$.

For the preceding paragraph to have a precise mathematical interpretation, the problem must be normalized so that as $m \to \infty$ there exist constants $\tilde{b}$, $\tilde{\eta}$, and μ, such that $b = \tilde{b}m^{1/2} \to \infty$, $\eta = \tilde{\eta}/m^{1/2} \to 0$, and $\theta - \theta_0 = \mu/m^{1/2} \to 0$. The details of this analysis are omitted.

As a simple example, consider a logistic model with independent 0-1 observations $x_1, x_2, \ldots, x_n, \ldots$ having

$$P_{\alpha,\beta}\{x_n = 1\} = e^{\alpha+\beta z_n}/(1 + e^{\alpha+\beta z_n}) = 1 - P_{\alpha,\beta}\{x_n = 0\}.$$

Here $z_1, z_2, \ldots$ are fixed covariates, and we may wish to test $H_0\colon \beta = 0$ against, say, $H_1\colon \beta > 0$ with α as a nuisance parameter. It is easy to see that

$$l(n;\alpha,\beta) = \alpha\sum_1^n x_i + \beta\sum_1^n z_i x_i - \sum_1^n \log(1 + e^{\alpha+\beta z_i}),$$

and by differentiation that

$$\dot{l}_\beta(n; \alpha; 0) = \Sigma z_i\left(x_i - \frac{e^\alpha}{1 + e^\alpha}\right),$$

$$\hat{\alpha}_{0,n} = \log[\bar{x}_n/(1 - \bar{x}_n)] \qquad \text{where } \bar{x}_n = \frac{1}{n}\sum_{i=1}^{n} x_i,$$

and

$$i^*(n; \alpha, 0) = \frac{e^\alpha}{(1 + e^\alpha)^2}\sum_{i=1}^{n}(z_i - \bar{z}_n)^2.$$

Hence the class of tests described above is based on the statistic $\dot{l}_\beta(n; \hat{\alpha}_{0,n}, 0) = \Sigma z_i(x_i - \bar{x}_n)$ plotted on a time scale determined by $i^*(n; \hat{\alpha}_{0,n}, 0) = \bar{x}_n(1 - \bar{x}_n)\Sigma(z_i - \bar{z}_n)^2$.

Problems

3.1.* (a) Use (3.15) and (3.16) to show that

$$E_\mu\{\exp(-\lambda\tau)\} = \exp\{-[(2\lambda + (\mu - \eta)^2)^{1/2} - (\mu - \eta)]b\}.$$

(No calculations are necessary.)

(b) Evaluate $E_\mu\{\exp(-\lambda\tau)\}$ by an application of Proposition 3.2.

(c) Evaluate $\int_0^\infty E_\mu e^{-\lambda\tau}\, d\lambda$ by (a) and calculus.

(d) Evaluate $E_\mu\{\exp[-\lambda(\tau \wedge m)]\}$ and hence give another derivation of (3.17).

3.2. Find the distribution of $\tilde{\tau} = \inf\{t: t \geq m_0, W(t) \geq b + \eta t\}$.

3.3.* Let $x_1, x_2, \ldots$ be independent with $P_p\{x_i = 1\} = p$, $P_p\{x_i = -1\} = q = 1 - p$, let $S_n = x_1 + \cdots x_n$, and for $b = 1, 2, \ldots$ let $\tau = \tau_b = \inf\{n: S_n = b\}$.

(a) Use the argument leading to (3.13) to evaluate $P_{1/2}\{\tau < m | S_m = b - j\}$, $j = 1, 2, \ldots$.

(b) Show that for $j = 1, 2, \ldots$

$$P_p\{\tau < m, S_m = b - j\} = \left(\frac{p}{q}\right)^b P_q\{S_m = b + j\}.$$

(c) Use the method of Problem 3.1 to evaluate $E_p(\tau \wedge m)$.

(d) Consider the symmetric sequential probability ratio test of (2.8) truncated at the nth observation. Use (b) to compute approximations for $P_p\{S_{N \wedge m} \geq c\}$ as in (3.36). (The final approximation involves three cumulative binomial probabilities. For moderately large m, the central limit theorem with continuity correction can be used to obtain a further approximation to the distribution of S_m.)

(e) Generalize the considerations of (a), (b), and (c) to allow for the possibility that $P\{x_i = 0\} = r > 0\, (p + q + r = 1)$. Use the results to study a truncated version of the sequential probability ratio test of Problem 2.3.

3.4.* (a) Verify the remark at the end of Section 4 concerning the disconnectedness of the set of points $(\tau \wedge m, W(\tau \wedge m))$ for which the lower confidence bound μ_* exceeds μ.

(b) Show that the lower confidence bound μ_* defined in (3.24) is *not* in general uniformly most accurate (cf. Cox and Hinkley, 1974, p. 212 or Lehmann, 1959, p. 78).

3.5. Derive (3.52).

3.6. Let $\eta = 0$, $b = m\zeta$, $\xi = m\zeta - x$. Define $\tau(b)$ by (3.7). Prove that $P_\xi^{(m)}\{\tau(b) < m\} = P_{-\xi}^{(m)}\{\tau(x) < m\} = P_{-\xi}\{\tau(x) < \infty\}$. The second equality seems rather surprising. Given an intuitive reason why it should be expected to hold in the limit as $m \to \infty$.

3.7. Let $c_+(t)$ $(c_-(t))$ be a smooth function taking on positive (negative) values for $t \geq 0$. Let $T = \inf\{t\colon W(t) \notin (c_-(t), c_+(t))\}$ and suppose that $\bar{G}(t,\xi)\,d\xi = P_0\{T > t, W(t) \in d\xi\}$ is known. Shown heuristically that $h_+(t)\,dt = P_0\{T \in dt, W(T) = c_+(T)\}$ can be evaluated by

$$h_+(t) = -\tfrac{1}{2}\frac{\partial \bar{G}(t,\xi)}{\partial \xi}\bigg|_{\xi = c_+(t)},$$

and use this result to obtain Theorem 3.42 from Corollary 3.43. Hint: Let $\tau_b = \inf\{t\colon W(t) \geq b\}$, and note that since a smooth curve is approximately a constant over a short interval, for small δ, ε

$$P_0\{t < T \leq t + \delta, W(T) = c_+(T)\} \cong \int_0^\varepsilon \bar{G}(t, c_+(t) - \xi) P_0\{\tau_\xi \leq \delta\}\,d\xi.$$

Observe that $\bar{G}(t, c_+(t)) = 0$ and use the explicit evaluation of $P_0\{\tau_\xi \leq \delta\}$ to complete the calculation.

3.8 Use the results of Section 9 to obtain an approximate sequential test for the parameter $\theta = \mu/\sigma$ of a normal population with mean μ and variance σ^2, where $\lambda = \sigma^2$ is a nuisance parameter.

3.9. Suppose that a Brownian $W_1(t)$ motion with drift μ_1 and variance 1 per unit time has been observed for m units of time. A second Brownian process $W_2(t)$ with drift μ_2 and variance 1 is to be observed sequentially in order to compare μ_1 and μ_2. Consider the test which continues sampling until

$$|W_2(t) - tm^{-1}W_1(m)| \geq b$$

and upon stopping asserts that $\mu_2 > \mu_1$ or $\mu_2 < \mu_1$ depending on whether the quantity within absolute value signs is positive or negative. Find the probability of asserting that $\mu_2 > \mu_1$ (a) conditionally given $m^{-1}W_1(m)$ and (b) unconditionally. Note that the unconditional probability depends on the parameter $\delta = \mu_1 - \mu_2$. Note also the formal similarity between the answer to (a) and (3.53), hence also between the answer to (b) and the unconditional probability (3.54). See Switzer (1983) for more information concerning this problem.

3.10 Consider τ_b defined by (3.7) with $\eta = 0$. Assume that $\mu > 0$. Use Proposition 3.3 to show that $\operatorname{var}(\tau_b) = b/\mu^3$. Show that as $b \to \infty$ $P_\mu\{(\tau_b - b\mu^{-1})/(b/\mu^3)^{1/2} \leq x\} \to \Phi(x)$, where Φ denotes the standard normal distribution.

Remark. Although it is possible to demonstrate this limiting relation by a direct calculation starting from (3.15), the following two possibilities are more instructive.

(i) Observe that for arbitrary $\varepsilon > 0$

$$\{W(m) \geq b\} \subset \{\tau_b \leq m\} \subset \{W(m) \geq b - \varepsilon m^{1/2}\} \cup \{\tau_b < m, W(m) < b - \varepsilon m^{1/2}\}.$$

Show that for $m = b\mu^{-1} + (b\mu^{-3})^{1/2}x$, $P_\mu\{\tau < m, W(m) < b - \varepsilon m^{1/2}\} \to 0$. This proves the relation $\lim P_\mu\{\tau_b \leq m\} = \lim P_\mu\{W(m) \geq b\}$.

(ii) Note that by the strong law of large numbers $b\tau_b^{-1} = W(\tau_b)/\tau_b \to \mu$ with probability 1 as $b \to \infty$ and apply Theorem 2.40 (generalized to continuous time).

3.11.† Generalize the asymptotic normality in Problem 3.10 to nonlinear boundaries. A possible formulation is as follows.

Let $c(t)$ be a smooth concave function with $c(0) > 0$, and for $\mu > 0$ assume there exists a value $\lambda = \lambda_\mu$ (necessarily unique) such that $\mu\lambda = c(\lambda)$. Let $T_m = \inf\{t\colon W(t) \geq mc(t/m)\}$. Use the fact that $W(T_m) = mc(T_m/m)$ and the strong law of large numbers to show that $m^{-1}T_m \to \lambda$ with probability one as $m \to \infty$. Use this result and Theorem 2.40 to show that T_m is asymptotically normally distributed with mean $m\lambda$ and variance $m\lambda^3/[c(\lambda) - \lambda c'(\lambda)]^2$. An alternative possibility is to generalize the method (i) of Problem 3.10.

3.12. Let T be any stopping time $\leq m$ such that $E_\xi^{(m)}[T/(m - T)] < \infty$. Show that

$$E_\xi^{(m)}\left[\frac{W(T) - m^{-1}\xi T}{m - T}\right] = 0$$

and

$$E_\xi^{(m)}\left\{\left[\frac{W(T) - m^{-1}\xi T}{m - T}\right]^2\right\} = m^{-1}E_\xi^{(m)}[T/(m - T)].$$

Remark. For an informal argument, differentiate the equation $P_\xi^{(m)}\{T < m\} = E_0^{(m)} l^{(m)}(T, W(T); \xi, 0)$ with respect to ξ. For a rigorous proof use the fact that $(W(t) - m^{-1}\xi t)/(m - t)$ and $(W(t) - m^{-1}\xi t)^2/(m - t)^2 - m^{-1}t/(m - t)$ are martingales under $P_\xi^{(m)}$.

3.13. Generalize Theorem 3.58 as follows (Hoeffding, 1960). Let $x_1, x_2, \ldots$ be independent with one of the probability densities f_0, f_1, f_2. Let $\tilde{T}$ be the stopping rule and α, β the error probabilities of a sequential test of H_0: The true density is f_0 against H_1: The true density is f_1, with $\alpha + \beta < 1$. Then

$$E_2(\tilde{T}) \geq \{[(\sigma/4)^2 - \mu\log(\alpha + \beta)]^{1/2} - \sigma/4\}^2/\mu^2,$$

where $\mu_i = E_2\{\log[f_2(x_1)/f_i(x_1)]\}$, $\mu = \max(\mu_0, \mu_1)$, and $\sigma^2 = \text{var}_2\{\log[f_1(x_1)/f_0(x_1)]\}$.

3.14.* Consider a Brownian motion process $W(t)$, $0 < t \leq m$, conditional on $W(m) = \xi$, and think of ξ as an unknown parameter. Then for $t < m$ $l^{(m)}(t, W(t); \xi, 0)$ is a likelihood function for the observations $W(s)$, $s \leq t$; and the observed Fisher information is

$$\mathscr{I}_t(\xi) = -\frac{\partial^2}{\partial\xi^2}\log l^{(m)}(t, W(t); \xi, 0) = t/m(m - t).$$

(a) Show that a sequential probability ratio test of ξ_0 against ξ_1 minimizes $E_{\xi_i}^{(m)}[\mathscr{I}_T(\xi_i)]$ for $i = 0, 1$ over all tests having no larger error probabilities at ξ_0 and ξ_1. Interpret this test in the context of predicting an election.

(b) Show that in terms of $W(t)$ the stopping boundary of the test in (a) is the same as the triangular test of Section 7. Use this observation to derive (3.53) directly from (2.11).

Hint: For (a), it may be helpful to consider a "one-sided" problem as a preliminary exercise (cf. Proposition 2.38 and Theorem 2.39).

3.15.* Let $\eta > 0$ and let $\tau = \inf\{t: W(t) \geq b + \eta t\}$ define a one-sided sequential probability ratio test of H_0: $\mu = 0$ against H_1: $\mu = 2\eta$ as discussed in II.4. Suppose that although one attempts to "stop" the process $W(t)$ at time τ, it runs on for some additional fixed length of time $\bar{m}$, so the final observed value is $W(\tau + \bar{m})$. Let $U = W(\tau + \bar{m})/(\tau + \bar{m})$. Show that if $U > \eta$ an attained significance level defined by large values of U equals

(3.61) $$\int_{-\infty}^{\infty} G[(b - \bar{m}U + \bar{m}^{1/2}z)/(U - \eta)]\phi(z)\,dz,$$

where $G(m)$ is the right hand side of (3.15) with $\mu = 0$. Given an asymptotic justification for approximating (3.61) by

(3.62) $$G[(b - \bar{m}U)/(U - \eta)] + \tfrac{1}{2}[\bar{m}/(U - \eta)^2]G''[(b - \bar{m}U)/(U - \eta)] + \cdots.$$

3.16. For τ defined by (3.7), use (3.16) to prove that for $\mu \geq \eta$ $[W(\tau) - \mu\tau]^2/\tau$ has a chi-square distribution with one degree of freedom. Hence the confidence interval $W(\tau)/\tau \pm z_\gamma/\tau^{1/2}$ fails to cover μ with exactly the nominal 2γ probability (although neither one-sided interval exactly achieves the nominal γ).

3.17.* As an alternative to the tests of Tables 3.1 and 3.2 take $\eta = 0$ and $b = c$. Computer power functions and expected sample size functions as in Tables 3.1 and 3.2 (a) for continuous observation and (b) for a group test with at most 7 groups of 7 observations each. (Use (3.28) in computing the power function of the group test.) Why is the choice $\eta = 0$ a natural one for increasing power in the neighborhood of $\mu = 0$?

CHAPTER IV

Tests with Curved Stopping Boundaries

1. Introduction and Examples

The stopping rules of Chapter II and III are defined by the crossing of linear boundaries by random walks (or Brownian motion). The linear boundaries arise naturally from sequential probability ratio tests of simple hypotheses against simple alternatives. For problems involving several parameters or composite hypotheses we shall want to consider curved stopping boundaries, which are more difficult to investigate; and only rarely can one obtain exact results—even for Brownian motion.

In this chapter we shall review some of the considerable progress which has been made—for the most part since 1976—in analyzing tests defined by curved stopping boundaries. Special attention is given to the so-called repeated significance tests and their application to clinical trials.

We begin with several examples. It will be helpful to think in terms of the following simplified problem. Patients enter a clinical trial sequentially and receive one of two treatments, A or B. The (immediate) responses of patients receiving treatment A are denoted by $y_1, y_2, \ldots$, those receiving treatment B by $z_1, z_2, \ldots$. To make a paired comparison of A and B we consider only the differences $x_n = y_n - z_n$ $(n = 1, 2, \ldots)$, and make the probabilistic assumption that $x_1, x_2, \ldots$ are independent and normally distributed with mean μ and variance σ^2. (We shall not discuss the manner in which patients are paired, which may include matching on prognostic factors.) To be specific, suppose that a large mean response is desirable. Then treatment A or B is preferred according as $\mu > 0$ or $\mu < 0$. The null hypothesis of no treatment difference is H_0: $\mu = 0$.

EXAMPLE 4.1. Test of Power One. This example is artificial, but theoretically fascinating; and the associated mathematical theory is somewhat simpler than that of repeated significance tests, which follow. Assume that σ is known and without loss of generality equals one. Suppose there is no cost of sampling and no requirement that a decision be made when H_0: $\mu = 0$ is true, i.e. as long as the data are consistent with H_0 we are content to observe $x_1, x_2, \ldots$ *ad infinitum*. On the other hand, suppose that under H_1: $\mu \neq 0$ we would like to stop sampling as soon as possible no matter what the value of μ. This leads to the

Definition. An α level test of power one of H_0: $\mu = 0$ against H_1: $\mu \neq 0$ is a stopping rule T such that $P_0\{T < \infty\} \leq \alpha$ and $P_\mu\{T < \infty\} = 1$ for all $\mu \neq 0$.

It is understood that to stop sampling is equivalent to rejecting H_0, i.e. the rejection region of the test is that part of the sample space where $T < \infty$. A good test of power one has a small expected sample size $E_\mu(T)$ (in some suitable sense) for all $\mu \neq 0$.

It is not obvious that tests of power one exist. Clearly the boundary of the continuation region can not consist of a finite number of line segments, for in that case $P_0\{T < \infty\} < 1$ implies $P_\mu\{T < \infty\} < 1$ for $|\mu|$ sufficiently small but positive.

To construct a test of power one let $Q = \int_{-\infty}^{\infty} P_\mu \phi(\mu)\,d\mu$ be a standard normal mixture of the probabilities $\{P_\mu\}$. The likelihood ratio of $x_1, \ldots, x_n$ under Q relative to P_0 (in the notation of Proposition 3.1) is

$$l_n = \int_{-\infty}^{\infty} l(n, S_n; \mu, 0)\phi(\mu)\,d\mu = \int_{-\infty}^{\infty} \exp[\mu S_n - n\mu^2/2]\phi(\mu)\,d\mu$$
$$= (n+1)^{-1/2}\exp[S_n^2/2(n+1)],$$

and a one-sided sequential probability ratio test of P_0 against Q is defined by the stopping rule (cf. (2.35))

$$(4.2) \quad T = \inf\{n: l_n \geq B\} = \inf\{n: |S_n| \geq \{(n+1)[\log(n+1) + 2\log B]\}^{1/2}\},$$

where, as before, $\inf \phi = +\infty$. The basic inequality (2.9) applies to show that $P_0\{T < \infty\} \leq B^{-1}$ for $B > 1$, and this can be made less than any pre-assigned α by taking $B \geq \alpha^{-1}$. Also by the strong law of large numbers $P_\mu\{n^{-1}S_n \to \mu\} = 1$, and since $n^{-1}\{(n+1)[\log(n+1) + 2\log B]\}^{1/2} \to 0$ as $n \to \infty$, we see by (4.2) that $P_\mu\{T < \infty\} = 1$ for all $\mu \neq 0$. Thus T defined by (4.2) is a test of power one of $\mu = 0$ against $\mu \neq 0$.

EXAMPLE 4.3. Repeated Significance Test. In practice if μ appears to be close to zero, one will eventually terminate a clinical trial with acceptance (non-rejection) of H_0: $\mu = 0$. For example, we might stop at $\min(T, m)$, where T is defined by (4.2), and reject H_0 if and only if $T \leq m$. The power and expected sample size are

$$(4.4) \qquad P_\mu\{T \leq m\}$$

and

$$E_\mu(T \wedge m). \tag{4.5}$$

Alternatively, there is now no obvious advantage to curved boundaries, so we might also consider the qualitatively similar test described in Remark 3.18.

An appealing *ad hoc* stopping rule is based on the observation that a conventional significance test of H_0: $\mu = 0$ with sample size n would reject H_0 if $|S_n| > bn^{1/2}$ for suitable b. Since we would like to reject H_0 after only a few observations (if rejection is appropriate), we consider the stopping rule

$$T = \inf\{n: n \geq m_0, |S_n| > bn^{1/2}\} \qquad (m_0 \geq 1) \tag{4.6}$$

and the sequential test: stop sampling at $\min(T, m)$ and reject H_0 if and only if $T \leq m$. Again the power is given by (4.4) and the expected sample size by (4.5). To obtain a given significance level α one requires a larger value of b in (4.6) when $m > m_0$ than for a fixed sample size significance test.

Alternative motivation for the definition (4.6) is as follows. To test H_0: $\mu = 0$ against the simple alternative $\mu = \mu_1$, one may use the test of III.3: stop sampling at $\min(\tau(\mu_1), m)$, where

$$\tau(\mu_1) = \inf\{n: l(n, S_n; \mu_1, 0) \geq B\}$$

and reject H_0 if and only if $\tau(\mu_1) \leq m$. According to II.4, for large m this test has approximately minimum expected sample size under μ_1 subject to having significance level no more than $P_0\{\tau(\mu_1) \leq m\} \cong P_0\{\tau(\mu_1) < \infty\} \leq B^{-1}$. It seems plausible that a reasonable test for the composite alternative H_1: $\mu \neq 0$ may be defined in terms of the stopping rule $\min_{\mu_1 \neq 0} \tau(\mu_1)$, or equivalently by

$$T = \inf\left\{n: \max_{\mu_1} l(n, S_n; \mu_1, 0) \geq B\right\} = \inf\{n: S_n^2/2n \geq a\}, \tag{4.7}$$

which is the same as (4.6) with $m_0 = 1$ and $b = (2a)^{1/2}$.

The interpretation of (4.7) as a repeated generalized likelihood ratio test will prove useful in analyzing (4.6) and in generalizing it to other parametric models.

EXAMPLE 4.8. Sequential t-tests. Suppose now that σ is an unknown nuisance parameter, and we would like to test H_0: $\mu = 0$ against H_1: $\mu \neq 0$ with a test similar to that of the preceding example. The log generalized likelihood ratio statistic is

$$\log\left\{\sup_{\mu,\sigma} \prod_{i=1}^n \sigma^{-1}\phi[(x_i - \mu)/\sigma]\right\} - \log\left\{\sup_{\sigma} \prod_{i=1}^n \sigma^{-1}\phi(x_i/\sigma)\right\}$$
$$= \tfrac{1}{2} n \log(1 + \bar{x}_n^2/v_n^2),$$

where

$$\bar{x}_n = n^{-1}\Sigma_1^n x_i \quad \text{and} \quad v_n^2 = n^{-1}\Sigma_1^n (x_i - \bar{x}_n)^2.$$

In analogy with (4.7) we define

$$T = \inf\{n: n \geq m_0, \tfrac{1}{2} n \log(1 + \bar{x}_n^2/v_n^2) > a\} \qquad (m_0 \geq 2) \tag{4.9}$$

and consider the test: stop sampling at $\min(T, m)$ and reject H_0 if and only if $T \leq m$.

Note that the joint distribution of $\{\bar{x}_n/v_n, n = 2, 3, \ldots\}$ depends on (μ, σ) only through the ratio $\theta = \mu/\sigma$. It follows that the distribution of T, hence the power and expected sample size function of this test are functions only of θ. Note also that in contrast to the case of known σ, here the introduction of stopping boundaries which are nonlinear relative to the sufficient statistics $(\Sigma_1^n x_i, \Sigma_1^n x_i^2)$ seems unavoidable.

EXAMPLE 4.10. Schwarz's Boundaries. The first person to investigate repeated likelihood ratio tests was Schwarz (1962), whose formulation was somewhat different than the one considered above. To test H_0: $\mu \leq 0$ against H_1: $\mu > 0$ (once again with σ known and equal to 1), let $\delta > 0$ and define the one-sided likelihood ratio statistics

$$\lambda_n^+ = \log\left\{\sup_{\mu > -\delta} l(n, S_n; \mu, -\delta)\right\}$$

and

$$\lambda_n^- = \log\left\{\sup_{\mu < \delta} l(n, S_n; \mu, \delta)\right\}.$$

Stop sampling at

$$T = \inf\{n\colon \max(\lambda_n^+, \lambda_n^-) \geq a\} = \inf\{n\colon |S_n| \geq (2an)^{1/2} - \delta n\}$$

and reject H_0 if and only if $S_T \geq 0$. This test may be considered a reasonable competitor to a (truncated) sequential probability ratio test or to Anderson's modification (III.7). Like the triangular version of Anderson's test, the stopping region can be truncated to reduce the maximum sample size with little effect on the power function or expected sample size.

The arrangement of the rest of this chapter is as follows. In Section 2 we develop approximations to the significance level, power, and expected sample size for repeated significance tests of the drift to Brownian motion. In Section 3 the corresponding discrete time results are stated, and some numerical examples are given to illustrate the accuracy of the approximations. A critical examination of these examples leads to modifications of the repeated significance tests, which are introduced in Section 4 and discussed in Sections 5 and 6. Additional theoretical results are contained in Sections 7 and 8.

The reader whose main interest is in applications may wish to turn directly to Section 3.

2. Repeated Significance Tests for Brownian Motion

Let $\{W(t), 0 \leq t < \infty\}$ be a Brownian motion with drift μ. Given $b > 0, m_0 > 0$ define

$$T = \inf\{t\colon t \geq m_0, |W(t)| \geq bt^{1/2}\}. \tag{4.11}$$

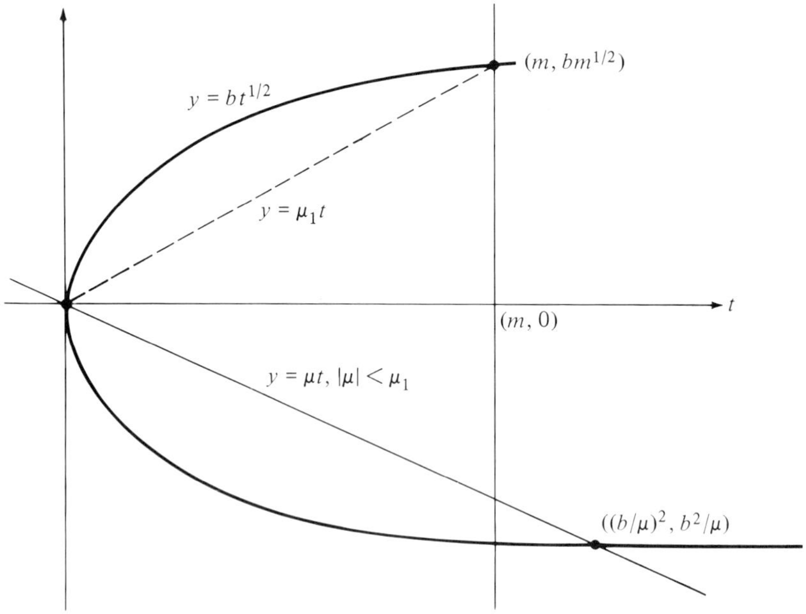

Figure 4.1.

In this section we consider in detail the test of H_0: $\mu = 0$ against H_1: $\mu \neq 0$ which stops sampling at $\min(T, m)$ $(m > m_0)$ and rejects H_0 if and only if $T \leq m$.

The situation is more complicated than in III.3 because we shall not obtain an exact expression for the joint distribution of T and $W(m)$, from which all else follows by straightforward calculus. Instead we shall obtain a number of asymptotic approximations, which will be useful under different relations among the crucial parameters b, m, and μ. In particular the approximation to $P_\mu\{T \leq m\}$ will be quite different under H_0: $\mu = 0$ than under H_1: $\mu \neq 0$.

We shall approximate the probability $P_\mu\{T \leq m\}$ and expected sample size $E_\mu(T \wedge m)$ asymptotically as $b \to \infty$, $m \to \infty$, and for some fixed $0 < \mu_1 < \infty$, $bm^{-1/2} = \mu_1$. Thus as both the parabolic stopping boundary scaled by b and the truncation m move towards $+\infty$, the slope of the line from $(0, 0)$ to $(m, bm^{1/2})$, where the boundary and truncation line meet, remains fixed at μ_1 (see Fig. 4.1).

By the law of large numbers the process $W(t)$ will travel close to the line $y = \mu t$, except for random fluctuations. Hence with probability one, for each $0 < \varepsilon < 1$, for all t sufficiently large

$$\mu(1 - \varepsilon) \leq t^{-1} W(t) \leq \mu(1 + \varepsilon).$$

It follows that for all b sufficiently large, $W(t)$ first crosses $bt^{1/2}$ between the points where $\mu t(1 - \varepsilon)$ and $\mu t(1 + \varepsilon)$ do, i.e., in the interval $([b/\mu(1 + \varepsilon)]^2, [b/\mu(1 - \varepsilon)]^2)$. Since ε is arbitrary, for each $\mu \neq 0$

(4.12) $$P_\mu\left\{\lim_{b\to\infty} b^{-2}T = \mu^{-2}\right\} = 1,$$

and in particular

(4.13) $$\begin{aligned} P_\mu\{T \le m\} &\to 0 \qquad \text{for } |\mu| < \mu_1 \\ &\to 1 \qquad \text{for } |\mu| > \mu_1. \end{aligned}$$

From (4.12) and (4.13) it follows that

$$\begin{aligned} E_\mu(T \wedge m) &\sim (b/\mu)^2 \qquad |\mu| > \mu_1 \\ &\sim m \qquad |\mu| < \mu_1. \end{aligned}$$

For μ in a neighborhood of μ_1 of size $O(m^{-1/2})$ the probability in (4.13) converges to a limit between 0 and 1, and a more interesting approximation to $E_\mu(T \wedge m)$ can be obtained (see Theorem 4.27 below).

The following fundamental result corresponds to (3.13) in the linear case. Recall the notation

$$P_\xi^{(m)}(A) = P_0(A \mid W(m) = \xi) \qquad (A \in \mathscr{E}_m).$$

Theorem 4.14. *Suppose $b \to \infty$, $m \to \infty$, $m_0 \to \infty$ in such a way that for some $0 < \mu_1 < \mu_0 < \infty$, $bm^{-1/2} = \mu_1$ and $bm_0^{-1/2} = \mu_0$. Let $\xi = m\xi_0$. For every closed subinterval K of $(\mu_1^2/\mu_0, \mu_1]$, uniformly for $\xi_0 \in K$*

$$P_\xi^{(m)}\{T < m\} \sim \mu_1 \xi_0^{-1} \exp\{-\tfrac{1}{2}m(\mu_1^2 - \xi_0^2)\}.$$

For all ξ_0

$$P_\xi^{(m)}\{T < m\} \le (m/m_0)^{1/2} \exp\{-\tfrac{1}{2}m(\mu_1^2 - \xi_0^2)\},$$

and uniformly for ξ_0 in closed subintervals of $[0, \mu_1^2/\mu_0)$

$$P_\xi^{(m)}\{T < m\} = o(\exp\{-\tfrac{1}{2}m(\mu_1^2 - \xi_0^2)\}).$$

The proof of Theorem 4.14 is preceded by several applications. Let $c \le b$. Since

(4.15) $$P_\mu\{T < m, |W(m)| < cm^{1/2}\} = \int_{|\xi| < cm^{1/2}} P_\xi^{(m)}\{T < m\} P_\mu\{W(m) \in d\xi\}$$

(cf. (3.10)), unconditioning Theorem 4.14 yields

Corollary 4.16. *Suppose in addition to the asymptotic normalization of Theorem 4.14 that $cm^{-1/2} = \gamma \in (\mu_1^2/\mu_0, \mu_1]$. Then uniformly for μ in closed subintervals of $(0, \infty)$*

(4.17) $$P_\mu\{T < m, |W(m)| < cm^{1/2}\} \sim \frac{\phi[m^{1/2}(\mu_1 - \mu)]}{\mu m^{1/2}} \mu_1 \gamma^{-1} e^{-m\mu(\mu_1 - \gamma)},$$

and

(4.18) $$P_0\{T < m, |W(m)| < cm^{1/2}\} \sim b\phi(b)\log(mc^2/m_0 b^2).$$

Corollary 4.19. *Under the asymptotic relations of Theorem* 4.14, *uniformly for* μ *in closed subintervals of* $(0, \infty)$

$$P_\mu\{T < m\} = 1 - \Phi[m^{1/2}(\mu_1 - \mu)] + \{\phi[m^{1/2}(\mu_1 - \mu)]/\mu m^{1/2}\}(1 + o(1))$$
(4.20)

as $b \to \infty$.

Corollary 4.19 follows at once from (4.17) and the obvious relation

$$P_\mu\{T < m\} = P_\mu\{|W(m)| > bm^{1/2}\} + P_\mu\{T < m, |W(m)| < bm^{1/2}\}.$$

It is interesting to note that $P_\mu\{T < m\}$ converges to 0 or 1 for fixed μ less than or greater than μ_1 and to a limit between 0 and 1 for $\mu = \mu_1 + \Delta m^{-1/2}$. This is a consequence of the behavior of the first probability on the right hand side of (4.20). The last term always converges to 0, but typically plays an important role in numerical calculations.

From (4.18) with $c = b$ one easily obtains an approximation for $P_0\{T < m\}$. A more detailed analysis, which is deferred to Chapter XI, yields the following more precise result.

Theorem 4.21. *Suppose* $b \to \infty, m \to \infty$, *and* $m_0 \to \infty$ *in such a way that for some* $0 < \mu_1 < \mu_0 < \infty$, $bm^{-1/2} = \mu_1$ *and* $bm_0^{-1/2} = \mu_0$. *Then for T defined by* (4.11) *as* $b \to \infty$

$$P_0\{T < m\} = (b - b^{-1})\phi(b)\log(m/m_0) + 4b^{-1}\phi(b) + o(b^{-1}\phi(b)).$$

Remarks (i) From the scaling property that $m^{-1/2}W(mt)$, $0 \le t < \infty$, is again a Brownian motion it follows that $P_0\{T < m\}$ is a function of b and the ratio m/m_0. Hence Theorem 4.21 holds for any m_0 and m whose ratio remains constant as $b \to \infty$. This is not the case for the corresponding results in discrete time.

Actually the condition that m/m_0 remain constant is more restrictive than necessary (see Jennen, 1985).

(ii) For numerical purposes we shall usually use an analogous discrete time result ((4.40) below). However, it is interesting to note that the approximation given by Theorem 4.21 is quite accurate, even when $P_0\{T \le m\}$ is not close to 0. An exact but complicated eigenfunction expansion of $P_0\{T \le m\}$ has been given by DeLong (1981), who also tabulated $P_0\{T \le m\}$ for a range of values of b and m/m_0. An indication of the accuracy of the approximation in Theorem 4.21 is given in Table 4.1.

Proof of Theorem 4.14. The proof is an elaboration of the idea leading to (3.13). One key idea is to think of time as running backwards from $t = m$ to $t = 0$. Let

Table. 4.1. Approximation and Exact Values of $P_0\{T \le m\}$

b	m/m_0	Approximation from (4.21)	Exact Value (DeLong)
2.9	10	.043	.044
3.4	20	.013	.013
2.4	5	.109	.113
1.9	50	.491	.501
1.4	100	.901	.866

$$P_{\lambda,\xi}^{(m)}(A) = P_0\{A \mid W(0) = \lambda, W(m) = \xi\} \qquad (A \in \mathscr{E}_m)$$

and define

$$T^* = \sup\{t\colon t \le m, |W(t)| \ge bt^{1/2}\},$$

so

$$P_{0,\xi}^{(m)}\{T < m\} = P_{0,\xi}^{(m)}\{T^* \ge m_0\}.$$

It is easy to see from (3.11) that the likelihood ratio of $\{W(s), t \le s \le m\}$ under $P_{\lambda,\xi}^{(m)}$ relative to $P_{0,\xi}^{(m)}$ is

$$\exp[\lambda W(t)/t - \lambda^2/2t - \lambda\xi/m + \lambda^2/2m].$$

Let

$$\tilde{P}_\xi^{(m)} = \int_{-\infty}^{\infty} P_{\lambda,\xi}^{(m)} \phi[(\lambda - \xi)/m^{1/2}]\, d\lambda/m^{1/2} \tag{4.22}$$

and note that under $\tilde{P}_\xi^{(m)}$, $\{W(t), 0 \le t \le m\}$ is just an ordinary driftless Brownian motion process with "initial" value $W(m) = \xi$ and time scale running backward from $t = m$ to $t = 0$ (see Figure 4.2).

By (4.22) the likelihood ratio of $\{W(s), t \le s \le m\}$ under $\tilde{P}_\xi^{(m)}$ relative to $P_{0,\xi}^{(m)}$ is

$$\begin{aligned} &\int_{-\infty}^{\infty} \exp[\lambda W(t)/t - \lambda^2/2t - \lambda\xi/m + \lambda^2/2m]\phi[(\lambda - \xi)/m^{1/2}]\, d\lambda/m^{1/2} \\ &\quad = (t/m)^{1/2} \exp[W^2(t)/2t - \xi^2/2m]. \end{aligned} \tag{4.23}$$

Let $\tilde{E}_\xi^{(m)}$ denote expectation under $\tilde{P}_\xi^{(m)}$. Since T^* is a stopping time in the reversed time scale, by Wald's likelihood ratio identity and (4.23)

$$\begin{aligned} P_{0,m\xi_0}^{(m)}\{T^* \ge m_0\} &= \tilde{E}_{m\xi_0}^{(m)}\{(m/T^*)^{1/2} \exp[-W^2(T^*)/2T^* + m\xi_0^2/2]; T^* \ge m_0\} \\ &= \exp[-m(\mu_1^2 - \xi_0^2)/2]\tilde{E}_{m\xi_0}^{(m)}\{(m/T^*)^{1/2}; T^* \ge m_0\}. \end{aligned} \tag{4.24}$$

To analyze this final expectation note that the $\tilde{P}_{m\xi_0}^{(m)}$ distribution of T^* is the same as the P_0 distribution of $(m - \tau^*)^+$, where

$$\tau^* = \inf\{t\colon |m\xi_0 + W(t)| \ge b(m - t)^{1/2}\}.$$

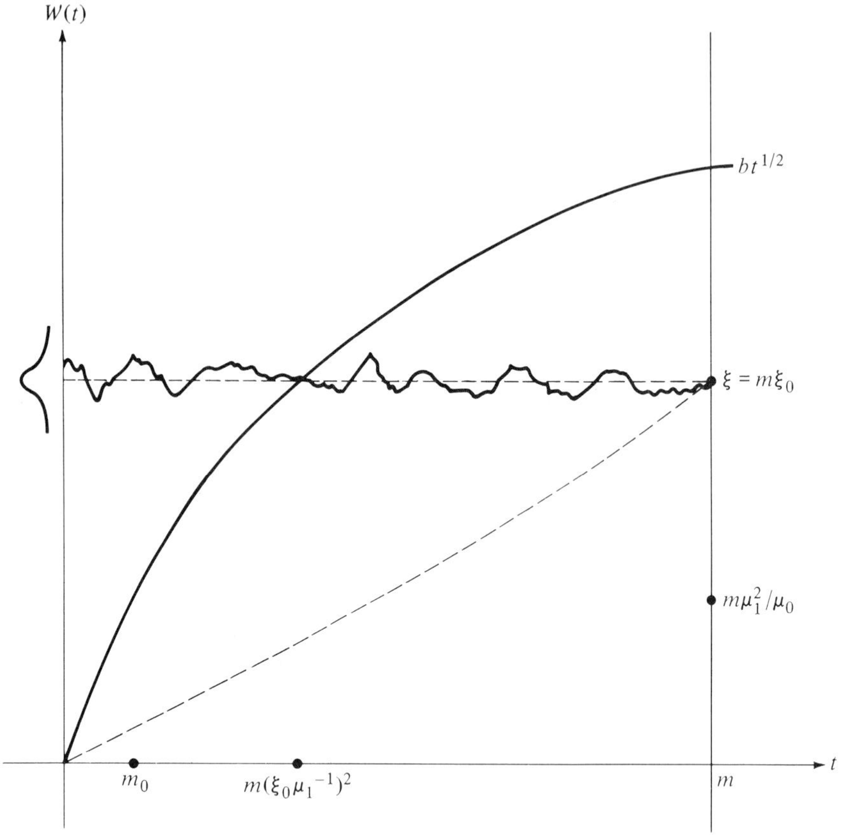

Figure 4.2.

Let $\varepsilon > 0$ and $\mu_1^2/\mu_0 < \xi_0 \leq \mu_1$. It is easy to see that

$$P_0\{\tau^* \leq \min(m - m_0, m[1 - \xi_0^2\mu_1^{-2}(1 - \varepsilon)])\} \to 1$$

uniformly in ξ_0 provided ξ_0 is bounded away from μ_1^2/μ_0. Also by (3.15) with $\mu = \eta = 0$,

$$P_0\{\tau^* \geq m[1 - \xi_0^2\mu_1^{-2}(1 + \varepsilon)]\} \to 1$$

uniformly in ξ_0 provided ξ_0 is bounded away from μ_1. The desired result for $\mu_1^2/\mu_0 < \xi_0 \leq \mu_1$ now follows from (4.24). Similar but easier arguments take care of the other cases. □

Remark 4.25. Variations of the preceding argument work for boundary curves other than $bt^{1/2}$. Introduction of the probability $\tilde{P}_\xi^{(m)}$ leads to the simple equation (4.24) facilitating the proof of Theorem 4.14, but it may not be appropriate for other boundaries. An instructive alternative possibility, which works quite generally, is to carry out a likelihood ratio calculation directly in terms of $P_{\lambda,\xi}^{(m)}$, in this case for $\lambda = \xi$. This has the effect of approximating the curved boundary by its tangent at $m(\xi_0/\mu_1)^2$. For fixed $\xi_0 < \mu_1$ rather elaborate analysis is required (see Siegmund, 1982). For ξ close to $bm^{1/2}$, specifically for $\xi = bm^{1/2} - x$ with x fixed, the details are much simpler. In this case crossing the curve has asymptotically the same (conditional) probability as crossing the tangent to the curve at $t = m$. For proving Corollary 4.19 it is only the values of ξ close to $bm^{1/2}$ which contribute (cf. Problems 4 and 14).

For a refinement of the proof of Theorem 4.14, which yields Theorem 4.21, see XI.1.

Now consider

$$
\begin{aligned}
E_\mu(T \wedge m) &= E_\mu(T) - E_\mu[(T-m); T > m] \\
&= m + E_\mu[(T-m); T < m].
\end{aligned}
\tag{4.26}
$$

For fixed $\mu > \mu_1$, $E_\mu[(T-m); T > m]$ will converge to 0, whereas for fixed $0 < \mu < \mu_1$, $E_\mu[(T-m); T < m]$ will converge to 0. Hence it seems reasonable to try to expand $E_\mu(T \wedge m)$ about $E_\mu(T)$ when $\mu > \mu_1$ and about m when $0 < \mu < \mu_1$. However, only for values of μ fairly close to μ_1, specifically for $\mu = \mu_1 + \Delta m^{-1/2}$, will these second order terms be large enough to play a significant role. In principal, for μ close to μ_1 one could develop equivalent expansions for $E_\mu(T \wedge m)$ starting from either of the expressions in (4.26). However, the choice of the starting expression will to some extent determine how well the result works when μ is not close to μ_1. Since large values of μ are much more important, we shall work only with the first expression in (4.26) and keep in mind that the result should not be expected to give good approximations when μ is small and $E_\mu(T \wedge m)$ about equal to m.

Theorem 4.27. *Suppose $b \to \infty$, $m \to \infty$, $m_0 \to \infty$ in such a way that for some $0 < \mu_1 < \mu_0 < \infty$, $bm^{-1/2} = \mu_1$ and $bm_0^{-1/2} = \mu_0$. Then for any Δ and $\mu = \mu_1 + \Delta m^{-1/2}$*, as $m \to \infty$

$$
\begin{aligned}
E_\mu(T \wedge m) = (b^2 - 1)/\mu^2 - (\mu - \tfrac{1}{2}\mu_1)^{-1}\{&m^{1/2}[\phi(\Delta) - \Delta\Phi(-\Delta)][1 - b^{-2}] \\
&- [\mu_1/8(\mu - \tfrac{1}{2}\mu_1)^2][\Phi(-\Delta)(1 + \Delta^2) - \Delta\phi(\Delta)] - b^{-1}\phi(\Delta)/\mu_1 \\
&+ \mu_1^{-1}b^{-1}[(2 + \Delta^2)\phi(\Delta) - \Delta(\Delta^2 + 3)\Phi(-\Delta)]\} + o(b^{-1}).
\end{aligned}
$$

The asymptotic expansion in Theorem 4.27 is rather complicated—involving terms of order m, $m^{1/2}$, 1, and $m^{-1/2}$. For moderately large m the terms of order $m^{-1/2}$ can be neglected to obtain a simpler approximation; but for group tests with a small value of m these terms often improve the approximation substantially (see Table 4.3). To simplify the exposition, Propositions 4.28 and 4.29 justify this approximation only up to terms which converge to 0 as $b \to \infty$. The origin of the terms of order b^{-1} is discussed in Problem 4.20.

Proposition 4.28. *Under the conditions of Theorem* 4.27 *on* b, m, *and* m_0, *for each* $0 < \mu < \mu_0$, *uniformly in closed subsets of* $(0, \mu_0)$,

$$E_\mu(T) = (b^2 - 1)/\mu^2 + o(1) \qquad (b \to \infty).$$

To justify (4.28) note that

$$W^2(T)/T = \mu^2 T + (W(T) - \mu T)^2/T + 2\mu(W(T) - \mu T)$$

and that on $\{T > m_0\}$, $W^2(T)/T = b^2$. Hence by Wald's identity (3.3) and some easy calculations

$$\mu^2 ET + E_\mu\{(W(T) - \mu T)^2/T\} = E_\mu(W^2(T)/T) = b^2 + o(1).$$

Equation (4.12) and Wald's identity (3.3) for the second moment lead to

$$\begin{aligned} E_\mu\{(W(T) - \mu T)^2/T\} &\sim \mu^2 b^{-2} E_\mu\{(W(T) - \mu T)^2\} \\ &= \mu^2 b^{-2} E_\mu(T) \to 1 \qquad (b \to \infty). \end{aligned}$$

Hence $\mu^2 E_\mu(T) + 1 = b^2 + o(1)$, which is equivalent to Proposition 4.28.

Proposition 4.29 below provides an asymptotic approximation for $E_\mu[(T - m); T > m]$ in (4.26), up to terms converging to 0 as $b \to \infty$.

Proposition 4.29. *Under the conditions of Theorem* 4.27, *for* T_+ *defined by* $T_+ = \inf\{t: t \geq m_0, W(t) \geq bt^{1/2}\}$ *and for* $\mu = \mu_1 + \Delta m^{-1/2}$

$$\begin{aligned} E_\mu\{E_\mu[T_+ - m | W(m)]; T_+ > m\} &= m^{1/2}(\mu - \tfrac{1}{2}\mu_1)^{-1}[\phi(\Delta) - \Delta\Phi(-\Delta)] \\ &\quad - \mu_1^{-2}[\Phi(-\Delta)(1 + \Delta^2) - \Delta\phi(\Delta)] + o(1) \end{aligned}$$

as $b \to \infty$.

To understand this result, re-write the expectation in Proposition 4.29 as

$$\int_{(0,\infty)} P_\mu\{W(m) \in \mu_1 m - dx\}(1 - P_\xi^{(m)}\{T_+ < m\}) E_\mu \tau_m(x), \tag{4.30}$$

where $\xi = \mu_1 m - x$ and

$$\tau_m(x) = \inf\{t: t > 0, W(t) = \mu_1 m^{1/2}[(m + t)^{1/2} - m^{1/2}] + x\}.$$

Since $\mu_1 m^{1/2}[(m + t)^{1/2} - m^{1/2}] \leq \frac{1}{2}\mu_1 t$, Wald's identity gives $\mu E_\mu \tau_m(x) = E_\mu W(\tau_m(x)) \leq \frac{1}{2}\mu_1 E_\mu \tau_m(x) + x$, or

$$E_\mu \tau_m(x) \leq x/(\mu - \tfrac{1}{2}\mu_1). \tag{4.31}$$

Split the range of integration in (4.30) into the three parts $(0, m^{1/8})$, $(m^{1/8}, m^{1/2} \log m)$, and $(m^{1/2} \log m, \infty)$. It is easy to see from (4.31) that the first and third integrals converge to 0. By Theorem 4.14, uniformly for $\xi \leq \mu_1 m - m^{1/8}$

$$P_\xi^{(m)}\{T_+ < m\} = o(m^{-1})$$

and hence by (4.30) and (4.31) it suffices to consider

$$m^{-1/2}\int_{m^{1/8}}^{m^{1/2}\log m}\phi[m^{1/2}(\mu_1-\mu)-xm^{-1/2}]E_\mu\tau_m(x)\,dx. \tag{4.32}$$

A Taylor series expansion of $(m+t)^{1/2}-m^{1/2}$ and Wald's identity yield

$$\mu E_\mu\tau_m(x)=x+\tfrac{1}{2}\mu_1 E_\mu\tau_m(x)-\mu_1 E_\mu\tau_m^2(x)/8m+0(E_\mu\tau_m^3/m^2).$$

In the range $m^{1/8}<x<m^{1/2}\log m$, by the law of large numbers the distribution of $x^{-1}\tau_m(x)$ converges to a point mass at its asymptotic expectation $2/\mu_1$, so $E_\mu\tau_m^i(x)\sim(2x/\mu_1)^i$ $(i=2,3)$ and hence

$$E_\mu\tau_m(x)=x/(\mu-\tfrac{1}{2}\mu_1)-x^2/m\mu_1^2+o(x^2/m). \tag{4.33}$$

Substituting this expression into (4.32) and integrating give the Proposition.

Remark 4.34. An alternative proof for Proposition 4.29 is to write

$$E_\mu(T-m;T>m)=\int_{(0,\infty)}P_\mu\{T-m>m^{1/2}x\}m^{1/2}\,dx$$

and use an appropriate modification of Corollary 4.19 (see Problem 4.8). Although this method is rather involved computationally, it has the advantage that it easily yields the term of order $m^{1/2}$ in Proposition 4.29 in much more complicated situations, where computation of the term of constant order is very difficult or impossible (see Problems 4.7 and 5.1).

Theorem 4.27 remains true if m_0 is fixed or grows more slowly than m as $b\to\infty$. The growth condition imposed in the statement of the Theorem makes its formulation consistent with Theorem 4.14 and simplifies the proof.

3. Numerical Examples for Repeated Significance Tests

The repeated significance test for a normal mean μ described in Example 4.3 in defined by the stopping rule of (4.6) or equivalently (4.7). Approximations to the error probabilities and expected sample size of this test are more complicated than for Brownian motion because of the excess over the boundary $S_T^2/2T-a$ (cf. (4.7)). Appropriate approximations are stated below, but their mathematical justification is more technical and is deferred to Chapter IX.

There it is shown that for all $\mu\neq 0$, $S_T^2/2T-a$ converges in law as $a\to\infty$, and the limiting distribution is evaluated. Important parameters of this limiting distribution are

$$\nu(\mu)=\lim_{a\to\infty}E_\mu\exp\{-(S_T^2/2T-a)\}\qquad(\mu\neq 0) \tag{4.35}$$

and

$$\rho(\mu) = \lim_{a\to\infty} E_\mu(S_T^2/2T - a) \qquad (\mu \neq 0). \tag{4.36}$$

It is shown in IX.3 that

$$v(\mu) = 2\mu^{-2}\exp\left[-2\sum_1^\infty n^{-1}\Phi(-\tfrac{1}{2}|\mu|n^{1/2})\right], \tag{4.37}$$

and a similar, numerically computable expression is given for ρ. For many purposes simple approximations suffice. These are

$$v(\mu) = e^{-\rho\mu} + o(\mu^2) \qquad (\mu \to 0) \tag{4.38}$$

and

$$\rho(\mu) = \rho\mu + \mu^2/8 + o(\mu^2) \qquad (\mu \to 0) \tag{4.39}$$

(cf. 10.37). Here $\rho \cong .583$ is the same constant which appears in the discrete time approximations of III.5. As an indication of the accuracy of (4.38), for $\mu = 2$, $e^{-2(.583)} = .3116$ compared to the correct value $v(2) = .3204$. The approximation (4.39) for $\rho(\mu)$ is slightly less accurate but good enough for most purposes. (Tables of $\rho(\mu)$ and of $\log(2/\mu^2 v(\mu))$ are given by Woodroofe, 1982, p. 33.)

We are now in a position to state useful approximations to $P_\mu\{T \le m\}$ and $E_\mu(T \wedge m)$ for the stopping rule T defined by (4.7). Let $\mu_1 = bm^{-1/2}$ and $\mu_0 = bm_0^{-1/2}$. Then

$$P_0\{T \le m\} \cong 2b\phi(b)\int_{\mu_1}^{\mu_0} x^{-1}v(x)\,dx + 2[1 - \Phi(b)]. \tag{4.40}$$

For $\mu \neq 0$

$$P_\mu\{T \le m\} \cong 1 - \Phi[(\mu_1 - |\mu|)m^{1/2}] + v(\mu_1)\phi[(\mu_1 - |\mu|)m^{1/2}]/|\mu|m^{1/2}. \tag{4.41}$$

For $\mu = \mu_1 + \Delta m^{-1/2}$

$$\begin{aligned} E_\mu[T \wedge m] \cong{} & [b^2 - 1 + 2\rho(\mu)]/\mu^2 \\ & - (\mu - \tfrac{1}{2}\mu_1)^{-1}\{m^{1/2}[\phi(\Delta) - \Delta\Phi(-\Delta)][1 - b^{-2}(1 + \rho(\mu_1))] \\ & + (2\mu - \mu_1)^{-1}\rho(2\mu - \mu_1)\Phi(-\Delta) \\ & - [\mu_1/8(\mu - \tfrac{1}{2}\mu_1)^2][\Phi(-\Delta)(1 + \Delta^2) - \Delta\phi(\Delta)] \\ & - b^{-1}v(\mu_1)\phi(\Delta)\mu_1^{-1}[1 + \rho(\mu_1)] \\ & + \mu_1^{-1}b^{-1}[(2 + \Delta^2)\phi(\Delta) - \Delta(\Delta^2 + 3)\Phi(-\Delta)]\} + o(b^{-1}). \end{aligned} \tag{4.42}$$

In (4.40)–(4.42) the simple approximations of (4.38) and (4.39) can be used in place of $v(\mu)$ and $\rho(\mu)$. Numerical values of the constant $(2/\pi)^{1/2}\int_\mu^\infty x^{-1}v(x)\,dx$ required for the evaluation of (4.40) are given in Table 4.2.

Table 4.2. $c(\mu) = (\frac{2}{\pi})^{1/2} \int_\mu^\infty x^{-1} \nu(x)\, dx$

μ	$c(\mu)$	μ	$c(\mu)$
.20	1.3817	1.00	.4109
.25	1.2252	1.05	.3893
.30	1.1011	1.10	.3694
.35	.9992	1.15	.3509
.40	.9134	1.20	.3337
.45	.8400	1.30	.3026
.50	.7761	1.40	.2755
.55	.7200	1.50	.2515
.60	.6703	1.60	.2304
.65	. 6259	1.70	.2116
.70	.5858	1.80	.1948
.75	.5497	2.00	.1663
.80	.5168	2.20	.1432
.85	.4869	2.60	.1085
.90	.4594	3.00	.0844
.95	.4342	3.40	.0671

Remarks. (i) If one were to put $\nu(\mu) \equiv 1$ and $\rho(\nu) \equiv 0$ in (4.41) and (4.42), these approximations would reduce to the corresponding results for Brownian motion (cf. (4.20) and Theorem 4.27). However, the relation between (4.40) and Theorem 4.21 is not so simple. If we set $\nu(\mu) \equiv 1$ in (4.40), we obtain the term of order $b\phi(b)$ in Theorem 4.21, but there is a discrepancy in the second order terms by the amount

$$b^{-1}\phi(b)[2 - \log(m/m_0)] + o(b^{-1}\phi(b)). \tag{4.43}$$

Woodroofe and Takahashi (1982, correction 1985) have calculated the actual second order term in an asymptotic expansion of $P_0\{T \le m\}$ as $b \to \infty$, but it is quite complicated and usually not more accurate than (4.40).

(ii) Recall from the discussion preceding Theorem 4.27 that one should not expect (4.42) to be particularly accurate for μ substantially smaller than μ_1. Experience shows (see below) that (4.42) is quite accurate for $\mu \ge \mu_1$, even for quite small m, but deteriorates rapidly when $\mu < \mu_1$. In this case (4.42) involves differences of terms which are individually large and hence is numerically unstable. Since the expected sample size for small μ is of relatively minor interest, alternative approximations have not been developed.

Table 4.3 illustrates the accuracy of (4.40)–(4.42). Numbers in parentheses are exact values computed numerically by Pocock (1977). The maximum sample size is nominally $m = 7$, but it is also interesting to regard this as a group sequential test, where the data are examined after every seventh observation up to a maximum of 7 groups or $7 \cdot 7 = 49$ observations. Then the test of Table 4.3 is directly comparable to that of Table 3.1. (See III.5 for a discussion of that test for grouped data.) To make such a comparison, the row with a particular value of μ in Table 3.1 should be compared with the row having the

Table 4.3. Repeated Significance Test: $b = 2.485$; $m_0 = 1$; $m = 7$; $\alpha \cong .049$ (.05) (Two-Sided)

μ	$\mu/7^{1/2}$	Power		$E_\mu(T \wedge m)$		$7 \cdot E_\mu(T \wedge m)$	
1.770	.67	.99	(.99)	2.56	(2.68)	18	(19)
1.587	.60	.97		3.01		21	
1.506	.57	.95	(.95)	3.25	(3.35)	23	(23)
1.125	.43	.76	(.75)	4.56	(4.63)	32	(32)
1.058	.40	.70		4.78		34	
.855	.32	.51	(.50)	5.56	(5.56)	39	(39)
.794	.30	.45		—			
.529	.20	.23		—			

Parenthetical entries computed numerically by Pocock (1977).

Table 4.4. Repeated Significance Test: $b = 2.66$; $m_0 = 7$; $m = 49$; $\alpha \cong .05$ (Two-Sided)

μ	Power	$E_\mu(T \wedge m)$
.67	.99	16
.60	.96	19
.57	.94	21
.40	.67	32
.30	.42	—
.20	.21	—

same numerical value entered under $\mu/7^{1/2}$ in Table 4.3. Similarly the expected sample size in Table 3.1 should be compared with seven times the expected sample size in Table 4.3. The repeated significance test in Table 4.3 has slightly greater power purchased at the price of a slightly larger expected sample size than the test of Table 3.1.

The rescaling argument of Remark 3.19 shows that for Brownian motion, the test of Table 4.3 would be exactly equivalent to one with $m_0 = 7$ and $m = 49$, up to a change of scale. Table 4.4 reports some results for such a test, where the data are inspected after each new observation. Compared to the test of Table 4.3, this test appears to have slightly less power and a slightly smaller expected sample size.

Table 4.5 and 4.6 below give similar information about two tests with a maximum sample size of $m = 144$. In Table 4.5, $m_0 = 16$, $m = 144$, and the data are examined after each observation. In Table 4.6 m_0 and m are nominally 1 and 9 respectively, but for comparative purposes one should think of this as a group test having a maximum of 9 groups with 16 observations per group for a maximum of $9 \times 16 = 144$ observations.

The relation of these tests is much the same as in Tables 4.3 and 4.4, so it appears unlikely that the relation is an accident of the specific choice of the test parameters b, m_0, and m.

Table 4.5 Repeated Significance Test: $b = 2.75$; $m_0 = 16$; $m = 144$; $\alpha \cong .050$

μ	Power	$E_\mu(T \wedge m)$	Power of Fixed Sample Test ($m = 144$)
.4	.99	44	1.00
.33	.93	62	.98
.25	.71	89	.85
.20	.50	112	.67
.15	.29	—	.44

Table 4.6. Repeated Significance Test: $b = 2.535$; $m_0 = 1$; $m = 9$; $\alpha \cong .050$

μ	$\mu/4$	Power	$E_\mu(T \wedge m)$	$16 \cdot E_\mu(T \wedge m)$
1.60	.4	.99	3.09	49
1.32	.33	.95	4.16	67
1.00	.25	.75	5.78	93
.80	.20	.55	6.79	112
.60	.15	.33	—	—

Remark 4.45. Although Table 4.3 indicates that the approximations (4.40)–(4.42) are quite good, occasionally one wants to check the accuracy or range of validity of a new approximation by simulation; or one wants to simulate a process for which no known approximation seems adequate. In some cases importance sampling can be very helpful for variance reduction (cf. Remark 2.26). An especially interesting case is the significance level of a repeated significance test. Let $Q = \int_{-\infty}^{\infty} P_\mu \phi(\mu)\, d\mu$, so that the likelihood ratio of $x_1, \ldots, x_n$ under Q relative to P_0 is $(n+1)^{-1/2} \exp[S_n^2/2(n+1)]$ (cf. Example 4.1). Then by Proposition 2.24, for any stopping rule N

$$P_0\{N \le m\} = E\{(N+1)^{1/2} \exp[-S_N^2/2(N+1)]; N \le m\}, \tag{4.46}$$

where E denotes expectation under the probability Q. Hence a possible scheme for simulating $P_0\{N \le m\}$ is to generate observations $x_1, x_2, \ldots$ according to Q and to use the average of independent realizations of $(N+1)^{1/2} \exp[-S_N^2/2(N+1)] I_{\{N \le m\}}$ for the estimate. This possibility seems particularly promising for the stopping rule T of a repeated significance test, because $\exp[-S_T^2/2(T+1)]$ is close to being the constant $e^{-b^2/2}$ on $\{T \le m\}$. Also $Q\{T \le m\}$ is typically large whereas $P_0\{T \le m\}$ is small, so a Q-realization of $x_1, x_2, \ldots, x_{T \wedge m}$ usually involves the generation of fewer random numbers then a P_0-realization.

For example, in checking the accuracy of the approximation to α given in Table 4.4, 2500 realizations of this scheme yielded the estimate .0504 with a standard error of .0011. About 40,000 P_0-realizations would be required for similar accuracy. The identity (4.46) is also a fruitful starting place for developing analytic approximations (see Section 8).

4. Modified Repeated Significance Tests

The differences among the various repeated significance tests of Tables 4.3–4.5 and between those tests and that of Table 3.1 fail to give a compelling reason for preferring one sequential test to the others. However, the differences between these sequential tests and corresponding fixed sample tests are considerable. As was noted in III.3 the price of the smaller expected sample size of the sequential tests is a loss of power compared to a fixed sample test of sample size m.

This loss of power is a serious disadvantage if, as is often the case in clinical trials, one anticipates a small treatment difference, which even a fixed sample test may have insufficient power to detect (Peto *et al.*, 1976). Hence it seems reasonable to try to modify the sequential tests to increase their power without losing completely their expected sample size advantage in case a large treatment effect does indeed exist.

One possibility is to use a stopping rule of the form (4.6) of a repeated significance test, but with a larger value of b, so that $P_0\{T \le m\}$ is considerably smaller than the desired overall significance level. Then we can add to the rejection region of a repeated significance test those points in the sample space where truncation occurs after m observations and $|S_m| \ge cm^{1/2}$, where c is chosen so that the overall type one error probability, $P_0\{T \le m\} + P_0\{T > m, |S_m| \ge cm^{1/2}\}$, achieves the desired level α. More formally, given b, $0 < c \le b$, m_0, and m, stop sampling at $\min(T, m)$, where T is defined by (4.6), and reject H_0 if either $T \le m$ or $T > m$ and $|S_m| \ge cm^{1/2}$. The power function is

$$P_\mu\{T \le m\} + P_\mu\{T > m, |S_m| \ge cm^{1/2}\}$$

$$= P_\mu\{|S_m| \ge cm^{1/2}\} + P_\mu\{T < m, |S_m| < cm^{1/2}\}. \tag{4.47}$$

We can regard this procedure as creating a family of tests interpolating the fixed sample ($b = \infty$) and repeated significance tests ($c = b$).

Calculating the power function of the modified test presents some new mathematical problems, which in the case of Brownian motion have already been discussed in (4.17) and (4.18). The corresponding discrete time results are given below, but for practical purposes simpler approximations often suffice. For example, if c is considerably smaller than b, then for $\mu \ne 0$ the term $P_\mu\{T < m, |S_m| < cm^{1/2}\}$ in (4.47) is usually quite small, and the power is approximately

$$P_\mu\{|S_m| \ge cm^{1/2}\}. \tag{4.48}$$

The neglected term is given approximately by

$$P_\mu\{T < m, |S_m| < cm^{1/2}\} \cong \frac{\phi[m^{1/2}(\mu_1 - |\mu|)]}{|\mu| m^{1/2}} \mu_1 \gamma^{-1} v(\mu_1^2 \gamma^{-1}) e^{-m|\mu|(\mu_1 - \gamma)}, \tag{4.49}$$

where $\mu_1 = bm^{-1/2}$, $\gamma = cm^{-1/2}$, and $\mu \ne 0$ (cf. (4.17)). This expression can be

Table 4.7. Modified Repeated Significance Test: $b = 3.12$; $c = 2.05$; $m_0 = 16$; $m = 144$; $P_0\{T \leq m\} \cong .018$; $\alpha \cong .05$

μ	Power	$E_\mu(T \wedge m)$	$P_\mu\{T \leq m\}$
.4	1.00	57	.97
.33	.97	79	.86
.25	.83	106	.57
.20	.65	—	.35
.15	.42	—	.17

Table 4.8. Modified Repeated Significance Test: $b = 2.91$; $c = 2.05$; $m_0 = 1$; $m = 9$; $P_0\{T \leq m\} \cong .018$; $\alpha \cong .05$

μ	$\mu/4$	Power	$E_\mu(T \wedge m)$	$16 \cdot E_\mu(T \wedge m)$	$P_\mu\{T \leq m\}$
1.6	.4	1.00	3.87	62	.98
1.32	.33	.97	5.14	82	.89
1.00	.25	.83	6.77	108	.61
.80	.20	.65	—	—	.39
.60	.15	.42	—	—	.20

used to confirm that (4.48) is a good approximation to (4.47) or to improve the approximation.

It is interesting to note that when $c = b$ the right hand side of (4.49) specializes to the second term on the right hand side of (4.41). However, in contrast to the case of Brownian motion there is a precise asymptotic justification for (4.49) only when $c < b$, and not when $c = b$ (cf. Corollary 9.55 and (9.64)–(9.66)).

For the significance level one can use the first term on the right hand side of (4.47) in conjunction with the approximation

$$P_0\{T < m, |S_m| < cm^{1/2}\} \cong 2b\phi(b) \int_{\mu_1^2/\gamma}^{\mu_0} x^{-1} v(x)\, dx, \tag{4.50}$$

where $\mu_1 = bm^{-1/2}$, $\gamma = cm^{-1/2}$, and $\mu_0 = bm_0^{-1/2}$ (see (4.18) and Corollary 9.55).

Tables 4.7 and 4.8 contain numerical examples. The tests of Tables 4.5 and 4.6 were modified by increasing b, so that $P_0\{T \leq m\} \cong .018$ for both tests and hence they can still be compared to each other with respect to expected sample size. Next c was determined so that (4.50) (in conjunction with (4.47)) indicated a significance level of .05. The power computations use (4.48), supplemented by (4.49) in the last two rows, where it makes a difference of about .01–.03.

The two tests have essentially the same power function, which is no longer substantially inferior to that of a fixed sample test (cf. Table 4.5). The continuously monitored sequential test has a slightly smaller expected sample size than the group sequential test. For the important larger values of $|\mu|$, the expected sample sizes are about 25–30% larger than in Tables 4.5 and 4.6, but they are still only $\frac{1}{3}$ to $\frac{1}{2}$ of the fixed sample size of 144.

This modification of the repeated significance test to increase its power is the same as the modification of the test of Table 3.1 to obtain that of Table 3.2. For linear boundaries there is another possibility, which we now consider. The boundaries in III.3 arise naturally from a stopping rule of the form

$$\tilde{\tau} = \inf\{n\colon l(n, S_n; \mu_1, 0) \geq B\},$$

where $l(n, S_n; \mu_1, \mu_0) = \exp[(\mu_1 - \mu_0)S_n - n(\mu_1^2 - \mu_0^2)/2]$ is the likelihood ratio of $x_1, \ldots, x_n$ under P_{μ_1} relative to P_{μ_0}. The slope η of the stopping boundary written in the form of (3.27), namely

$$\tilde{\tau} = \inf\{n\colon S_n \geq b + \eta n\}, \tag{4.51}$$

is $\eta = \mu_1/2$. For the composite alternative hypothesis $\mu > 0$, the choice of μ_1, hence of η, is somewhat arbitrary; and the power of the test can be increased by using small values of η. Consideration of the locally most powerful test statistic, $(\partial/\partial\mu)l(n, S_n; \mu, 0)|_{\mu=0}$ suggests the specific choice $\eta = 0$.

To be more precise and make the present discussion consistent with the two-sided repeated significance tests, let $b > 0$, $-\infty < \eta < \infty$, and $m = 1, 2, \ldots$ ($m \leq -b/\eta$ if $\eta < 0$). Define

$$N = \inf\{n\colon |S_n| \geq b + \eta n\}. \tag{4.52}$$

Consider the test of $H_0\colon \mu = 0$ against $H_1\colon \mu \neq 0$ which stops sampling at $\min(N, m)$ and rejects H_0 if and only if $N \leq m$. The power of the test is $P_\mu\{N \leq m\}$, which can often be approximated by the results of III.5 (cf. Remark 3.18). In particular the significance level is

$$\begin{aligned} P_0\{N \leq m\} &\leq 2P_0\{\tilde{\tau} \leq m\} \\ &= 2[P_0\{S_m \geq b + \eta m\} + P_0\{\tilde{\tau} < m, S_m < b + \eta m\}], \end{aligned} \tag{4.53}$$

where $\tilde{\tau}$ is defined by (3.27) and the last term of (4.53) is given approximately by (3.28). For conventional values of the significance level, except for μ close to 0, the power and expected sample size can be approximated by

$$P_\mu\{\tilde{\tau} \leq m\} \tag{4.54}$$

and

$$E_\mu(\tilde{\tau} \wedge m), \tag{4.55}$$

for which III.5 suggests approximations. If necessary, one can use results of III.6 to evaluate and improve upon the use of (4.54) and (4.55).

For a given significance level α, the fixed sample test of sample size m can be obtained as a limiting case as $b \to \infty$ and $\eta \to -\infty$.

Table 4.9 gives a numerical example for the case $\eta = 0$. The stopping rule (4.52) with $\eta = 0$ is just that of a sequential probability ratio test for testing $\mu < 0$ against $\mu > 0$, although the point of view is much different than in Chapter II or in III.6. The most noticeable difference between this test and the modified repeated significance test of Tables 4.7 and 4.8 is the larger expected sample size for large values of μ.

Table 4.9 Truncated Sequential Probability Ratio Test: $b = 26.3$; $m = 144$; $\alpha \cong .050$

μ	Power	$E_\mu(N \wedge m)$
.4	1.00	67
.33	.97	81
.25	.83	102
.20	.65	116
.15	.42	128

In addition to increasing the power of sequential tests, the modifications proposed in this section make the sequential tests compare more favorably with fixed sample tests in regard to attained significance level and estimation, which are discussed next.

5. Attained Significance Level and Confidence Intervals

Similarly to III.4, attained significance levels and confidence intervals can be defined for repeated significance tests. We shall see below that they provide some additional evidence in favor of modified repeated significance tests with b substantially larger than c.

We continue to assume that $x_1, x_2, \ldots$ are independent and normally distributed with mean μ and variance 1. Let $S_n = x_1 + \cdots + x_n$ and let T be defined by (4.6).

For the two-sided sequential test of H_0: $\mu = 0$ against H_1: $\mu \neq 0$ which stops sampling at $\min(T, m)$, the discussion of III.4 suggests defining the (two-sided) attained significance level as follows:

(i) if $T = m_0$ and $|S_{m_0}| = zm_0^{1/2}$, the attained level is $2[1 - \Phi(z)]$;
(ii) if $T = n \in (m_0, m]$ the attained level is $P_0\{T \leq n\}$;
(iii) if $T > m$ and $|S_m| = cm^{1/2}$, the attained level is

$$P_0\{T \leq m\} + P_0\{T > m, |S_m| \geq cm^{1/2}\}$$
$$= P_0\{|S_m| \geq cm^{1/2}\} + P_0\{T \leq m, |S_m| < cm^{1/2}\}.$$

For the Brownian motion process this definition would correspond to ordering sample points according to the value of $|W(T \wedge m)/T \wedge m|$ with large values providing more evidence against H_0: $\mu = 0$. In discrete time this relation is only approximate, and there is some loss of information in not distinguishing between different values of $|S_{T \wedge m}/T \wedge m|$ except insofar as T changes. For group sequential tests with a small number of large groups the information loss may be significant enough to indicate the advisability of a somewhat different definition, but here only this simple one is considered (see IX.5).

By Theorem 4.21 (cf. also (4.40)) $P_0\{T \le n\}$ is roughly proportional to $\log n$, and hence only for extremely small n is $P_0\{T \le n\}$ appreciably smaller than $P_0\{T \le m\}$. For example, for the repeated significance test of Table 4.6, which has $m = 9$ and $\alpha = P_0\{T \le 9\} \cong .05$, we find that $P_0\{T = 1\} = .011$ and $P_0\{T \le 3\} \cong .027$. Hence the evidence against H_0 as measured by the attained significance level is not as strong as the drastic action of curtailing the test after a relatively small number of observations would seem to indicate.

The situation is somewhat better for a modified repeated significance test with b chosen large enough that $P_0\{T \le m\}$ is substantially smaller than the overall significance level of the test. For example, for the modified test of Table 4.8, for which $m = 9$ and $P_0\{T \le 9\} \cong .018$, $P_0\{T = 1\} = .0036$ and $P_0\{T \le 3\} \cong .0091$. Hence by taking b large, we can be assured that early termination of the test will lead to small p-values, but of course the price is an increase in the expected sample size for large $|\mu|$.

It is also straightforward to define confidence intervals for repeated significance tests similarly to III.4. For example, for $m_0 < n \le m$ define

$$\mu_{*1}(n) = \inf\{\mu: P_\mu\{T \le n, S_T > 0\} \ge \gamma\} \tag{4.56}$$

(cf. (3.21)) and

$$\mu_{*3}(n) = \inf\{\mu: P_\mu\{T \ge n\} + P_\mu\{T < n, S_T > 0\} \ge \gamma\}. \tag{4.57}$$

For $-bm^{1/2} < x < bm^{1/2}$ let

$$\mu_{*2}(x) = \inf\{\mu: P_\mu\{T \le m, S_T > 0\} + P_\mu\{T > m, S_m \ge x\} \ge \gamma\} \tag{4.58}$$

(cf. (3.22)). A slightly more complicated argument than that given in III.4 (cf. Woodroofe, 1982, p. 100 ff. for details) shows that

$$\begin{aligned} & = \mu_{*1}(T) && \text{if } T \le m \text{ and } S_T > 0 \\ \mu_*(T \wedge m, S_{T \wedge m}) & = \mu_{*2}(S_m) && \text{if } T > m \\ & = \mu_{*3}(T) && \text{if } T \le m \text{ and } S_T < 0 \end{aligned} \tag{4.59}$$

is a $(1 - \gamma)100\%$ lower confidence bound for μ. A $(1 - \gamma)100\%$ upper confidence bound μ^* can be defined similarly, and $[\mu_*, \mu^*]$ is a $(1 - 2\gamma)100\%$ confidence interval.

The following heuristic observations provide a different and useful perspective in our further study of these confidence intervals. Suppose that we would like to estimate μ by $\bar{x}_n = n^{-1}S_n$ with prescribed proportional accuracy, i.e. so that for given $0 < \delta < 1$ and $0 < \gamma < \frac{1}{2}$,

$$P_\mu\{|\bar{x}_n - \mu| \le \delta|\mu|\} \ge 1 - 2\gamma \tag{4.60}$$

for all $\mu \ne 0$. Condition (4.60) is satisfied provided n is chosen so that

$$\delta n^{1/2}|\mu| \ge z_\gamma, \tag{4.61}$$

Table 4.10 90% Confidence Intervals: $b = 3.12$; $m_0 = 16$; $m = 144$

Data	μ_*	μ^*
$T = 39, S_T > 0$	.16	.74
$T = 61, S_T > 0$	.13	.59
$T > 144, m^{-1}S_m = .25$	.08	.38
$T > 144, m^{-1}S_m = .20$	.04	.33

where $1 - \Phi(z_\gamma) = \gamma$. In ignorance of μ it is impossible to choose a fixed sample size n fulfilling (4.61), but it is possible to estimate μ by $\bar{x}_n$ and sample sequentially until (4.61) appears to be true, i.e. until $\delta n^{1/2}|\bar{x}_n| \geq z_\gamma$. This gives the stopping rule (4.6) with $b = z_\gamma/\delta$.

In terms of proportional accuracy a reasonable requirement might be to estimate μ to within 50% of its value 90% of the time. The heuristic argument of the preceding paragraph suggests putting $\delta = .5$ and $z_\gamma = 1.645$, so $b = 3.29$. This value of b is somewhat larger than those considered in IV.3.

Table 4.10 gives some examples of 90% intervals computed from hypothetical data. The test parameters are $b = 3.12$, $m_0 = 16$, and $m = 144$ from Table 4.7. The approximation (4.49) was used for the required probability calculations.

Because of the heuristic discussion of proportional accuracy given above, it is interesting to consider

$$\tilde{\mu} = \tfrac{1}{2}(\mu_* + \mu^*)$$

as a point estimator of μ and note that the ratio $(\mu^* - \mu_*)/\tilde{\mu}$ is remarkably constant in the first three rows of Table 4.10, although the constant is somewhat larger than our heuristic analysis would suggest. One should also note that in the first three rows, $\tilde{\mu}$ and the interval $[\mu_*, \mu^*]$ are shifted to the left with respect to $\hat{\mu} = S_{T \wedge m}/T \wedge m$ and the naive interval $\hat{\mu} \pm 1.645/(T \wedge m)^{1/2}$, even if we assume that $S_T = bT^{1/2}$ whenever $T \leq m$. The discrepancy is greatest at the lower endpoint of the confidence interval. In the last row of Table 4.10, where the final value S_m is not particularly close to the stopping boundary, the confidence interval is essentially the same as the naive interval, as one expects from (4.58) and (4.59).

It is interesting to compare the intervals in Table 4.10 with the corresponding intervals for the test of Table 4.8, considered as a group sequential test with a maximum of 9 groups of 16 observations each, for a maximum of $9 \times 16 = 144$ observations. Data leading to $T = 61$ and $S_T > 0$ in Table 4.10 would presumably lead to 64 observations (4 groups) for the test of Table 4.8 and hence to the 90% confidence interval [.12, .65]. Data leading to $T = 39$ and $S_T > 0$ in Table 4.10 would probably lead to 48 observations (3 groups) but might possibly lead to 32 observations (2 groups) in Table 4.8. The confidence intervals would be [.14, .79] or [.18, .89] respectively. There appears to be a slight loss of efficiency in the case of the group sequential test, which might be recovered by taking into account the actual value of $S_{T \wedge m}/T \wedge m$ in the definition of the confidence interval (cf. IX.5 and Problem 9.15).

It is also interesting to compare the confidence interval defined above with the naive interval $\hat{\mu} \pm z_\gamma/(T \wedge m)^{1/2}$, by computing the actual coverage probability for the naive interval. The calculation is a difficult one, involving the joint distribution of T and S_T. An *ad hoc* approximation is given by Siegmund (1978). Here we consider the simpler case of Brownian motion, which yields essentially the same insights.

Let $W(t)$, $0 \le t < \infty$, denote Brownian motion with drift μ and let

$$T_+ = \inf\{t: t \ge m_0, W(t) \ge bt^{1/2}\}.$$

(For simplicity we are ignoring the lower part of the stopping boundary, but except for very small values of μ this has a negligible effect on subsequent calculations.) Let $\mu_\pm = W(T_+ \wedge m)/T_+ \wedge m \pm z_\gamma/(T_+ \wedge m)^{1/2}$ denote the end points of the naive interval. It is easy to see that

$$\begin{aligned} P_\mu\{\mu_- > \mu\} &= P_\mu\{T_+ = m_0, W(m_0) - m_0\mu > z_\gamma m_0^{1/2}\} \\ &\quad + P_\mu\left\{m_0 < T_+ < m \wedge \left(\frac{b - z_\gamma}{\mu}\right)^2\right\} \\ &\quad + P_\mu\{T_+ > m, W(m) - m\mu > z_\gamma m^{1/2}\} \end{aligned} \tag{4.62}$$

and

$$\begin{aligned} P_\mu\{\mu_+ < \mu\} &= P_\mu\{T_+ = m_0, W(m_0) - m_0\mu < -z_\gamma m_0^{1/2}\} \\ &\quad + P_\mu\left\{\left(\frac{b + z_\gamma}{\mu}\right)^2 < T_+ \le m\right\} \\ &\quad + P_\mu\{T_+ > m, W(m) - m\mu < -z_\gamma m^{1/2}\}. \end{aligned}$$

In most cases these formulas can be simplified. For example, if $m_0 < (b - z_\gamma)^2/\mu^2 < m$, the sum of the first two terms on the right hand side of (4.62) is just $P_\mu\{T_+ < (b - z_\gamma)^2/\mu^2\}$ and the third term is 0. Hence insofar as (4.20) can be used to provide an approximation (the m in (4.20) is here set equal to $(b - z_\gamma)^2/\mu^2$), we find that

$$P_\mu\{\mu_- > \mu\} \cong 1 - \Phi(z_\gamma) + \phi(z_\gamma)/(b - z_\gamma), \tag{4.63}$$

which does not depend on μ and is considerably larger than the nominal value of $\gamma = 1 - \Phi(z_\gamma)$. For $\gamma = .05$ and $b = 3.2$, (4.63) gives .116. The probability of the event $\{\mu_+ < \mu\}$ is always smaller than the nominal value, γ, but the relation is more complicated than (4.63) for interesting values of μ. For $b = 3.2$, $m = 144$, and $z = 1.645$, for $\mu = .3$, .2, and .15, Corollary 4.16 yields the approximations $P_\mu\{\mu_+ < \mu\} \cong .048, .049$, and .043 respectively. Hence the actual probability that the naive interval fails to cover μ is much larger than the nominal amount at the lower endpoint and slightly smaller at the upper endpoint. The total non-coverage probability is substantially larger than the nominal amount.

Note that in contrast to the linear boundary of III.4, the set of points $(T_+ \wedge m, W(T_+ \wedge m))$ where $\mu_- > \mu$ (or where $\mu_+ < \mu$) is a connected set.

Similarly one can define attained significance levels and confidence intervals for the truncated sequential probability ratio test discussed at the end of Section 4. In contrast to repeated significance tests, early termination of this test can lead to very small attained significance levels and to confidence intervals which are fairly close to the naive interval, at least in the case of continuously inspected data. This is not surprising in light of the somewhat larger sample sizes of the truncated sequential probability ratio test for large $|\mu|$. For the test of Table 4.9 applied to Brownian motion, the probability that the naive 90% interval has a lower endpoint exceeding μ is about .07 when μ is about .2–.4.

6. Discussion

The results of Section 4 indicate that although repeated significance tests have somewhat less power than corresponding fixed sample tests, they can be modified to achieve essentially the same power function while maintaining a much smaller sample size for large $|\mu|$. This modification also yields more accurate estimates and smaller attained significance levels when the test is terminated after a small number of observations.

For the modified repeated significance test, continuous examination of the data seems to yield essentially the same power function as a group test and a slightly smaller expected sample size. (This comparison is predicated on the assumption that the ratio of maximum to minimum sample size, m/m_0, is the same for the two tests.) There are undoubtedly occasions when group sequential tests are feasible to administer, but continuous inspection is not. However, continuous inspection may not be as onerous as is sometimes argued (e.g. Pocock, 1977). It is not really necessary to examine the data continuously, but only to be prepared to adapt the inspection schedule to the data if they appear to be approaching a stopping boundary. In fact it may be psychologically difficult to stick to the schedule of a group sequential test if freehand extrapolation of the data already available suggests that the test will cross a stopping boundary well in advance of the next inspection point.

A competitor to the modified repeated significance test, which has essentially the same power and expected sample size for a broad range of values of μ, is a truncated sequential probability ratio test or a version of that test for grouped data (cf. Table 4.9). Although it has a larger expected sample size for large $|\mu|$, it compensates for this by yielding smaller attained significance levels and shorter confidence intervals when the tests terminate after a small number of observations. There does not seem to be a persuasive scientific reason for preferring one test to the other in general.

Some secondary reasons for preferring repeated significance tests are as follows:

(i) Their interpretation as repeated likelihood ratio tests (cf. (4.7) and Example 4.8) makes it easy to generalize them to multiparameter problems where all reasonable stopping rules involve non-linear functions of the sufficient statistics, so the sequential probability ratio test loses its comparative simplicity.

(ii) Since $P_0\{T \leq m\}$ changes very slowly with m and is only a fraction of the overall significance level, modified repeated significance tests are particularly flexible in adjusting to a change in the anticipated maximum sample size once an experiment has begun. In other words, the stopping boundary can be selected without knowing the maximum sample size precisely, and this may be a distinct advantage in the early stages of an experiment.

(iii) For small $|\mu|$ the experiment is more likely to terminate with the maximum sample size m. In clinical trials μ usually represents only the most important end point in a multidimensional parameter space. In the absence of strong evidence in this coordinate in favor of one treatment, one often wants to have the largest possible sample size in order to explore other coordinates.

Finally we note that the tests discussed in this chapter can easily be modified for early stopping when H_0 appears to be true (cf. III.8 and Problem 4.1).

In discussing clinical trials the viewpoint of this book has been one of acceptance of classical criteria for statistical inference: significance, power, and expected sample size. The principal issues have been how to choose reasonable procedures within this framework.

Anscombe (1963) has criticized this approach because it is not Bayesian, and, more importantly for the present discussion, because it is not decision oriented. He argues that the primary goal of a clinical trial is not to make inferences about a treatment effect, but to decide which treatment to use in the future (since *some* treatment must be used). Hence for Anscombe termination leads to two possible decisions: decide in favor of Treatment A or in favor of Treatment B. One is not permitted to say that H_0: $\mu = 0$ is a tenable hypothesis and therefore neither treatment is strongly endorsed. An important practical consequence of this viewpoint, which can be discussed in classical terms without reference to explicit loss functions or prior distributions, is that instead of demanding evidence that $\mu \neq 0$ before terminating a trial and deciding in favor of one treatment, it is only necessary to show in some sense that one treatment may be better and almost certainly is not distinctly inferior to the other in order to justify termination.

More specifically, if one were to use the stopping rule of a truncated, symmetric sequential probability ratio test for a clinical trial, it should not be used as in Table 4.9 to test H_0: $\mu = 0$, but rather should be used to decide whether $\mu < 0$ or $\mu > 0$ as in Table 3.6. There, in effect an indifference zone exists near $\mu = 0$ where an incorrect decision is made about 50% of the time, but is regarded as unimportant. And when μ is large a decision is made with fewer observations, because it is not necessary to infer that μ is positive, but only what is tantamount to saying that μ exceeds the lower boundary of the indifference zone. (In fact, Table 3.8 shows that the test of Table 3.6 essentially never yields the inference that $\mu \neq 0$ at level .05.)

I believe that a reasonable reply to this particular criticism is the following. The actual decision which treatment to use often involves many imponderables which make it practically difficult, even if theoretically possible, to formulate a reasonable loss function incorporating all the required ingredients. Hence the statistician's job is to present evidence which assists the decision makers. Sometimes that evidence will be in the form of a strong recommendation: the data favor Treatment A to such an extent that the experimenters considered it advisable to terminate the experiment in order that all future patients could receive Treatment A. In other cases the data seem fairly persuasive by the end of the experiment although they did not seem sufficiently so during the experiment to warrant an early termination. And often the evidence will contain enough ambiguity that a decision which treatment to use may be based on secondary considerations: side effects, cost, special characteristics of the patient, etc., which may not be a part of the formal model.

Although it is obviously not optimal in any decision theoretic framework, a modified repeated significance test is oriented towards achieving essentially a large fixed sample size in those ambiguous cases where decision making is likely to be based on subtle advantages of one treatment relative to the other. At the same time it allows experimenters the flexibility to terminate an experiment early in the fortunate circumstance that one treatment appears so much better than the second in an important performance index that other considerations are insignificant in comparison. The results of this chapter demonstrate that to a considerable extent it succeeds in achieving these goals simultaneously, and hence is a considerable improvement over a fixed sample design.

7. Some Exact Results

The methods developed in Section 2 for studying repeated significance tests for Brownian motion are quite general in the sense that they allow one to make similar approximations for a wide class of stopping boundaries. For a special set of (non-linear) boundaries they yield exact results. This section illustrates the possibilities by discussing two examples in some detail.

Recall from Section 2 the notation

$$P_{\lambda,\xi}^{(m)}(A) = P\{A \mid W(0) = \lambda, W(m) = \xi\} \qquad (A \in \mathscr{E}_m).$$

Again it will be convenient to introduce a new probability $\tilde{P}_{\xi}^{(m)}$ defined by (4.22). Recall also that the likelihood ratio of $\{W(s), t \leq s \leq m\}$ under $\tilde{P}_{\xi}^{(m)}$ relative to $P_{0,\xi}^{(m)}$ is given by $l^{(m)*}(t, W(t); \xi)$ in (4.23). Now define

$$T^* = \sup\{t \colon t \leq m, l^{(m)*}(t, W(t); \xi) \geq B\}. \tag{4.64}$$

Table 4.11 Test Defined by (4.66): $a = 3.42$; $m = 49$; $\alpha = .050$

μ	Power	$E_\mu(T \wedge m)$
.67	.99	17
.60	.96	21
.40	.64	34

For all $B > 1$ Wald's likelihood ratio identity yields (cf. 4.24)

$$P_{0,\xi}^{(m)}\{T^* > 0\} = \tilde{E}_\xi^{(m)}[1/l^{(m)*}(T^*, W(T^*); \xi); T^* > 0] = B^{-1}\tilde{P}_\xi^{(m)}\{T^* > 0\}. \tag{4.65}$$

For notational convenience define a by $B = m^{-1/2}\exp[\frac{1}{2}(a^2 - \xi^2/m)]$, where $|\xi| < [m(a^2 - \log m)]^{1/2}$ and $m < e^{a^2}$. It is easy to see that $\tilde{P}_\xi^{(m)}\{T^* > 0\} = 1$, and to rewrite (4.65) in terms of

$$T = \inf\{t: |W(t)| \geq [t(a^2 - \log t)]^{1/2}\} \tag{4.66}$$

as

$$P_{0,\xi}^{(m)}\{T < m\} = m^{1/2}\exp[-\tfrac{1}{2}(a^2 - \xi^2/m)]. \tag{4.67}$$

By calculus one can now obtain for T defined by (4.66)

$$P_\mu\{T \leq m\} = P_\mu\{|W(m)| \geq [m(a^2 - \log m)]^{1/2}\} + 2\exp[-\tfrac{1}{2}(a^2 + m\mu^2)]\sinh\{\mu[m(a^2 - \log m)]^{1/2}\}/(2\pi)^{1/2}\mu \tag{4.68}$$

and

$$E_\mu(T \wedge m) = mP_\mu\{T > m\} + \phi(a)\int_0^m \{t(a^2 - \log t)\}^{1/2}\cosh\{\mu[t(a^2 - \log t)]^{1/2}\}e^{-\mu^2 t/2}\,dt \tag{4.69}$$

for $m < e^{a^2}$.

Table 4.11 contains a numerical example of the stopping rule T of (4.66) used to test H_0: $\mu = 0$ with the rejection region $\{T \leq m\}$. Except for the slightly larger expected sample size for large $|\mu|$ this test behaves much like the repeated significance test of Table 4.4. (It is even more similar to the corresponding test for Brownian motion, for which $b = 2.79$.)

A test defined by (4.66) seems attractive because it does not require the somewhat arbitrary minimum sample size m_0 of a repeated significance test, and for Brownian motion one can calculate the operating characteristics exactly. However, for practical purposes a minimum sample size is probably advisable, and in discrete time one must still resort to approximations.

Remark. For $\mu > 0$ one discovers an interesting comparison between (4.68) and Corollary 4.19 by putting

(4.70) $$\mu_1 m = [m(a^2 - \log m)]^{1/2},$$

so (4.68) becomes

(4.71) $$P_\mu\{T \le m\} = 1 - \Phi[m^{1/2}(\mu_1 - \mu)] + \phi[m^{1/2}(\mu_1 - \mu)]/m^{1/2}\mu + O(\phi[m^{1/2}(\mu_1 + \mu)]/m^{1/2}) \qquad (m \to \infty).$$

The normalization (4.70) makes $[t(a^2 - \log t)]^{1/2}$ and the boundary $bt^{1/2}$ of Corollary 4.19 equal at $t = m$, and (4.71) says that up to a very small error the approximation given for the boundary $bt^{1/2}$ in Corollary 4.19 is exactly correct for T defined by (4.66). This is perhaps not surprising because the two curves are very close to each other for t near to m, which is the critical range of values for the approximation of Corollary 4.19. Since the stopping boundary in (4.66) is fairly close to parabolic over most of the range $0 \le t \le m$, (4.71) also helps to explain why the approximation of Corollary 4.19 gives much more accurate numerical results than the proof itself leads one to expect.

The key to the calculation (4.67) is introduction of the measure $\tilde{P}_\xi^{(m)}$ by mixing the $P_{\lambda,\xi}^{(m)}$ and then letting the likelihood ratio $l^{(m)*}(t, W(t); \xi)$ "define" the stopping rule as in (4.64). In principle one can use almost arbitrary mixtures of the $P_{\lambda,\xi}^{(m)}$ to define a variety of boundaries for which exact crossing probabilities can be calculated.

If one chooses a mixing distribution degenerate at a fixed point, i.e. one works directly with $P_{\lambda,\xi}^{(m)}$, this argument is essentially equivalent to the derivation of (3.13) and produces a linear boundary. Therefore, to produce a two-sided boundary similar to the symmetric triangular case of Remark 3.50 it seems reasonable to consider a (symmetric) distribution concentrated on two points. Specifically, let $\tilde{P}_\xi^{(m)}$ be of the form

$$\int_{-\infty}^{\infty} P_{\lambda,\xi}^{(m)} \phi[m^{-1/2}(\lambda - \xi)]\, dG(\lambda) \bigg/ \int_{-\infty}^{\infty} \phi[m^{-1/2}(\lambda - \xi)]\, dG(\lambda),$$

where now G gives probability $\frac{1}{2}$ to the points $\pm\lambda_0$. It is easy to show that the likelihood ratio of $\{W(s), t \le s \le m\}$ under $\tilde{P}_\xi^{(m)}$ relative to $P_{0,\xi}^{(m)}$ is

(4.72) $$l^{(m)*}(t, W(t); \xi) = \cosh[\lambda_0 W(t)/t] \exp(-\lambda_0^2/2t + \lambda_0^2/2m)/\cosh(\lambda_0 \xi/m).$$

Again defining T^* by (4.64) but with $l^{(m)*}$ now given by (4.72), putting B in the convenient form $b \exp(\lambda_0^2/2m)/\cosh(\lambda_0 \xi/m)$, and reasoning as above, we obtain for the stopping rule

(4.73) $$T = \inf\{t\colon \cosh(\lambda_0 W(t)/t) \ge b \exp(\lambda_0^2/2t)\}$$

the basic result

(4.74) $$P_{0,\xi}^{(m)}\{T < m\} = b^{-1} \cosh(\lambda_0 \xi/m) \exp(-\lambda_0^2/2m),$$

provided that the right hand side of (4.74) is less than one. The stopping boundaries defined implicitly by (4.73) are linear near $t = 0$; and for $b < 1$ they are closed at $t_1 = -\lambda_0^2/2 \log b$, near which they behave like $\pm(t_1 - t)^{1/2}$. Using

(4.74) one can easily calculate the operating characteristics of a test which for suitable λ_0, b, and m closely resembles the symmetric triangular test of III.6. See Daniels (1982) for details.

8. The Significance Level of Repeated Significance Tests for General One-Parameter Families of Distributions

In this section we discuss an alternative method for approximating the significance level of repeated significance tests. In some important respects it is less general than the method of Theorem 4.14. For example, it does not seem especially well suited for dealing with stopping boundaries which are too "far" from $t^{1/2}$. However, it easily yields approximations for the significance level in a number of cases for which the calculations of Theorem 4.14 are very difficult to carry out.

To illustrate the method we first provide an alternate derivation of the dominant term in Theorem 4.21 and then give a heuristic discussion of a much more general result.

Proof that $P_0\{T \le m\} \sim b\phi(b)\log(m/m_0)$. (Here T is defined in (4.11) and the asymptotic relations $bm^{-1/2} = \mu_1 < \mu_0 = bm_0^{-1/2}$ are assumed to hold.)

Let $Q = \int_{-\infty}^{\infty} P_\mu \phi(\mu)\, d\mu$. By Proposition 3.1 the likelihood ratio of $\{W(s), s \le t\}$ under Q relative to P_0 is

$$\begin{aligned} L_t &= \int_{-\infty}^{\infty} l(t, W(t); \mu, 0)\phi(\mu)\, d\mu = \int_{-\infty}^{\infty} \exp[\mu W(t) - \mu^2 t/2]\phi(\mu)\, d\mu \\ &= (t+1)^{-1/2} \exp\{W^2(t)/2(t+1)\}. \end{aligned}$$

Observe that $P_0\{T = m_0\} = P_0\{|W(m_0)| \ge bm_0^{1/2}\} \sim 2b^{-1}\phi(b)$ as $b \to \infty$, and hence it suffices to consider $P_0\{m_0 < T \le m\}$. By a continuous time version of Proposition 2.24

$$P_0\{m_0 < T \le m\} = E\{(T+1)^{1/2} \exp(-W^2(T)/2(T+1)); m_0 < T \le m\},$$

where E denotes expectation with respect to Q. Since $|W(T)| = bT^{1/2}$ whenever $m_0 < T \le m$, this becomes

$$P_0\{m_0 < T \le m\}$$

$$= be^{-b^2/2} \int_{-\infty}^{\infty} E_\mu\left\{\left[\frac{T+1}{b^2}\right]^{1/2} \exp\left[\frac{b^2}{2(T+1)}\right]; m_0 < T \le m\right\} \phi(\mu)\, d\mu. \tag{4.75}$$

The argument leading to (4.13) shows that

$$P_\mu\{m_0 < T \le m\} \to \begin{cases} 1 & \text{if } \mu_1 < |\mu| < \mu_0 \\ 0 & \text{if } |\mu| \notin [\mu_1, \mu_0]. \end{cases}$$

Hence by (4.12)

$$E_\mu\left\{\left[\frac{T+1}{b^2}\right]^{1/2}\exp\left[\frac{b^2}{2(T+1)}\right]; m_0 < T \le m\right\} \to |\mu|^{-1}e^{\mu^2/2} \quad \text{if } \mu_1 < |\mu| < \mu_0$$

$$\to 0 \qquad \text{if } |\mu| \notin [\mu_1, \mu_0].$$

Substituting these results into (4.75) yields

$$(4.76)\qquad \begin{aligned} P_0\{m_0 < T \le m\} &\sim b\phi(b)2\int_{\mu_1}^{\mu_0}|\mu|^{-1}\,d\mu \\ &= b\phi(b)2\log(\mu_0/\mu_1) \\ &= b\phi(b)\log(m/m_0). \end{aligned}$$

Remark. The preceding argument in conjunction with (4.35) provides one justification for (4.40).

Now suppose that $x_1, x_2, \ldots, x_n, \ldots$ are independent with probability density function $f(x;\theta)$, where for simplicity θ is assumed to be real. Let $l_n(\theta) = \sum_1^n \log f(x_k;\theta)$ denote the log likelihood function,

$$\dot{l}_n(\theta) = \frac{d}{d\theta}l_n(\theta)$$

the efficient score function, and

$$-\ddot{l}_n(\theta) = -\frac{d^2}{d\theta^2}l_n(\theta)$$

the Fisher information of $x_1, \ldots, x_n$. Also let $I(\theta) = -E_\theta[\ddot{l}_1(\theta)]$ denote the expected Fisher information of a single observation.

Consider the following test of H_0: $\theta = \theta_0$ against H_1: $\theta \ne \theta_0$ based on the likelihood ratio statistic $\Lambda_n = \sup_\theta l_n(\theta) - l_n(\theta_0)$. Define the stopping rule

$$(4.77)\qquad T = \inf\{n\colon n \ge m_0, \Lambda_n \ge a\},$$

stop sampling at $\min(T, m)$, and reject H_0 if and only if $T \le m$. For the special case of normal data with mean θ, variance 1, and $\theta_0 = 0$, it is easily seen that $l_n(\theta) = -\frac{1}{2}\sum_1^n (x_k - \theta)^2 + \text{const.}$ and (4.77) reduces to (4.7).

The significance level of this test is $P_{\theta_0}\{T \le m\} = P_{\theta_0}\{\Lambda_{m_0} \ge a\} + P_{\theta_0}\{m_0 < T \le m\}$. The following heuristic discussion shows that the argument given above in the case of Brownian motion provides the basic ideas for approximating $P_{\theta_0}\{m_0 < T \le m\}$ under quite general conditions.

Assume that for each n there exists a unique maximum likelihood estimator $\hat{\theta}_n$ which can be obtained by solving the equation $\dot{l}_n(\zeta) = 0$, so

$$\Lambda_n = l_n(\hat{\theta}_n) - l_n(\theta_0)$$

and

$$(4.78)\qquad l_n(\zeta) = l_n(\hat{\theta}_n) + \tfrac{1}{2}(\zeta - \hat{\theta}_n)^2\ddot{l}_n(\hat{\theta}_n) + \cdots.$$

Define $Q = \int P_\zeta \, d\zeta/(2\pi)^{1/2}$, where the integral extends over the entire parameter space. In general Q is not a probability measure, but one can nevertheless let L_n denote the likelihood ratio of $x_1, \ldots, x_n$ under Q relative to P_{θ_0}, so by an obvious generalization of Proposition 2.24

$$P_{\theta_0}\{m_0 < T \le m\} = \int_{\{m_0 < T \le m\}} L_T^{-1} \, dQ \tag{4.79}$$
$$= \int E_\theta\{L_T^{-1}; m_0 < T \le m\} \, d\theta/(2\pi)^{1/2}.$$

By the definition of Q

$$L_n = \int \exp[l_n(\zeta) - l_n(\theta_0)] \, d\zeta/(2\pi)^{1/2}.$$

By the law of large numbers with P_θ probability one, $-n^{-1}\ddot{l}_n(\theta) \to I(\theta)$, and hence under additional regularity conditions

$$-n^{-1}\ddot{l}_n(\hat{\theta}_n) \to I(\theta). \tag{4.80}$$

In particular $\ddot{l}_n(\hat{\theta}_n)$ diverges to $-\infty$. It follows from (4.78) that with P_θ probability one

$$L_n \sim \exp[l_n(\hat{\theta}_n) - l_n(\theta_0)] \int \exp[\ddot{l}_n(\hat{\theta}_n)(\zeta - \hat{\theta}_n)^2/2] \, d\zeta/(2\pi)^{1/2}$$
$$\sim \exp(\Lambda_n)/|\ddot{l}_n(\hat{\theta}_n)|^{1/2}.$$

Hence by (4.79)

$$P_{\theta_0}\{m_0 < T \le m\}$$
$$\sim a^{1/2} e^{-a} \int E_\theta\{a^{-1/2}|\ddot{l}_T(\hat{\theta}_T)|^{1/2} e^{-(\Lambda_T - a)}; m_0 < T \le m\} \, d\theta/(2\pi)^{1/2}. \tag{4.81}$$

From the expansion (4.78) and the consistency of $\{\hat{\theta}_n\}$ it follows that to a first approximation the behavior of Λ_n under P_θ for $\theta \ne \theta_0$ is similar to the behavior of the random walk $l_n(\theta) - l_n(\theta_0)$ with mean value

$$nJ(\theta, \theta_0) = nE_\theta\{l_1(\theta) - l_1(\theta_0)\}.$$

(J is the Kullback–Leibler information.) In particular, for $\theta \ne \theta_0$

$$P_\theta\left\{\lim_n n^{-1}\Lambda_n = J(\theta, \theta_0)\right\} = P_\theta\left\{\lim_n n^{-1}[l_n(\theta) - l_n(\theta_0)] = J(\theta, \theta_0)\right\} = 1,$$

and hence

$$P_\theta\left\{\lim_a a^{-1} T = 1/J(\theta, \theta_0)\right\} = 1. \tag{4.82}$$

If m_0, m, and a are related by

$$m^{-1}a = \eta_1 < m_0^{-1}a = \eta_0,$$

it follows from (4.82) that

$$(4.83)\qquad \begin{aligned} P_\theta\{m_0 < T \le m\} &\to 1 \qquad \text{if } \eta_1 < J(\theta,\theta_0) < \eta_0 \\ &\to 0 \qquad \text{if } J(\theta,\theta_0) \notin [\eta_1,\eta_0]. \end{aligned}$$

Also by (4.80) and (4.82)

$$(4.84)\qquad P_\theta\left\{\lim_a a^{-1}[-\ddot{l}_T(\hat{\theta}_T)] = I(\theta)/J(\theta,\theta_0)\right\} = 1.$$

Substituting (4.83) and (4.84) into (4.81) yields

$$(4.85)\qquad \begin{aligned} &P_{\theta_0}\{m_0 < T < m\} \\ &\sim a^{1/2}e^{-a}\int_{\{\theta:\eta_1 < J(\theta,\theta_0) < \eta_0\}} [I(\theta)/J(\theta,\theta_0)]^{1/2} E_\theta\{e^{-(\Lambda_T - a)}\}\, d\theta/(2\pi)^{1/2}. \end{aligned}$$

In the case of Brownian motion, by the continuity of the sample paths $\Lambda_T - a = 0$ whenever $T > m_0$, and simple algebra shows that (4.85) reduces to (4.76). For the discrete time normal case, substitution of (4.35) into (4.85) yields the first term on the right hand side of (4.40).

A consequence of Theorem 9.12 is that under quite general conditions

$$\tilde{v}(\theta) = \lim_{a\to\infty} E_\theta e^{-(\Lambda_T - a)}$$

exists, and the results of Chapters VIII and IX provide algorithms for its computation. Substituting this limit into (4.85) yields the approximation

$$(4.86)\qquad \begin{aligned} &P_{\theta_0}\{m_0 < T \le m\} \\ &\sim (2\pi)^{-1/2}a^{1/2}e^{-a}\int_{\{\theta:\eta_1 < J(\theta,\theta_0) < \eta_2\}} [I(\theta)/J(\theta,\theta_0)]^{1/2}\tilde{v}(\theta)\, d\theta. \end{aligned}$$

For applications of (4.86) which would be substantially more difficult to derive by the method of proof of Theorem 4.14, see Theorems 5.29 and 9.69.

Problems

4.1.* Let $x_1, x_2, \ldots$ be independent $N(\mu, 1)$. Consider a "one-sided" repeated significance test for H_0: $\mu = 0$ against H_1: $\mu > 0$, modified along the lines of III.8 for early termination when H_0 appears to be true by introducing a "lower" stopping boundary $-\tilde{b}n^{1/2} + \delta n$ or $-\tilde{b} + \delta n$. Give approximations for the error probabilities and expected sample size in the spirit of those developed in III.6 for truncated sequential probability ratio tests (cf. especially (3.36) and (3.37)). Find choices of the test parameters which define a test having error probabilities at $\mu = 0$ and $\mu = .6$ and also maximum sample size about the same as the test of Table 3.11. Make additional appropriate numerical comparisons (see Problem 4.19).

Remark. A repeated significance test with $b = 2.95$, $c = 2.07$, $m_0 = 7$, $m = 49$ is roughly comparable to the test of Table 3.2.

4.2. Show that a repeated significance test for a normal mean is asymptotically optimal in the following sense. Let N_m be a family of stopping rules indexed by m. Suppose $P_0\{N_m \le m\} \to 0$ and for some $\mu_1 > 0$ and all $|\mu| > \mu_1$, $mP_\mu\{N_m > m\} \to 0$. (For a repeated significance test, interpret these conditions in terms of significance level and power.) Show that for each μ with $|\mu| > \mu_1$

$$\tfrac{1}{2}\mu^2 E_\mu(N_m \wedge m) \ge [-\log P_0\{N_m \le m\}](1 + o(1))$$

and that for T defined by (4.6) with $b = \mu_1 m^{1/2}$

$$\tfrac{1}{2}\mu^2 E_\mu(T \wedge m) = -\log P_0\{T < m\} + 0(\log m).$$

4.3.* How should the definitions of attained significance level and confidence interval given in Section 5 be modified if one wants to take into account the amount by which S_n jumps over the boundary $\pm bn^{1/2}$? Why might this modification be important for a group sequential test? (For a discussion of one possible definition and the required probability calculation, see IX.5.)

4.4. Suppose that ξ in Theorem 4.14 is of the form $\xi = bm^{1/2} - z$, where $z > 0$ is fixed as $b \to \infty$. Show that Theorem 4.14 in effect says that $P_\xi^{(m)}\{T < m\}$ asymptotically equals the (conditional) probability of crossing the tangent at $t = m$ to the curve $bt^{1/2}$ during the time interval $(0, m)$.

4.5.† Let $T = \inf\{t: |W(t)| \ge [(t + 1)(\log(t + 1) + 2\log B)]^{1/2}\}$ (cf. (4.2)). By modifying the argument given in support of Proposition 4.28 show that as $b = \log B \to \infty$

$$\mu^2 E_\mu(T) = 2b + \log(2b/\mu^2) + \mu^2 - 1 + o(1).$$

4.6. For T defined as in Problem 4.5, use the argument of Section 8 to show that $P_0\{T < \infty\} = B^{-1}$ $(B > 1)$.

4.7.† Under the conditions of Theorem 4.27, show that T is asymptotically normally distributed (cf. Problems 3.10 and 3.11). Use this result to derive heuristically the terms of order m and of order $m^{1/2}$ in Theorem 4.27. Extend these results to the more general curved boundaries of Problem 3.11 and a general class of random walks having positive mean and finite variance.

4.8. Under the conditions of Proposition 4.29, use Corollary 4.19 to show that $P_\mu\{T_+ > m + m^{1/2}x\} = 1 - \Phi(\Delta + \tfrac{1}{2}\mu_1 x) - m^{-1/2}\phi(\Delta + \tfrac{1}{2}\mu_1 x)(\tfrac{1}{2}\Delta x + \mu_1 x^2/4 + \mu_1^{-1})(1 + o(1))$. Integrate this expression as in Remark 4.34 to obtain another demonstration of Proposition 4.29.

4.9.† Show that for the stopping rule (4.11), for $\mu \ne 0$,

$$E_\mu[W(T)/T] = \mu(1 + 2b^{-2}) + o(b^{-2}) \qquad (b \to \infty).$$

Hint: Expand $T^{-1/2}$ about $(b/\mu)^2$; use Proposition 4.28 and Problem 4.7—assuming that the quadratic term is uniformly integrable.

4.10. Show that for the stopping rule (4.11)

$$P_0\{T \le m\} \le 2[1 - \Phi(b)] + b\phi(b)[\log(m/m_0) + 2b^{-2}].$$

Hint: Consider the argument on page 34 of Ito and McKean (1965) modified by taking into account the minimum sample size m_0 (as in Problem 3.2) and by approximating the curved boundary by secants.

4.11. Let $x_1, x_2, \ldots$ be random variables whose joint distribution depends on a parameter θ. A $(1 - \alpha)$ 100% *confidence sequence* is a sequence of random sets $A_n = A_n(x_1, \ldots, x_n)$ such that for every θ

$$P_\theta\{\theta \in A_n \text{ for all } n\} \geq 1 - \alpha.$$

Use the results of Example 4.1 to give a $(1 - \alpha)$ 100% confidence sequence for a normal mean under repeated sampling from a normal population with known variance. Note that if (A_n) is a $(1 - \alpha)$ 100% confidence sequence, then for every stopping rule T, A_T is a $(1 - \alpha)$ 100% confidence set for θ.

4.12. Show that the stopping rule (4.2) can be obtained by the following Bayesian considerations. Suppose there is a prior probability p that $\mu = 0$, and conditional on $\mu \neq 0$, μ has a $N(0, 1)$ distribution. Then for some $0 < c < 1$, T in (4.2) is of the form "stop as soon as the posterior probability that $\mu = 0$ is $\leq c$" (see Cornfield, 1966).

4.13.† Show that the method of Section 8 can be used to prove (4.18). Show heuristically how (4.18) implies Theorem 4.14. *Hint:* Start from the Wald-like identity

$$\begin{aligned} &P_0\{m_0 < T \leq m, |W(m)| < cm^{1/2}\} \\ &\quad = E\{(T + 1)^{1/2} \exp[-W^2(T)/2(T + 1)] \\ &\qquad \times P_0\{|W(m)| < cm^{1/2} | T, W(T)\}; m_0 < T \leq m\}, \end{aligned}$$

where $E(\cdot) = \int_{-\infty}^{\infty} E_\mu(\cdot)\phi(\mu)\, d\mu$. In discrete time this identity follows easily from Theorem 2.24.

4.14.† Prove the result of Problem 4.4 by the method hinted at in Remark 4.25, and show how to obtain Corollary 4.19 as a consequence.

4.15.† Let $\rho > 0$ and $T = \inf\{t\colon t \geq m_0, |W(t)| \geq bt^{1/2} + \rho\}$. Show that under the conditions of Theorem 4.14, for $\mu_1^2/\mu_0 < \xi_0 < \mu_1$

$$P_\xi^{(m)}\{T < m\} \sim \mu_1 \xi_0^{-1} \exp\{-\mu_1^2 \xi_0^{-1} \rho - \tfrac{1}{2} m(\mu_1^2 - \xi_0^2)\}. \tag{4.87}$$

Verify that using (4.87) to obtain an approximation for $P_\mu\{T < m, |W(m)| < cm^{1/2}\}$ yields the same expression as replacing $v(x)$ by $\exp(-\rho x)$ in (4.49) (if $\mu \neq 0$) or (4.50) (if $\mu = 0$) (recall also (4.38)). How does this problem relate to (3.28)?

4.16. Give in terms of the exponential integral an approximation for $P_\mu\{T < m, |W(m)| < cm^{1/2}\}$ (T defined in (4.11)) which is valid for all μ. Modify this approximation for discrete time.

4.17. Let $W(t) = (W_1(t), W_2(t))'$ be two-dimensional Brownian motion with mean vector $(\mu_1 t, \mu_2 t)'$, var $W_i(t) = \sigma_i^2 t$, and $\operatorname{cov}[W_1(t), W_2(t)] = \rho\sigma_1\sigma_2 t$. Let T be the stopping rule (4.11) or (3.7) defined in terms of W_1, so the approximate bias of $W_1(T)/T$ as an estimator of μ_1 is given by Problem 4.9 or (3.20) respectively (in the case $\sigma_1 = 1$). What is the approximate bias of $W_2(T)/T$ as an estimator of μ_2? *Hint:* Recall that $W_2(t) - \mu_2 t - \rho\sigma_2[W_1(t) - \mu_1 t]/\sigma_1$ and $W_1(t)$ are independent.

4.18. Consider a Brownian motion process $W(t), 0 \le t \le m$, conditional on $W(m) = \xi$, as in Problem 3.14. Devise a test of power one (cf. Example 4.1) of $H_0\colon \xi = 0$ against $H_1\colon \xi \neq 0$. Develop approximations for $E_\xi^{(m)}[T/(m - T)]$ and $E_\xi^{(m)}(T)$ (cf. Problem 3.12).

4.19. Assuming the asymptotic scaling of Theorem 4.21, use the method of Section 8 to show that

$$E_0(T \wedge m) = m - b^3\phi(b)[\mu_1^{-2}\log(m/m_0) - \mu_1^{-2} + \mu_0^{-2}] + o(b^3\phi(b)).$$

Indicate how this approximation (or better yet a discrete time version) would be useful in a discussion of Problem 4.1.

4.20.† Derive heuristically the terms of order b^{-1} in the asymptotic expansion of Theorem 4.27.

Remarks. One of these terms comes from $P_\xi^{(m)}\{T_+ < m\}$ in (4.30), for ξ close to $bm^{1/2}$. The rest result from expanding $E_\mu \tau_m(x)$ to the next order of precision beyond that given in (4.33). It seems possible also to use the method suggested in Remark 4.34 after obtaining the next term in (4.20) (for $\mu = \mu_1 + \Delta/m^{1/2}$). That the $o(1)$ in Proposition 4.28 is actually $o(b^{-1})$ is easy to see heuristically and has been proved by Jennen (1985).

CHAPTER V

Examples of Repeated Significance Tests

1. Introduction

We consider below a number of examples of repeated significance tests in order to discover the extent to which the methods of Chapter IV can be adapted to a variety of more difficult problems. Often precise development of an asymptotic theory analogous to that obtained in Chapter IV for normally distributed data of known variability is complicated, and hence we shall concentrate on the use of normal approximations to obtain a rough idea of the probabilities and expected sample sizes involved. For continuously monitored tests Brownian motion provides simple approximations. For group tests one can use a discrete time normal approximation to take into account the sometimes substantial excess over the stopping boundary.

Tests for Bernoulli data and some applications are described in Section 2. Section 3 is concerned with the problem of comparing more than two treatments. The discussion of confidence regions is particularly interesting because the "exact" sequential regions are not only shifted towards the origin as in one dimension, but also have a different shape than the conventional fixed sample regions. Section 4 discusses tests for a normal mean with unknown variance. Here there are no new conceptual problems, but only the technical one of providing reasonable approximations to various operating characteristics. Section 5 takes up the problem of censored survival data analyzed by the proportional hazards model. An appropriate Brownian motion approximation is proposed and informally justified. The results of a Monte Carlo study and some applications are presented in Section 6.

The various sections of this chapter are independent and can be read in any order.

2. Bernoulli Data and Applications

This section is concerned with some examples involving categorical data—especially paired comparisons and testing the equality of two binomial populations.

For the theoretically simpler case of paired comparisons, suppose experimental units are matched in pairs with one unit in each pair receiving one of the two possible treatments. Let $x_{ij} \in \{0, 1\}$ denote the response (failure or success) for the ith treatment in the jth pair, $i = 1, 2; j = 1, 2, \ldots$. The x_{ij} are assumed to be independent random variables, and the success probabilities

$$p_{ij} = P\{x_{ij} = 1\} \qquad (q_{ij} = 1 - p_{ij})$$

are assumed to satisfy the log-linear relation

$$\log(p_{ij}/q_{ij}) = \alpha_i + \beta_j \qquad (i = 1, 2; j = 1, 2, \ldots).$$

We are interested in comparing the treatment effects α_1 and α_2; the unit effects $\beta_1, \beta_2, \ldots$ are nuisance parameters.

Note that

$$P\{x_{1j} = 1 | x_{1j} \neq x_{2j}\} = \frac{p_{1j}q_{2j}}{p_{1j}q_{2j} + p_{2j}q_{1j}} = \frac{e^{\alpha_1 - \alpha_2}}{1 + e^{\alpha_1 - \alpha_2}},$$

which does not depend on β_j. Hence by considering only those pairs in which $x_{1j} \neq x_{2j}$, and in the nth such pair letting $y_n = 1$ or 0 according as $x_{1j} = 1$ and $x_{2j} = 0$ or $x_{1j} = 0$ and $x_{2j} = 1$, we obtain a sequence $y_1, y_2, \ldots$ of independent Bernoulli variables with

$$P\{y_n = 1\} = p = \frac{e^{\alpha_1 - \alpha_2}}{1 + e^{\alpha_1 - \alpha_2}}, \qquad P\{y_n = 0\} = q = 1 - p. \tag{5.1}$$

In this way problems of inference about $\alpha_1 - \alpha_2$ become problems of inference about the single Bernoulli parameter p. Of course, the effect of the nuisance parameters is still present, since in general several pairs (x_{1j}, x_{2j}) must be observed to produce a single y_n. In the special case $\beta_1 = \beta_2 = \cdots$, so $p_{1j} = p_1$ and $p_{2j} = p_2$ for all j, the expected number of pairs (x_{1j}, x_{2j}) which must be observed in order to generate a single y is $(p_1q_2 + p_2q_1)^{-1}$. In general the relation between the number of variables y_n observed and the actual sample size is very complicated.

A repeated likelihood ratio test of H_0: $\alpha_1 = \alpha_2$ or equivalently $p = \frac{1}{2}$ against H_1: $p \neq \frac{1}{2}$ is defined as follows (cf. (4.7), Example 4.8, and (4.77)). Let $H(x) = x \log x + (1 - x)\log(1 - x) + \log 2$ and $S_n = \sum_1^n y_j$. Given $0 < m_0 < m$ and $0 < d \leq a$, let

$$T = \inf\{n: n \geq m_0, nH(n^{-1}S_n) \geq a\}. \tag{5.2}$$

Stop sampling at $\min(T, m)$ and reject H_0 if either $T \leq m$ or $T > m$ and $mH(m^{-1}S_m) \geq d$. As usual we are interested in the power,

$$(5.3)\qquad \begin{aligned} &P_p\{T \le m\} + P_p\{T > m, mH(m^{-1}S_m) \ge d\} \\ &\quad = P_p\{mH(m^{-1}S_m) \ge d\} + P_p\{T < m, mH(m^{-1}S_m) < d\} \end{aligned}$$

and expected sample size $E_p(T \wedge m)$.

In general there are two ways to approximate these quantities: (a) to adapt the theory developed in Chapter IV for normal data to the problem at hand, or (b) use the results of Chapter IV as a rough central limit type approximation. The first approach is theoretically more satisfying, but involves a greater mathematical investment. The second seems adequate for most practical purposes and is the path taken here. (See IX.3 for a more precise analysis with more extensive numerical examples.)

Let

$$(5.4)\qquad \tilde{S}_n = n\{2H(n^{-1}S_n)\}^{1/2}\,\mathrm{sgn}(n^{-1}S_n - \tfrac{1}{2}),$$

so (5.2) can be rewritten in the form

$$(5.5)\qquad T = \inf\{n\colon n \ge m_0, |\tilde{S}_n| \ge bn^{1/2}\},$$

where $b = (2a)^{1/2}$. Also put $c = (2d)^{1/2}$. A Taylor series expansion of $[H(x)]^{1/2}$ about $x = p$, yields

$$\tilde{S}_n \cong \{n[2H(p)]^{1/2} + (S_n - np)H'(p)/[2H(p)]^{1/2}\}\,\mathrm{sgn}(n^{-1}S_n - \tfrac{1}{2}),$$

which suggests approximating $\tilde{S}_n$ by a Brownian motion process with drift and variance

$$(5.6)\qquad \tilde{\mu} = 2H(p)^{1/2}\,\mathrm{sgn}(p - \tfrac{1}{2}) \text{ and } \tilde{\sigma}^2 = pq[H'(p)]^2/2H(p).$$

For $b = 3.15$, $c = 2.05$, $m_0 = 10$, and $m = 49$ this approximation in conjunction with the results of Chapter IV gives $P_{1/2}\{T \le 49\} \cong .013$, $P_{.7}\{T \le 49\} \cong .50$, $E_{.7}(T \wedge 49) \cong 37$, and $P_{1/2}\{T \le 49\} + P_{1/2}\{T > 49, |\tilde{S}_{49}| > 2.05(49)^{1/2}\} \cong .047$. Monte Carlo estimates (10,000 trials) of .012, .49, 39, and .050 respectively indicate that the approximation is acceptably accurate for these values of b, c, m_0, and m.

It is worth noting that there are two somewhat different mathematically precise interpretations of the preceding approximation. One is that for any fixed p

$$\tilde{\sigma}^{-1}n^{-1/2}(\tilde{S}_{[nt]} - [nt]\tilde{\mu}), \qquad 0 \le t \le t',$$

with $\tilde{\mu}$ and $\tilde{\sigma}$ given by (5.6), converges to Brownian motion with 0 drift. The second is that if p depends on n and converges to $\frac{1}{2}$ in such a way that $n^{1/2}(p - \frac{1}{2}) \to \frac{1}{2}\xi$ for some $-\infty < \xi < \infty$, then $n^{-1/2}\tilde{S}_{[nt]}$, $0 \le t \le t'$, converges to Brownian motion with drift ξ. The fact that a single asymptotic approximation has some validity for fixed p and for local alternatives $p \to \frac{1}{2}$ is a consequence of the square root transformation in (5.4). The corresponding asymptotic results for $nH(n^{-1}S_n)$ itself do not fit together so smoothly. (See Problem 5.3 for a generalization.)

An alternative model for comparing two treatments is a two population test.

Suppose that for $i = 1, 2, x_{i1}, x_{i2}, \ldots$ are independent Bernoulli variables with success probability p_i, and assume that the $\{x_{1j}\}$ and $\{x_{2k}\}$ are independent. Let $\bar{x}_{in} = n^{-1} \sum_{j=1}^{n} x_{ij}$ ($i = 1, 2, n = 1, 2, \ldots$). The likelihood ratio statistic for testing H_0: $p_1 = p_2$ against H_1: $p_1 \neq p_2$ based on the observations $x_{11}, \ldots, x_{1n_1}, x_{21}, \ldots, x_{2n_2}$ is

$$n_1 H(\bar{x}_{1n_1}) + n_2 H(\bar{x}_{2n_2}) - (n_1 + n_2) H\left(\frac{n_1 \bar{x}_{1n_1} + n_2 \bar{x}_{2n_2}}{n_1 + n_2}\right). \tag{5.7}$$

To discuss the behavior of the process (5.7) it is convenient to make some assumptions about the relation of n_1 to n_2. For a large class of experiments it is reasonable to assume that n_1 increases proportionally to n_2 as sampling proceeds. The simplest and most important case is that of balanced treatment allocation, so n_1 and n_2 are approximately equal. For simplicity we shall assume that $n_1 = n_2 = n$, say, although the following discussion is easily modified to handle the case where n_1 is proportional to n_2. For the possibility that n_1/n_2 changes in an irregular fashion, see Chapter VI—especially Problem 6.3.

For $n_1 = n_2 = n$, the statistic in (5.7) becomes

$$Z_n = nI(\bar{x}_{1n}, \bar{x}_{2n}),$$

where $I(x, y) = H(x) + H(y) - 2H[(x + y)/2]$. A repeated likelihood test stops sampling at $\min(T, m)$, where

$$T = \inf\{n: n \geq m_0, Z_n \geq a\},$$

and rejects H_0: $p_1 = p_2$ if either $T \leq m$ or $T > m$ and $Z_m > d$. Because we are now concerned with a composite null hypothesis in a multi-parameter family of distributions, asymptotic theory analogous to that developed in Chapter IV is considerably more complicated. See Woodroofe (1982), Lalley (1983), or Hu (1985) for a precise treatment. A Brownian motion approximation can be based on

$$\tilde{S}_n = n[2I(\bar{x}_{1n}, \bar{x}_{2n})]^{1/2} \operatorname{sgn}(\bar{x}_{1n} - \bar{x}_{2n}),$$

and a Taylor series expansion of $I(x, y)$ about (p_1, p_2) indicates that an appropriate mean and variance are

$$\tilde{\mu} = [2I(p_1, p_2)]^{1/2}$$

and

$$\tilde{\sigma}^2 = [p_1 q_1 \log^2(p_1 \bar{q}/q_1 \bar{p}) + p_2 q_2 \log^2(p_2 \bar{q}/q_2 \bar{p})]/2I(p_1, p_2),$$

where $\bar{p} = (p_1 + p_2)/2$ and $\bar{q} = 1 - \bar{p}$. Siegmund and Gregory (1980) give numerical evidence that this approximation is quite satisfactory.

As an example consider the clinical trial described by Brown *et al.* (1960) to test the value of a large dose of antitoxin in cases of clinical tetanus. At the time of the study the value of antitoxin once symptoms had already appeared was uncertain, the antitoxin itself sometimes had very severe side effects (including death), and the antitoxin was expensive and of limited availability in Ibadan and Jamaica, where the study was conducted. Since the outcome of interest was life or death within a relatively short time following treatment, a sequential trial presented obvious advantages for early detection of a large treatment effect, should one be present.

To compare antitoxin with placebo it was decided to use a matched pair analysis and to choose a sequential design which for testing H_0: $p = \frac{1}{2}$ against H_1: $p \neq \frac{1}{2}$ had significance level .05 and power .95 at $p = .75$ (p is the parameter defined in (5.1)). The test actually used by Brown *et al.* was defined by straight line boundaries (cf. Problem 5.6), but for illustrative purposes we consider a repeated likelihood ratio test. Some numerical trial and error using the Brownian motion approximation with mean and variance given by (5.7) indicates that a repeated likelihood ratio test defined by the parameters $m_0 = 9, m = 63$, and $b = (2a)^{1/2} = 2.79$ has approximately the desired error probabilities. Also, the expected sample size when $p = .75$ is approximately 26. A Monte Carlo experiment (10,000 trials) gave estimates of .044, .950, and 28 for the significance level, power, and expected sample size to show that the crude Brownian motion approximation is reasonably accurate in this case.

Let $y_n = 1$ or 0 according as in the nth untied pair there is success (life) with antitoxin and failure (death) with placebo, or vice versa. The data in Brown *et al.* were $y_1 = \cdots = y_6 = 1$, $y_7 = 0$, $y_8 = 1$, $y_9 = 0$, $y_{10} = y_{11} = 1$, $y_{12} = 0$, $y_{13} = \cdots = y_{18} = 1$, which for the test described above would lead to stopping and rejecting H_0 in favor of antitoxin at $T = 18$. (The test actually used by Brown *et al.* also resulted in stopping at the 18th untied pair.)

It is interesting to calculate the attained significance of this result and a confidence interval for p. Because of the small numbers involved it is easy to perform exact computations. For example, direct enumeration of all possibilities shows that if $T \leq 18$, then $T \in \{9, 11, 14, 18\}$ and $P_p\{T \leq 18, T^{-1}S_T > \frac{1}{2}\} = p^9 + 9p^{10}q + 54p^{12}q^2 + 318p^{15}q^3$. Hence the attained significance is $P_{1/2}\{T \leq 18\} = .022$, and a 90% confidence interval for p is (.59, .92). The midpoint of this interval is .75, which is slightly less than the 0% "interval," .79. Both values are adjusted down from the empirical frequency of success, .83, and are remarkably close to the .75 of the original design considerations.

There was a difference in prognosis for different patients in the trial, but because of the need to administer treatment immediately upon arrival this difference was not accounted for in the randomization nor in the sequential analysis. The actual pairing, for example, was based on chronology, not on prognosis. As a consequence, the results were somewhat less clear cut than the preceding discussion suggests. See Brown *et al.* for details and Vakhil *et al.* (1968) for subsequent developments. In retrospect it appears that the test was terminated earlier than now seems advisable.

A similar design with similar results was used by Freireich *et al.* (1963) to compare 6-Mercaptopurine with placebo for prolonging the duration of a remission of leukemia. After being judged to be in a state of partial or complete remission, a patient was paired with a second patient in the same state. One patient in each pair received the treatment and the other a placebo. Success (failure) occurred in the ith pair if the time to relapse or censoring for the patient on 6-Mercaptopurine (placebo) exceeded the time to relapse for the patient on placebo (6-Mercaptopurine). (Other pairs were presumably to be discarded, although it appears that there were no such pairs.)

Let p denote the probability of success. It was decided to use a sequential test of H_0: $p = \frac{1}{2}$ against H_1: $p \neq \frac{1}{2}$ which had significance level .05 and power .95 at $p = .75$. These are in fact the requirements of the preceding test, and hence that same test is one possibility. Moreover, after 18 pairs the pattern of success (1) and failure (0) was 1, 0, 1, 1, 1, 0, 1, 1, 1, 1, 1, 1, 1, 0, 1, 1, 1, 1 for a total of 15 successes and 3 failures. Hence if the same design as that discussed above were used, stopping with rejection of H_0 would occur again after 18 pairs with the same attained significance level, confidence interval for p, etc. In actuality the data were reviewed every three months. At the third review the eighteenth pair had not yet been observed. By the fourth review there were already 21 pairs, the last three of which were also successes, so the test was stopped with the rejection of H_0. For additional discussion of this trial see Section 6.

A third example is the sequential clinical trial reported by Fogel *et al.* (1982) for testing the efficacy of vasopressin in stopping upper gastrointestinal bleeding. The circumstances leading to the decision to conduct this trial sequentially are interesting. At the time of the study vasopressin had been in use for about 25 years for controlling acute upper gastrointestinal bleeding. Nevertheless, the evidence in support of vasopressin was ambiguous, and it seemed desirable to consider a new trial. Since there was considerable variability in the opinions and practice of the cooperating physicians with regard to use of vasopressin, a sequential design was selected so that the trial could be terminated in the event that vasopressin proved as effective as some earlier studies had claimed (as high as a 70% success rate compared to 28% for placebo). Although there were several endpoints of interest, for the purpose of the sequential design success was defined as a termination of bleeding within six hours. The data were modeled as independent Bernoulli outcomes with probabilities p_1 for success with vasopressin and p_2 for success with placebo. A horizon of approximately 45 patients on each treatment over a three year period was anticipated. Although there was no formal provision for early termination if H_0: $p_1 = p_2$ appeared to be true, an *ad hoc* decision was made to terminate the trial after thirty months, because none of the endpoints of interest (except for the relatively unimportant one of total transfusion requirement) showed anything approximating a significant effect. Twenty-nine patients had received vasopressin with thirteen successes; thirty-one had received placebo resulting in eleven successes. For these data the value of χ^2 for testing $p_1 = p_2$ is .55. Analysis of other endpoints and stratification of the study group according to the cause of bleeding failed to suggest a sizeable treatment effect. In particular, the mortality rate was 35% in the control group and 28% in the experimental group.

In this experiment the importance of the sequential design was its (in fact unrealized) potential for early termination which made it ethically permissible to initiate a new trial to reexamine a treatment which was already in use.

3. Comparing More than Two Treatments

This section describes repeated significance tests for comparing more than two treatments, with special attention to the case of three treatments. Simple examples in clinical trails might be a comparison of one drug at a high dosage level, the same drug at a low dosage level, and placebo, or treatment A, treatment B, and a synthesis of treatments A and B used together.

To simplify the situation as much as possible, assume that observations are made sequentially on vectors $x_n = (x_{1n}, \ldots, x_{rn})'$ $(n = 1, 2, \ldots)$, where r denotes the number of treatments and x_{ij} is the response of the jth experimental unit assigned to treatment i. With minor modifications the results can be applied to randomized schemes of allocating treatments which approximate this idealization of perfect balance or to (data independent) allocation schemes which assign different proportions of experimental units to the different treatments. The x_{ij} are assumed to be independently and normally distributed with mean μ_i $(i = 1, 2, \ldots, r)$ and common known variance $\sigma^2 = 1$.

The log likelihood ratio statistic for testing the hypothesis of no treatment differences H_0: $\mu_1 = \cdots = \mu_r$ against H_1: not all μ_i equal is

$$Z_n = \frac{1}{2} n \sum_{i=1}^{r} (\bar{x}_{i\cdot} - \bar{x}_{\cdot\cdot})^2, \tag{5.8}$$

where $\bar{x}_{i\cdot} = n^{-1} \sum_{j=1}^{n} x_{ij}$ and $\bar{x}_{\cdot\cdot} = (rn)^{-1} \sum_{i,j} x_{ij}$. If H_0 is true, $2Z_N$ has a chi-square distribution with $r - 1$ degrees of freedom. A repeated significance test of H_0 may be defined as follows (cf. (4.7)); given $m_0 < m$ and $0 < d \le a$, stop sampling at $\min(T, m)$, where

$$T = \inf\{n: n \ge m_0, Z_n \ge a\} \tag{5.9}$$

and reject H_0 if either $T \le m$ or $T > m$ and $Z_m \ge d$. Like the fixed sample likelihood ratio test, this is in effect a test of the parameter $\theta = [\sum_i (\mu_i - \bar{\mu}.)^2]^{1/2}$, where $\bar{\mu}. = r^{-1} \sum_i \mu_i$, in the sense that its power function

$$P_\theta\{T \le m\} + P_\theta\{T > m, Z_m \ge d\}$$

and expected sample size

$$E_\theta(T \wedge m)$$

depend on the vector $\mu = (\mu_1, \ldots, \mu_r)'$ only through the value of θ. For this reason the test does not in itself provide information about the relative merits of the different treatments.

Let C be an orthogonal $r \times r$ matrix with $c_{rj} = r^{-1/2}$ $(j = 1, 2, \ldots, r)$, and consider

$$(\theta_1, \ldots, \theta_r)' = C\mu. \tag{5.10}$$

Then $\theta_r = r^{1/2}\bar{\mu}.$, and knowledge of the treatment effect vector $(\mu_1 - \bar{\mu}., \ldots, \mu_r - \bar{\mu}.)$ is equivalent to knowledge of the vector

$$(\theta_1,\ldots,\theta_{r-1},0)' = C(\mu_1 - \bar{\mu}.,\ldots,\mu_r - \bar{\mu}.)'$$

of length $\theta = (\sum_1^{r-1}\theta_i^2)^{1/2} = [\sum_1^r(\mu_i - \bar{\mu}.)^2]^{1/2}$. One reasonable way of making quantitative comparative statements about the treatment effects is to give a joint confidence region for θ and the angle ω which the vector $(\theta_1,\ldots,\theta_{r-1})'$ makes with some convenient direction in $r-1$ dimensional space.

Suppose, for example, that $r = 3$. Let $(c_{11},c_{12},c_{13}) = (2^{-1/2},-2^{-1/2},0)$ and $(c_{21},c_{22},c_{23}) = (6^{-1/2},6^{-1/2},-2\cdot 6^{-1/2})$, so that $\theta_1 = (\mu_1-\mu_2)/2^{1/2}$ and $\theta_2 = (\mu_1+\mu_2-2\mu_3)/6^{1/2}$. In this case ω may be taken to be the angle that $(\theta_1,\theta_2)'$ makes with the positive θ_1 axis in the (θ_1,θ_2) plane.

Even more specifically, consider the case where Treatment 1 is a drug at a high level of dosage, Treatment 2 is the same drug at a low level of dosage, and Treatment 3 is a placebo. Then under H_1, subhypotheses of considerable interest might be that Treatment 1 is effective but Treatments 2 and 3 are equally ineffective, or alternatively that Treatments 1 and 2 are about equally effective and better than Treatment 3. These subhypotheses can be stated more precisely as H_{11}: $\mu_1 > \mu_2 = \mu_3$ or H_{12}: $\mu_1 = \mu_2 > \mu_3$ and are easily seen to be equivalent to H_{11}: $\omega = \pi/6$ or H_{12}: $\omega = \pi/2$. An experiment of this sort is potentially of interest when the drug may have severe side effects and it is conceivable that Treatment 1 is actually harmful, while Treatment 2 is beneficial or perhaps no better than placebo. For example, the hypothesis of "disaster," H_{13}: $\mu_1 < \mu_2 = \mu_3$ is equivalent to H_{13}: $\omega = 7\pi/6$. Consequently a confidence region for (θ,ω) can be informative about the superiority of Treatment 1, the advisability of eliminating Treatment 3 and performing a new experiment to compare Treatments 1 and 2, etc.

Let C be the matrix of (5.10) and define for $i = 1,2,\ldots,r-1$, $j = 1,2,\ldots,n$

$$y_{ij} = \sum_k c_{ik}x_{kj} = \sum_k c_{ik}(x_{kj} - \bar{x}._j).$$

Let $S_{in} = y_{il} + \cdots + y_{in}$ and $S_n = (S_{1n},\ldots,S_{r-1,n})'$. The log likelihood ratio statistic in (5.8) is just $\|S_n\|^2/2n = (\sum_{i=1}^{r-1}S_{i,n}^2)/2n$, so an equivalent description of the stopping rule (5.9) and repeated significance test is as follows: given $m_0 < m$ and $0 < c \le b$, stop sampling at $\min(T,m)$, where

$$T = \inf\{n\colon n \ge m_0,\ \|S_n\| \ge bn^{1/2}\} \tag{5.11}$$

and reject H_0: $\mu_1 = \cdots = \mu_r$ if either $T \le m$ or $T > m$ and $\|S_m\| \ge cm^{1/2}$.

The significance level, power, and expected sample size of this test may be computed approximately by means of the following generalizations of (4.40)–(4.44), (4.49), and (4.50). Let $r_1 = r-1$ be the dimension of the random walk $\{S_n, n = 1,2,\ldots\}$. Let $\theta = [\sum_{i=1}^r(\mu_i - \bar{\mu}.)^2]^{1/2} = \|E_\mu S_1\|$, $\gamma_1 = bm^{-1/2} < \gamma_0 = bm_0^{-1/2}$, and $\gamma = cm^{-1/2} \le \gamma_1$.
Then

$$\begin{aligned} &P_\mu\{T \le m\} + P_\mu\{T > m,\ \|S_m\| \ge cm^{1/2}\} \\ &\quad = P_\mu\{\|S_m\| \ge cm^{1/2}\} + P_\mu\{T < m,\ \|S_m\| < cm^{1/2}\}, \end{aligned} \tag{5.12}$$

where for $\mu \neq 0$

Table. 5.1 Repeated Significance Test with Three Treatments: $b = 3.5$; $c = 2.5$; $m_0 = 7$; $m = 49$

θ	Power	$P_\theta\{T \le m\}$	$E_\mu(T \wedge m)$	Power of Fixed Sample Test ($m = 49$)
.7	.995	.95	23	.995
.5	.88	.64	35	.89
.3	.44	.18	—	.46
0	.051	.018	—	.05

(5.13)
$$P_\mu\{T < m, \|S_m\|\} < cm^{1/2}\} \cong \nu(\gamma_1^2\gamma^{-1})\frac{\phi[m^{1/2}(\gamma_1 - \theta)]}{(m\theta\gamma)^{1/2}} e^{-m\theta(\gamma_1 - \gamma)} \frac{\gamma_1^{r_1}}{(\theta\gamma)^{r_1/2}}$$

and

(5.14)
$$P_0\{T < m, \|S_m\| < cm^{1/2}\} \cong 2^{-(r_1-2)/2}[\Gamma(r_1/2)]^{-1} b^{r_1} e^{-b^2/2} \int_{\gamma_1^2/\gamma}^{\gamma_0} x^{-1}\nu(x)\,dx.$$

For $m_0 \ge 1$ and $\theta = \gamma_1 + \Delta m^{-1/2}$

(5.15)
$$\begin{aligned} E_\mu(T \wedge m) \cong {} & [m\gamma_1^2 + 2\rho(\theta) - r_1]/\theta^2 - \{m^{1/2}(\theta - \tfrac{1}{2}\gamma_1)^{-1}[\phi(\Delta) - \Delta\phi(-\Delta)]\} \\ & - \tfrac{1}{2}(\theta - \tfrac{1}{2}\gamma_1)^{-2}\Phi(-\Delta)\rho(2\theta - \gamma_1) + \gamma_1^{-2}\{(r_1 - 1)\Phi(-\Delta) \\ & + (1 + \Delta^2)\Phi(-\Delta) - \Delta\phi(\Delta)\}. \end{aligned}$$

Here ν and ρ are as in (4.35)–(4.39).

For the most part the derivations of these expressions follow from modifications of the calculations in the one-dimensional case (see Problem 5.1 and XI.2). Note, however, that (5.15) is to one order less precision asymptotically than (4.42). The terms of order $m^{-1/2}$ in an expansion of $E_\mu(T \wedge m)$ are dimension dependent and perhaps difficult to determine. From a practical point of view it is convenient that no new numerical tables are required.

Table 5.1 gives a numerical example. A Monte Carlo experiment with 2500 replications using importance sampling as described in Remark 4.46 estimated $P_0\{T \le 49\}$ to be $.0184 \pm .0005$. Monte Carlo estimates of other entries in Table 5.1 differed from the theoretical approximations by at most one unit to the accuracy given.

To find a confidence region for $(\theta_1, \ldots, \theta_{r_1})'$, first observe that the arguments of III.4 and IV.5 yield a confidence interval for $\theta = (\sum_1^{r_1} \theta_i^2)^{1/2} = [\sum_1^r (\mu_i - \bar{\mu}.)^2]^{1/2}$, say

(5.16)
$$[\theta_*(T \wedge m, \|S_{T \wedge m}\|), \theta^*(T \wedge m, \|S_{T \wedge m}\|)].$$

Let ω denote the angle which the vector $(\theta_1, \ldots, \theta_{r_1})'$ makes with the positive θ_1 axis, and let ω_n denote the corresponding (random) angle made by the vector

S_n, $n = 1, 2, \ldots$. We shall find a confidence region for $(\theta_1, \ldots, \theta_{r_1})'$ by combining the confidence interval (5.6) for θ with a confidence region for ω for each value of θ. Suppose for simplicity that $r = 3$, so $r_1 = 2$.

To begin, assume that θ is known and consider the problem of finding a confidence interval for ω. Since $(T \wedge m, \|S_{T\wedge m}\|)$ is ancillary, i.e. its distribution does not depend on ω, it seems reasonable to consider the conditional distribution of $\omega_{T\wedge m}$ given $(T \wedge m, \|S_{T\wedge m}\|)$. Obviously under $P_{0,0}$ the distribution is uniform on $[-\pi, \pi)$, and a likelihood ratio calculation reduces the general case to this one as follows. The likelihood ratio of $y_1, \ldots, y_n$ under $P_{\theta,\omega}$ relative to $P_{0,0}$ is

$$\exp(\theta_1 S_{1n} + \theta_2 S_{2n} - n(\theta_1^2 + \theta_2^2)/2) = \exp[\theta \|S_n\| \cos(\omega_n - \omega) - n\theta^2/2]. \tag{5.17}$$

Hence by a direct calculation, for any stopping rule τ defined in terms of $\|S_n\|$, $n = 1, 2, \ldots$, and in particular for $T \wedge m$,

$$P_{\theta,\omega}\{\omega_\tau \in dw \mid \tau, \|S_\tau\|\} = \exp[\theta \|S_\tau\| \cos(w - \omega)]\, dw / 2\pi I_0(\theta \|S_\tau\|). \tag{5.18}$$

Here $I_0(v) = \int_{-\pi}^{\pi} \exp(v \cos w)\, dw/2\pi$ is the usual modified Bessel function, and the distribution in (5.18) is the von Mises distribution on $[-\pi, \pi)$. For any fixed $\gamma_2 \in (0, 1)$ there exists a function $u(x)$ such that

$$P_{\theta,\omega}\{|\omega_{T\wedge m} - \omega| \le u(\theta \|S_{T\wedge m}\|) \mid T \wedge m, \|S_{T\wedge m}\|\} = 1 - \gamma_2, \tag{5.19}$$

which defines a confidence interval for ω for each fixed θ.

Let the confidence coefficient associated with (5.16) be $1 - \gamma_1$. Since θ_* and θ^* depend only on $T \wedge m$ and $\|S_{T\wedge m}\|$, by (5.19)

$$\begin{aligned} P_{\theta,\omega}\{\theta_* \le \theta \le \theta^*, |\omega_{T\wedge m} - \omega| \le u(\theta \|S_{T\wedge m}\|)\} \\ = E_{\theta,\omega}[P_{\theta,\omega}\{|\omega_{T\wedge m} - \omega| \le u(\theta\|S_{T\wedge m}\|) \mid T \wedge m, \|S_{T\wedge m}\|\}; \theta_* \le \theta \le \theta^*] \\ = (1 - \gamma_2) P_{\theta,\omega}\{\theta_* \le \theta \le \theta^*\} \ge (1 - \gamma_2)(1 - \gamma_1), \end{aligned}$$

so the region

$$\theta_* \le \theta \le \theta^*, \; \omega_{T\wedge m} - u(\theta\|S_{T\wedge m}\|) \le \omega \le \omega_{T\wedge m} + u(\theta\|S_{T\wedge m}\|) \tag{5.20}$$

defines a $(1 - \gamma_2)(1 - \gamma_1)$ confidence region for (θ, ω).

To get a better feeling for the shape of the region (5.20) note that as $v \to \infty$, a simple Taylor series expansion of the von Mises density $\exp[v \cos(w - \omega)]\, dw/2\pi I_0(v)$ indicates that it can be approximated by a normal density with mean ω and variance v^{-1}. Numerical investigation suggests using a normal distribution with variance $(v - 1/2)^{-1}$ (Mardia, 1972, p. 64). Hence $u(x) \cong z_{\gamma_2}/(x - 1/2)^{1/2}$, where z satisfies $1 - \Phi(z_\gamma) = \gamma/2$. With this approximation the confidence region (5.20) becomes approximately

$$\begin{aligned} \theta_* \le \theta \le \theta^*, \; \omega_{T\wedge m} - z_{\gamma_2}/(\theta\|S_{T\wedge m}\| - 1/2)^{1/2} \\ \le \omega \le \omega_{T\wedge m} + z_{\gamma_2}/(\theta\|S_{T\wedge m}\| - 1/2)^{1/2}. \end{aligned} \tag{5.21}$$

These relations are displayed in Figure 5.1.

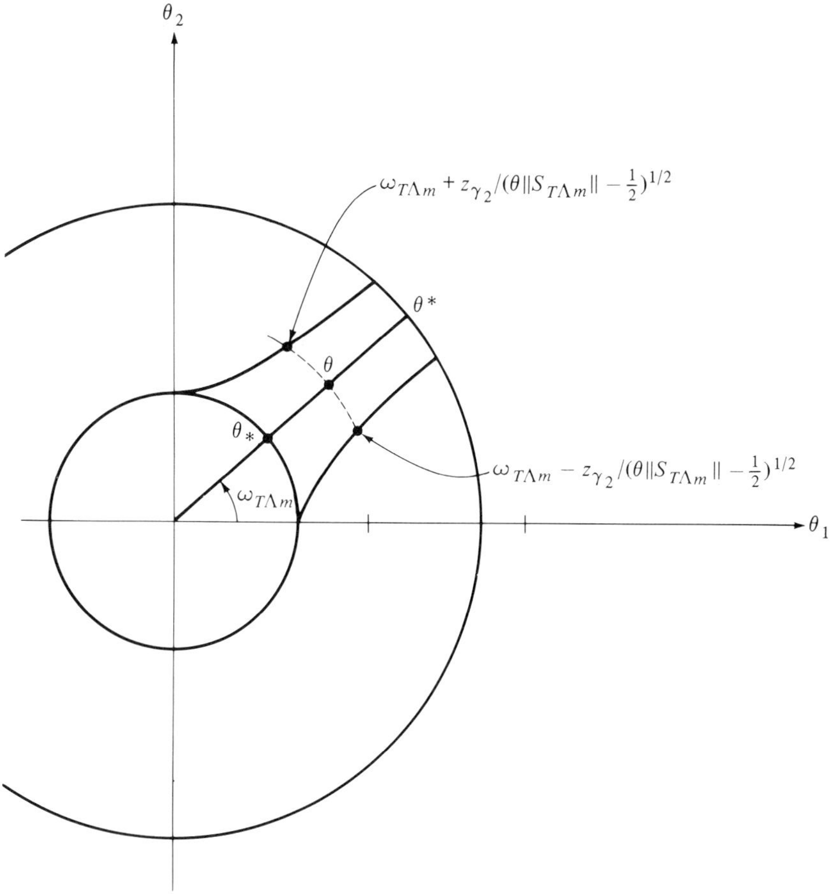

Figure 5.1.

Suppose that θ is so large that $P_\theta\{T \leq m\} \cong 1$, and hence $P_\theta\{T \wedge m = T\} \cong 1$. By the definition of T and (the obvious generalization of) (4.12) as $b \to \infty$

$$\|S_T\| \sim bT^{1/2} \sim b^2/\theta$$

with probability close to one, so a crude approximation to (5.21) for large θ is

$$\begin{aligned} \theta_* \leq \theta \leq \theta^*, \omega_{T \wedge m} - z_{\gamma_2}/(b^2 - 1/2)^{1/2} \\ \leq \omega \leq \omega_{T \wedge m} + z_{\gamma_2}/(b^2 - 1/2)^{1/2}. \end{aligned} \tag{5.22}$$

Although one would presumably use (5.20) or (5.21) in practice, the approximation (5.22) may be useful as a design consideration to help in selecting an appropriate value of b. For example, if $\gamma_2 = .05$, so $z_{\gamma_2} = 1.96$, and $b = 3.5$ as in Table 5.1, the angular width of the ω-interval in (5.22) is 1.14. Our earlier discussion indicated the possibility of a particular interest in the subhypotheses $\omega = \pi/6$ and $\omega = \pi/2$, which are separated by $\pi/2 - \pi/6 = 1.05$. It seems likely from (5.22) that at least for large θ the confidence region (5.20) will not simultaneously include $\omega = \pi/6$ and $\omega = \pi/2$, but if distinguishing these subhypotheses is particularly important one may wish to use a larger value of b.

Much of the preceding discussion applies to any spherically symmetric stopping rule. Only the approximation (5.22) used specific properties of T defined in (5.9). On the other hand, it is quite convenient that the width of the ω interval in (5.22) depends neither on θ nor on the data, so this might be interpreted as a point in favor of the parabolic stopping boundaries of repeated significance tests.

4. Normal Data with Unknown Variance

The models discussed so far have focused on the case of normal data of known variability. Although this is a useful conceptual model and can arise in practice as a large sample approximation to more complex models, a more common practical situation when observations are approximately normally distributed is that the variability is unknown. This section discusses the simplest such model, although the methods are applicable more generally.

Assume that $x_1, x_2, \ldots$ are independent and normally distributed with mean μ and unknown variance σ^2, and that we are interested in testing H_0: $\mu = 0$ against H_1: $\mu \neq 0$. Assume also that if σ were known we would use a repeated significance test or modification thereof as described in Chapter IV, so our present goal is simply to extend the theory developed there to cover the case of unknown σ. Example 4.8 suggests a possible solution: let $\bar{x}_n = n^{-1}\sum_1^n x_i$ and $v_n^2 = n^{-1}\sum_1^n (x_i - \bar{x}_n)^2$; stop sampling at $\min(T, m)$ where

$$T = \inf\{n\colon n \geq m_0, \tfrac{1}{2}n\log(1 + \bar{x}_n^2/v_n^2) \geq a\} \qquad (m_0 \geq 2)$$

and reject H_0 if and only if $T \leq m$. Since the joint distribution of $\bar{x}_n/v_n$, $n = 2, 3, \ldots$ depends on (μ, σ) only through the value of $\theta = \mu/\sigma$, the operating characteristics of this test depend only on θ. The power and expected sample size are

$$P_\theta\{T \leq m\} \tag{5.23}$$

and

$$E_\theta(T \wedge m). \tag{5.24}$$

(In this regard note also that the methods of III.4 and IV.5 produce a confidence interval for θ, but not for μ by itself. However, for measurig the extent to which a normal population is shifted away from one centered at 0, the ratio $\theta = \mu/\sigma$ is often easier to interpret than μ alone.)

It is, of course, possible to modify this test as in IV.4 by choosing a second constant $0 < d < a$ and taking as the rejection region the event $\{T \leq m\} \cup$

$\{T > m, Z_m \geq d\}$, where $Z_m = \frac{1}{2}m \log(1 + \bar{x}_m^2/v_m^2)$. Now the power is

$$
\begin{aligned}
P_\theta\{T \leq m\} &+ P_\theta\{T > m, Z_m \geq d\} \\
&= P_\theta\{Z_m \geq d\} + P_\theta\{T < m, Z_m < d\}.
\end{aligned} \tag{5.25}
$$

We begin by discussing approximate computation of (5.23) and (5.24). Even if we ultimately focus on a modified repeated significance test, (5.23) is of interest for $\theta = 0$ as an attained significance level and for $\theta \neq 0$ as a key ingredient in calculating confidence intervals for θ (see IV.5). Again there are two ways of approaching these calculations: (a) to adapt the methods developed in Chapter IV or (b) to use the results of Chapter IV as a rough central limit type approximation. Both methods are illustrated below.

Consider first the approximate computation of (5.23) when $\theta = 0$ by an adaptation of the argument in IV.8. The key idea in that argument involved introduction of a measure Q, such that the likelihood ratio of the data under Q relative to P_0 was a simple function of the statistic defining the stopping rule T. The problem of testing H_0: $\mu = 0$ against H_1: $\mu \neq 0$ is invariant under the transformation which sends $(x_1, x_2, \ldots, x_n)$ into $(cx_1, \ldots, cx_n)$ for an arbitrary constant $c \neq 0$. See Cox and Hinkley (1974), p. 157 ff. or Lehmann (1959), Chapter VI. Under this group of transformations,

$$
(y_2, \ldots, y_n) = (x_1^{-1}x_2, x_1^{-1}x_3, \ldots, x_1^{-1}x_n)
$$

is maximal invariant, and since the statistic of interest, $Z_n = \frac{1}{2}n \log(1 + \bar{x}_n^2/v_n^2)$, is also invariant, a reasonable starting place is calculation of the likelihood ratio of $(y_2, \ldots, y_n)$ under P_θ relative to P_0. (It follows from general theory or the following calculation that the joint distribution of $y_2, \ldots, y_n$ depends only on θ.) Conditioning on x_1 and integrating leads easily to the likelihood ratio

$$
\begin{aligned}
&\tilde{l}(n, y_2, \ldots, y_n; \theta, 0) \\
&= \frac{e^{-n\theta^2/2} \int_{-\infty}^{\infty} |x|^{n-1} \exp\left[-\frac{1}{2}x^2 \sum_1^n y_i^2 + \theta x \sum_1^n y_i\right] dx}{\int_{-\infty}^{\infty} |x|^{n-1} \exp\left[-\frac{1}{2}x^2 \sum_1^n y_i^2\right] dx}
\end{aligned} \tag{5.26}
$$

where $y_1 = x_1^{-1}x_1 = 1$. In analogy with the argument of IV.8 let $Q = \int_{-\infty}^{\infty} P_\theta \, d\theta/(2\pi)^{1/2}$, where P_θ is regarded as a probability on the space of sequences $(y_2, y_3, \ldots)$. Straightforward calculation shows that the likelihood ratio of $y_2, \ldots, y_n$ under Q relative to P_0 is

$$
\begin{aligned}
\int_{-\infty}^{\infty} \tilde{l}(n, y_2, \ldots, y_n; \theta, 0) \, d\theta/(2\pi)^{1/2} &= n^{-1/2} \left(\frac{\sum_1^n y_i^2}{\sum_1^n y_i^2 - n^{-1}\left(\sum_1^n y_i\right)^2} \right)^{n/2} \\
&= n^{-1/2}[(1 + \bar{x}_n^2/v_n^2)]^{n/2} \\
&= n^{-1/2} \exp\{\tfrac{1}{2}n \log(1 + \bar{x}_n^2/v_n^2)\}.
\end{aligned}
$$

This likelihood ratio plays the same role as $(t + 1)^{-1/2} \exp\{W^2(t)/2(t + 1)\}$ in IV.8. In analogy with (4.75) a likelihood ratio identity yields the representation

$$P_0\{T \le m\} = P_0\{Z_{m_0} \ge a\} + P_0\{m_0 < T \le m\}$$

$$= P_0\{Z_{m_0} \ge a\} + \int_{-\infty}^{\infty} E_\theta\{T^{1/2} e^{-Z_T}; m_0 < T \le m\}\, d\theta/(2\pi)^{1/2},$$

(5.27)

where $Z_n = \frac{1}{2} n \log(1 + \bar{x}_n^2/v_n^2)$.

Starting from (5.27) one can go step by step through the argument of IV.8 making minor changes to arrive at an asymptotic evaluation of $P_0\{m_0 < T \le m\}$. Of course, it is necessary to be able to calculate

$$v^*(\theta) = \lim_{a \to \infty} E_\theta\{\exp[-(Z_T - a)]\} \qquad (\theta \ne 0), \tag{5.28}$$

the existence of which is guaranteed by Theorem 9.12 and Proposition 9.30 (cf. (4.35)). The result is summarized as

Theorem 5.29. *Suppose* $a \to \infty$, $m_0 \to \infty$, *and* $m \to \infty$ *in such a way that for some* $0 < \theta_1 < \theta_0 < \infty$ $(e^{2a/m_0} - 1)^{1/2} = \theta_0$, $(e^{2a/m} - 1)^{1/2} = \theta_1$. *Let* $b = (2a)^{1/2}$. *Then*

$$P_0\{m_0 < T \le m\} \sim 2b\phi(b) \int_{\theta_1}^{\theta_0} [\log(1 + x^2)]^{-1/2} v^*(x)\, dx, \tag{5.30}$$

where v^* *is defined by* (5.28).

Evaluation of (5.28) is discussed in VIII.4 and Chapter IX. Unfortunately, in contrast to Table 4.2, which is also applicable for group tests and multivariate problems, in the case of unknown σ these variations lead to different integrals. Hence one should probably regard the integral in (5.30) as something to be computed numerically by a small program which is valid for a number of similar situations rather than as tabled constants.

The first term on the right hand side of (5.27) is easily computed exactly from tables of Student's t. It is not hard to show that this term is of order $\phi(b)/b$ under the conditions of Theorem 5.29, and hence is of smaller order of magnitude than (5.30). However, for numerical purposes it seems reasonable to approximate $P_0\{T \le m\}$ by adding the exact value of $P_0\{Z_{m_0} \ge a\}$ to (5.30). Not only do the Monte Carlo values given below support this suggestion, but Theorem 4.21 shows that for Brownian motion the analogous *ad hoc* device would give quite reasonable numerical results for m/m_0 in the important range of about 5 to 10.

Table 5.2 gives some examples of the approximation of $P_0\{T \le m\}$ by the right hand side of (5.27) with $P_0\{m_0 < T \le m\}$ approximated by (5.30). For comparison the results of a 3600 replication Monte Carlo experiment are included. Importance sampling as described in Remark 4.46 was used for variance reduction. Also included is the Brownian approximation given by

Table 5.2 Significance Level of Repeated t-Test

b	m_0	m	Analytic Approx.	Monte Carlo	Brownian Approx.
2.73	7	28	.057	.053 ± .001	.046
2.80	9	49	.051	.048 ± .001	.044
2.84	10	71	.050	.047 ± .001	.044
2.89	10	111	.051	.049 ± .001	.046
3.45	10	148	.012	.011 ± .0001	.010

Theorem 4.21 for the probability that a Brownian motion without drift crosses the boundary $bt^{1/2}$ ($b = (2a)^{1/2}$) in the indicated range. The theoretical approximation developed above is quite good. The Brownian approximation gives answers which are consistently too small, but surprisingly accurate. (Recall that for the case of known variance the Brownian approximation is too large and not nearly so accurate.) Because of its comparative simplicity one is tempted to use the Brownian approximation, which, however, should be checked more thoroughly before regarding these figures as more than a happy accident.

Approximating (5.23) when $\theta = 0$ is reasonably simple because invariance arguments lead to a one-dimensional problem. Analogous approximations for (5.23) and (5.24) when $\theta \neq 0$ seem to be substantially more difficult because apparently one must study the two-dimensional random walk $(\sum_1^n x_i, \sum_1^n x_i^2)$, n = 1, 2, ... in detail. As an alternative we consider the possibility of using Brownian motion to provide a crude but simple approximation.

Consider $\tilde{S}_n = n\{\log(1 + \bar{x}_n^2/v_n^2)\}^{1/2} \operatorname{sgn}(\bar{x}_n)$ in terms of which the stopping rule (4.9) can be expressed

$$T = \inf\{n: n \geq m_0, |\tilde{S}_n| \geq b\sqrt{n}\} \tag{5.31}$$

with $b = (2a)^{1/2}$. It is useful to think of $\tilde{S}_n$ as having the form

$$\tilde{S}_n = ng\left(n^{-1}\sum_1^n x_i, n^{-1}\sum_1^n x_i^2\right),$$

where

$$g(\xi, \eta) = \{-\log(1 - \xi^2/\eta)\}^{1/2} \operatorname{sgn}(\xi).$$

A linear Taylor series expansion of g about $(\mu, \sigma^2 + \mu^2)$ yields

$$\tilde{S}_n = n\{\log(1 + \theta^2)^{1/2} \operatorname{sgn}(\theta) + \theta(S_{1n} - \tfrac{1}{2}\theta S_{2n})/(1 + \theta^2)\{\log(1 + \theta^2)\}^{1/2} + \cdots,$$

where $S_{1n} = \sigma^{-1}\sum_1^n (x_i - \mu)$ and $S_{2n} = \sigma^{-2}\sum_1^n (x_i - \mu)^2 - n$. This suggests approximating $\tilde{S}_n$ by a Brownian motion process with mean

$$\tilde{\mu} = \{\log(1 + \theta^2)\}^{1/2} \operatorname{sgn}(\theta) \tag{5.32}$$

and variance

$$\tilde{\sigma}^2 = \theta^2(1 + \tfrac{1}{2}\theta^2)/\{(1 + \theta^2)^2 \log(1 + \theta^2)\}. \tag{5.33}$$

Table 5.3. Power and Expected Sample Size of Repeated t-Test

				$P_\theta\{T \le m\}$		$E_\theta(T \wedge m)$	
b	θ	m_0	m	Approx.	Monte Carlo	Approx.	Monte Carlo
2.73	.8	7	28	.936	.933 ± .004	13	14
2.73	.6	7	28	.708	.724 ± .013	18	19
2.80	.6	9	49	.935	.939 ± .004	22	23
2.80	.4	9	49	.593	.580 ± .018	35	35
2.84	.5	10	71	.938	.940 ± .004	34	33
2.84	.4	10	71	.772	.770 ± .010	45	43

As in Section 2 there are two rather different precise mathematical interpretations of this approximation. One is that for any $\theta \neq 0$ as $n \to \infty$

$$\tilde{\sigma} n^{-1/2}(S_{[nt]} - [nt]\tilde{\mu}), \qquad 0 \le t \le \tilde{t},$$

converges to Brownian motion with 0 drift. The second is that if θ depends on n and converges to 0 in such a way that for some fixed ξ, $-\infty < \xi < \infty$, $\theta \sim \xi n^{-1/2}$ as $n \to \infty$, then $n^{-1/2}\tilde{S}_{[nt]}$ converges to Brownian motion with drift ξ. That a single asymptotic approximation is valid both for fixed alternatives $\theta \neq 0$ and for local alternatives $\theta \to 0$ is a consequence of the square root transformation applied to the log likelihood ratio statistic. The corresponding asymptotic results for $\frac{1}{2}n\log(1 + \bar{x}_n^2/v_n^2)$ itself do not fit together so conveniently (see Problem 5.3).

If we replace S_n by a Brownian motion with mean and variance given by (5.32) and (5.33), the stopping rule (5.31) is of the form studied in Chapter IV. After making minor adjustments to account for the variance parameter $\tilde{\sigma}^2$, we easily obtain approximations for (5.23) and (5.24) when $\theta \neq 0$.

Table 5.3 reports some numerical examples and the results of a 900 replication Monte Carlo experiment to check the accuracy of this approximation. Importance sampling was used for variance reduction in estimating (5.23).

The Brownian motion approximation appears to be quite accurate. For $P_\theta\{T \le m\}$ there is no simple mathematical reason that this should be the case. With regard to $E_\theta(T \wedge m)$ it is interesting to note that the suggestion in Problem 4.7 applied to the present case gives an approximation of the form $K_1 m + K_2 m^{1/2}$ with the same K_1 ad K_2 as the proposed Brownian approximation (see Problem 5.2). The Brownian approximation also contains terms of constant order and of order $m^{-1/2}$ as $m \to \infty$, which cannot be exactly correct for the t-statistic. Neglecting them would lead to approximations in Table 5.3 which are consistently slightly too large.

Consider now a modified repeated t-test with power function given by (5.25). Although it is fairly complicated to give precise approximations to (5.25), one can usually piece together practically useful results. For example, when $\theta = 0$, one can compute the first term on the right hand side of (5.25) exactly and use Brownian motion to give a crude approximation to the second. This should be fairly accurate when d is small compared to a and hence the second term is small compared to the first. When $\theta \neq 0$ one can use a Brownian motion approximation; or one can compute the first term on the right hand side of (5.25) exactly from tables of the non-central t-distribution and use Brownian motion only to approximate the comparatively unimportant second term.

5. Survival Analysis—Theory

Most of this book is concerned with very simple models, primarily involving the normal distribution, with the implicit understanding that the central limit theorem frequently permits the approximate reduction of more complex models to simpler ones. Here we focus on one particularly important special problem concerning the proportional hazards model of survival analysis and obtain some results along the lines suggested in III.9 in a different context. A complete mathematical development contains several difficult points which have been omitted from the following heuristic discussion. The results of a Monte Carlo experiment are included to illustrate the insights and limitations of the approximate theory.

A number of authors, starting with Cox (1975), have discussed central limit theory for the proportional hazards model. Their results are concerned with the case that survival is measured from a common point in time for all items on a test. This theory is adequate for engineering studies of reliability in which items are simultaneously put on test or for epidemiological studies of survival after some fixed historical event. It also suffices for controlled clinical trials in which patients enter over a period of time and survival is measured from the time of entry—*provided* the analysis is conducted at one fixed point in time or after some fixed number of deaths. However, for reasons explained below, sequential analysis of survival data with staggered entry times requires a somewhat different approach. The one followed here is due to Sellke and Siegmund (1983).

Suppose that $k \le \infty$ patients enter a clinical trial at nonrandom times y_1, $y_2, \ldots, y_k$, with only finitely many entries in each finite time interval. (Equivalently one can assume that entry occurs according to a point process which is independent of everything else, and that all statements are made conditional on a realization of the process.) Associated with the ith patient is a triple (z_i, x_i, c_i), where z_i is a covariate (possibly vector-valued), x_i is the survival time after entry into the trial, and c_i is a censoring variable. Thus the ith person is on test until time $y_i + x_i \wedge c_i$. If $x_i \le c_i$, he dies while on test, and x_i is recorded. Otherwise that observation is censored, and it is known only that $x_i > c_i$. Of course, at any particular time t, there is in effect a second censoring variable $(t - y_i)^+$ in the sense that the time on test prior to time t for patient i is $x_i \wedge c_i \wedge (t - y_i)^+$. We shall call x_i, c_i, and $(t - y_i)^+$ the "age" of the ith patient at death, at censoring, and at time t, respectively.

Our basic stochastic model is that (z_i, x_i, c_i), $i = 1, 2, \ldots, k$ are independent and identically distributed, that the z_i are uniformly bounded, and that given

z_i, x_i, and c_i are conditionally independent with x_i having a continuous cumulative hazard function of the form

$$d\Lambda_i(s) = \exp(\beta z_i)\, d\Lambda(s), \tag{5.34}$$

for some unknown parameter β and baseline cumulative hazard function Λ. In the case of vector covariates βz_i denotes an inner product, but for simplicity of presentation we consider primarily the case of scalar z_i, which then usually denotes an indicator of the treatment used for the ith patient.

It is desirable to be able to relax the stochastic assumptions about the z's and c's and to treat them as ancillary statistics. In the case of identical arrival times $y_1 = y_2 = \cdots = y_k$ this does not pose serious problems (e.g. Sellke, 1982). For an arbitrary arrival process, however, removal of these assumptions requires a substantially deeper analysis (cf. Sellke, 1985).

It is convenient to introduce the notation

$$N_i(t, s) = I_{\{y_i + x_i \le t,\, x_i \le c_i,\, x_i \le s\}} \qquad (s \le t) \tag{5.35}$$

to indicate a patient who arrived and died before time t, was uncensored at the time of death, and whose age at death was less than or equal to s. We also define the set of patients at risk at time t and age s by

$$R(t, s) = \{i\colon y_i \le t - s,\ x_i \wedge c_i \ge s\} \qquad (s \le t). \tag{5.36}$$

Suppose that at time t, $v = v(t)$ deaths have been observed, and denote the ordered ages at death by $\tau_{(1)}(t) < \tau_{(2)}(t) < \cdots < \tau_{(v)}(t)$. Hence $\tau_{(1)}(t) = x_{i_1}$ if $x_{i_1} = \min\{x_j\colon\ y_j + x_j \le t,\ x_j \le c_j\}$, $\tau_{(2)}(t) = x_{i_2}$ if $x_{i_2} = \min\{x_j\colon\ x_j > \tau_{(1)}(t),\ y_j + x_j \le t,\ x_j \le c_j\}$, etc. Let $z_{(j)} = z_{(j)}(t)$ denote the covariate of the patient dying at age $\tau_{(j)}(t)$. Then for fixed t the likelihood of $(z_{(j)}, \tau_{(j)})$, $j = 1, 2, \ldots, v$ is

$$L(t, \beta) = \prod_{j=1}^{v} \left[\exp\left\{ -\int_{\tau_{(j-1)}}^{\tau_{(j)}} \sum_{i \in R(t,s)} e^{\beta z_i}\, d\Lambda(s) \right\} e^{\beta z_{(j)}}\, d\Lambda(\tau_{(j)}) \right]$$
$$\times \exp\left\{ -\int_{\tau_{(v)}}^{t} \sum_{i \in R(t,s)} e^{\beta z_i}\, d\Lambda(s) \right\},$$

where $\tau_{(0)} = 0$. For each j in the indicated product, the exponential factor gives the probability of no death at ages between $\tau_{(j-1)}$ and $\tau_{(j)}$; and the second two factors give the probability (density) that the person with covariate $z_{(j)}$ dies at age $\tau_{(j)}$. $L(t, \beta)$ is easily rewritten

$$L(t, \beta) = \exp\left\{ -\int_0^t \sum_{i \in R(t,s)} e^{\beta z_i}\, d\Lambda(s) \right\}$$
$$\prod_{j=1}^{v} \left[\sum_{i \in R(t, \tau_{(j)})} e^{\beta z_i}\, d\Lambda(\tau_{(j)}) \right] \prod_{j=1}^{v} \left(\frac{e^{\beta z_{(j)}}}{\sum_{i \in R(t, \tau_{(j)})} e^{\beta z_i}} \right). \tag{5.37}$$

The last factor in (5.37) does not depend on Λ and is the conditional probability that those persons dying at ages $\tau_{(1)}, \ldots, \tau_{(v)}$ have respectively the covariates $z_{(1)}, \ldots, z_{(v)}$ given that deaths occur at $\tau_{(1)}, \ldots, \tau_{(v)}$. It is called the partial likelihood by Cox (1975), who recommends that it be treated like an ordinary likelihood function for making inferences about β in the presence of the nuisance parameter Λ. With the aid of the notation (5.35), the log partial likelihood may be rewritten as

$$l(t,\beta) = \sum_{i=1}^{k} \int_{[0,t]} \left[\beta z_i - \log\left(\sum_{j \in R(t,s)} e^{\beta z_j}\right)\right] N_i(t, ds). \tag{5.38}$$

Differentiating (5.38) with respect to β gives the score function

$$\dot{l}(t,\beta) = \sum_{i=1}^{k} \int_{[0,t]} [z_i - \tilde{\mu}(t,s,\beta)] N_i(t, ds), \tag{5.39}$$

where

$$\tilde{\mu}(t,s,\beta) = \sum_{j \in R(t,s)} z_j \exp(\beta z_j) \Big/ \sum_{j \in R(t,s)} \exp(\beta z_j). \tag{5.40}$$

The maximum partial likelihood estimator of β is the solution $\hat{\beta}(t)$ of

$$\dot{l}(t,\beta) = 0.$$

Tests of H_0: $\beta = \beta_0$ can be based on $\hat{\beta}$, $\sup_\beta l(t,\beta) - l(t,\beta_0)$, or directly on the score function $\dot{l}(t,\beta_0)$. In any case the standard Taylor series approximation

$$0 = \dot{l}(t,\hat{\beta}(t)) = \dot{l}(t,\beta) + [\hat{\beta}(t) - \beta]\ddot{l}(t,\beta^*)$$

indicates that the behavior of these tests is intimately linked with that of $\dot{l}(t,\beta)$, which we now consider.

The important structural feature of $\dot{l}(t,\beta)$, $0 \le t < \infty$, is that it can be approximated uniformly in t by a martingale to which the central limit theorem applies. Certain simplifications and certain technicalities appear because of the continuous time setting of the problem. To make maximum use of the simplifications and duck the technicalities, the following discussion proceeds by analogy with discrete time martingale theory, which for these purposes is summarized in Appendix 3. For a complete development of the appropriate foundations in continuous time, see Jacobsen (1982).

According to Cox (1975), the efficient score (logarithmic derivative) of a partial likelihood function is itself a martingale under quite general conditions, but the appropriate interpretation of this statement in regard to $\dot{l}(t,\beta)$ is rather subtle.

For $s \ge 0$ let $\tilde{\mathscr{E}}_s$ denote the class of events determined by conditions on y_i, c_i, z_i, $I_{\{x_i \le s\}}$, and $x_i I_{\{x_i \le s\}}$, $i = 1, 2, \ldots, k$. Since

$$\begin{aligned} P\{s < x_i \le s + \delta | \tilde{\mathscr{E}}_s\} &= \{1 - \exp[\Lambda_i(s) - \Lambda_i(s+\delta)]\} I_{\{x_i > s\}} \\ &= \{\Lambda_i(s+\delta) - \Lambda_i(s) + o[\Lambda_i(s+\delta) - \Lambda_i(s)]\} I_{\{x_i > s\}}, \end{aligned}$$

it follows that $I_{\{x_i < s\}} - \Lambda_i(x_i \wedge s)$ is a martingale with respect to $\tilde{\mathscr{E}}_s$. (This is the continuous time analogue of (A.9) in Appendix 3.) For any fixed $t \ge 0$, since $c_i \wedge (t - y_i)^+$ is a stopping time with respect to $\tilde{\mathscr{E}}_s$,

(5.41) $$I_{\{x_i \le s \wedge c_i \wedge (t-y_i)^+\}} - \Lambda[x_i \wedge s \wedge c_i \wedge (t-y_i)^+]$$

is also an $\tilde{\mathscr{E}}_s$-martingale (cf. (A.18)). Now let $\mathscr{E}_{t,s}$ be the class of events which have been observed by time t and involve ages $\le s$, i.e. the events determined by

$$I_{\{y_i \le t\}},\ y_i I_{\{y_i \le t\}},\ I_{\{x_i \le s \wedge c_i \wedge (t-y_i)^+\}},\ x_i I_{\{x_i \le s \wedge c_i \wedge (t-y_i)^+\}},$$

$$I_{\{c_i \le s \wedge x_i \wedge (t-y_i)^+\}},\ c_i I_{\{c_i \le s \wedge x_i \wedge (t-y_i)^+\}},$$

and $z_i I_{\{y_i \le t\}}$. Since $\mathscr{E}_{t,s} \subset \tilde{\mathscr{E}}_s$, it follows that (5.41) is also an $\mathscr{E}_{t,s}$-martingale in s for each fixed t (cf. (A.8)). It is convenient to put $A_i(t, ds) = I_{\{i \in R(t,s)\}}\, d\Lambda_i(s)$, so by (5.36)

$$A_i(t,s) = \Lambda_i[s \wedge x_i \wedge c_i \wedge (t-y_i)^+];$$

and with the notation (5.35), (5.41) becomes

(5.42) $$N_i(t,s) - A_i(t,s),$$

which is an $\mathscr{E}_{t,s}$-martingale in s for fixed t.

Now let

(5.43) $$\dot{l}(t,s,\beta) = \sum_i \int_{[0,s]} [z_i - \tilde{\mu}(t,u,\beta)][N_i(t,du) - A_i(t,u)]$$

so by (5.39) and (5.40)

$$\dot{l}(t,t,\beta) = \dot{l}(t,\beta).$$

Moreover, since the integrands in (5.43) are uniformly bounded and $\mathscr{E}_{t,s}$ predictable, the indefinite stochastic integral inherits the martingale property of (5.42), i.e. $\dot{l}(t,s,\beta)$ is an $\mathscr{E}_{t,s}$ martingale in s for each fixed t (cf. (A.10)).

The martingale property in s of $\dot{l}(t,s,\beta)$ provides the possibility of very simple proofs of the asymptotic normality of $\dot{l}(t,\beta) = \dot{l}(t,t,\beta)$ at one fixed value of t. (A similar approach allows one to consider $\dot{l}(t,\beta)$ at the time of the mth death for some fixed value of m.) It is interesting to note that by the preceding analysis a problem with staggered entry and censoring variables c_i becomes equivalent to a problem with simultaneous entry and censoring variables $c_i \wedge (t-y_i)^+$. Some reflection reveals a much more complicated situation if one is interested in the joint distribution of $\dot{l}(t,\beta)$ as a process in t. In particular $\dot{l}(t,\beta)$ need not be a martingale in t (see Problem 5.4).

Let $N_i(t) = N_i(t,t)$, $A_i(t) - A_i(t,t)$, and $\mathscr{E}_t = \mathscr{E}_{t,t}$. Then $N_i(t)$ is an indicator that person i is observed to die before time t, and $A_i(t)$ is the cumulative hazard to which person i is exposed while on test before time t. Since (5.41) is symmetric in t and s, it follows that (5.41) (hence also (5.42)) is an $\mathscr{E}_{t,s}$ martingale in t for fixed s. Hence by a simple calculation

(5.44) $$\{N_i(t) - A_i(t), \mathscr{E}_t\}$$

is a martingale in t.

A change of variable in (5.39) yields

$$(5.45) \qquad \dot{l}(t,\beta) = \sum_i \int_{[0,t]} \{z_i - \tilde{\mu}(t, s - y_i, \beta)\}\{N_i(ds) - A_i(ds)\}.$$

If the integrand in (5.45) did not depend on t, then the indicated stochastic integral would inherit the martingale property of (5.44) (cf. (A.10)). This suggests that one approximate $\tilde{\mu}(t,s,\beta)$ by something not involving t and use the resulting martingale as an approximation to $\dot{l}(t,\beta)$.

If $k \to \infty$ and $\sum_{i=1}^k I_{\{y_i \le t-s\}} \to \infty$, the law of large numbers shows that $\tilde{\mu}(t,s,\beta)$ defined by (5.40) is approximately

$$(5.46) \qquad \mu(s) = \frac{E\{z_1 \exp(\beta z_1); x_1 \wedge c_1 \ge s\}}{E\{\exp(\beta z_1); x_1 \wedge c_1 \ge s\}}.$$

Hence one possibility for a martingale to approximate (5.45) is

$$(5.47) \qquad Q(t) = \sum_i \int_{[0,t]} \{z_i - \mu(s - y_i)\}\{N_i(ds) - A_i(ds)\}.$$

Let

$$(5.48) \quad N(t,s) = \sum_i N_i(t,s), \qquad A(t,s) = \sum_i A_i(t,s), \qquad D(t) = EN(t,t),$$

and

$$\begin{aligned} r(t) &= \dot{l}(t,\beta) - Q(t) \\ (5.49) \qquad &= -\sum_i \int_{[0,t]} \{\tilde{\mu}(t,s,\beta) - \mu(s)\}\{N_i(t,ds) - A_i(t,ds)\} \\ &= -\int_{[0,t]} \{\tilde{\mu}(t,s,\beta) - \mu(s)\}\{N(t,ds) - A(t,ds)\}. \end{aligned}$$

Theorem 5.53 below shows that in a certain sense $r(t)$ is small compared to $[D(t)]^{1/2}$, and this permits conclusions about the behavior of $\dot{l}(t,\beta)$ to be inferred from knowledge of the behavior of the martingale $Q(t)$.

According to III.9 (see also Appendix 3) a natural means of normalizing $\dot{l}(t,\beta)$ involves the observed Fisher information

$$-\ddot{l}(t,\beta) = -\partial^2 l(t,\beta)/\partial\beta^2,$$

which by (5.39) equals

$$(5.50) \qquad -\ddot{l}(t,\beta) = \int_{[0,t]} \tilde{\sigma}^2(t,s,\beta) N(t,ds),$$

where

$$(5.51) \quad \tilde{\sigma}^2(t,s,\beta) = \sum_{j \in R(t,s)} [z_j - \tilde{\mu}(t,s,\beta)]^2 \exp(\beta z_j) \Big/ \sum_{j \in R(t,s)} \exp(\beta z_j).$$

On the other hand the natural mechanism for normalizing $Q(t)$ is its predictable quadratic variation $\langle Q \rangle(t)$ (cf. Appendix 3), which by the independence of the different terms in (5.47) is

$$\begin{aligned}\langle Q\rangle(t) &= \sum_i \int_{[0,t]} [z_i - \mu(s - y_i)]^2 A_i(ds) \\ &= \sum_i \int_{[0,t]} [z_i - \mu(s)]^2 A_i(t, ds).\end{aligned} \tag{5.52}$$

The following result shows that $\dot{l}$ is close to Q and $-\ddot{l}$ is close to $\langle Q\rangle$ uniformly in t and the arrival process.

Theorem 5.53. *For all sufficiently small* $\varepsilon > 0$, *uniformly in the arrival process defined by* $\{y_i, i = 1, 2, \ldots, k\}$

$$\lim_{K\to\infty} P\{|\dot{l}(t,\beta) - Q(t)| \geq K + [D(t)]^{1/2-\varepsilon} \text{ for some } t\} = 0 \tag{5.54}$$

and

$$\lim_{K\to\infty} P\{|\ddot{l}(t,\beta) + \langle Q\rangle(t)| \geq K + [D(t)]^{1-2\varepsilon} \text{ for some } t\} = 0, \tag{5.55}$$

where $D(t) = EN(t,t)$ *is the expected number of deaths observed in* $[0, t]$.

For a complete proof of Theorem 5.53 see Sellke and Siegmund (1983).

To formulate an appropriate limit theorem, suppose now that we are given a sequence of arrival processes indexed by n: $\{y_{n1}, y_{n2}, \ldots, y_{nk_n}\}$. The assumptions on (z_i, x_i, c_i), $i = 1, 2, \ldots$ are the same as above. For $v > 0$ define

$$\tau_n(v,\beta) = \inf\{t\colon -\ddot{l}(t,\beta) \geq nv\}.$$

(The dependence of $-\ddot{l}$ on n is suppressed in the notation.) Our goal is to formulate conditions on the sequence of arrival processes and the joint distribution of (z_1, x_1, c_1) which permit the conclusions that

$$P\{\tau_n(v,\beta) < \infty\} \to 1 \qquad (n \to \infty) \tag{5.56}$$

and on $[0, v]$, in law

$$\lim_{n\to\infty} n^{-1/2}\dot{l}[\tau_n(\cdot,\beta),\beta] = W(\cdot), \tag{5.57}$$

where W is standard (driftless) Brownian motion (see III.9).

The conditions given by Sellke and Siegmund (1983) have the desirable property that they do not distinguish between a high rate of entry over a fixed time interval or a fixed rate of entry over a long time interval. They do have some technical aspects which makes one suspect that they are not optimal. The less technical of Sellke and Siegmund's two sets of conditions is that

$$k_n < \infty, \qquad \lim_{n\to\infty} n^{-1}k_n = \infty \qquad \text{and } \lim_{n\to\infty} (n\log n)^{-1}k_n = 0. \tag{5.58}$$

The condition that k_n grow faster than n is obviously necessary for (5.56), because by (5.50) $-\ddot{l}(t,\beta) = O(N(t,t)) = O(k_n)$. The condition that k_n grow more slowly than $n \log n$ is a technical one. It can always be fulfilled by the experimenter, who can decide to halt the intake of new patients into the study.

Theorem 5.59. *If* (5.58) *is satisfied, then* (5.56) *and* (5.57) *hold for every* $v > 0$.

PROOF. By (5.52), (5.58), and the law of large numbers

$$n^{-1}\langle Q\rangle(\infty) = n^{-1}\sum_i \int_{[0,\infty]} [z_i - \mu(s)]^2 A_i(\infty, ds) \to \infty \tag{5.60}$$

in probability as $n \to \infty$. Since by (5.58) $n^{-1}[D(t)]^{1-2\varepsilon} \to 0$ uniformly in t as $n \to \infty$, (5.60) and (5.55) imply that (5.56) holds for all $v > 0$. Since the z's are bounded, so is $\tilde{\sigma}^2(t, s, \beta)$ defined by (5.51). Hence by (5.50) and (5.56)

$$-n^{-1}\ddot{l}[\tau_n(v, \beta), \beta] \to v$$

in probability as $n \to \infty$. Another application of (5.55) shows that

$$n^{-1}\langle Q\rangle[\tau_n(v, \beta)] \to v,$$

for all v, so by a martingale central limit theorem (cf. Appendix 3 for references) $n^{-1/2}Q[\tau_n(\cdot, \beta)]$ converges in law to a Brownian motion process on every bounded interval $[0, v]$. Finally (5.54) and (5.58) show that

$$n^{-1/2}Q[\tau_n(\cdot, \beta)] \text{ and } n^{-1/2}\dot{l}[\tau_n(\cdot, \beta), \beta]$$

have the same limiting behavior, so the theorem follows. □

For testing H_0: $\beta = \beta_0$ using the statistic $\dot{l}(t, \beta_0)$, Theorem 5.59 has the interpretation that if $\dot{l}(t, \beta_0)$ is plotted as a function of $-\ddot{l}(t, \beta_0)$, the process has approximately (under H_0) the same behavior as zero drift Brownian motion plotted as a function of time. If one wanted to compare $|W(t)|$ with $bt^{1/2}$ for $t \in [m_0, m]$ to obtain a repeated significance test, the analogous comparison here would be between $\dot{l}(t, \beta_0)$ and $b[-\ddot{l}(t, \beta_0)]^{1/2}$ for $-\ddot{l}(t, \beta_0) \in [m_0, m]$. See III.9 for a more precise statement.

To get some idea of how $\dot{l}(t, \beta_0)$ behaves when $\beta \neq \beta_0$, note that as $\beta \to \beta_0$, by (5.40) and (5.51)

$$\tilde{\mu}(t, s, \beta) = \tilde{\mu}(t, s, \beta_0) + (\beta - \beta_0)\tilde{\sigma}^2(t, s, \beta_0) + o(\beta - \beta_0).$$

Hence by (5.39) and (5.50)

$$\dot{l}(t, \beta_0) = \dot{l}(t, \beta) + (\beta - \beta_0)[-\ddot{l}(t, \beta_0)] + o(\beta - \beta_0), \tag{5.61}$$

so by Theorem 5.59, for β close to β_0, $\dot{l}(t, \beta_0)$ behaves like a Brownian motion with drift $\beta - \beta_0$ in the time scale of $-\ddot{l}(t, \beta_0)$ (see Figure 5.2).

The conclusion of the preceding paragraph is formalized in the following theorem.

Theorem 5.62. *Suppose that the assumptions of Theorem 5.59 are satisfied, and that β varies with n in such a way that for some fixed values β_0 and Δ, the true*

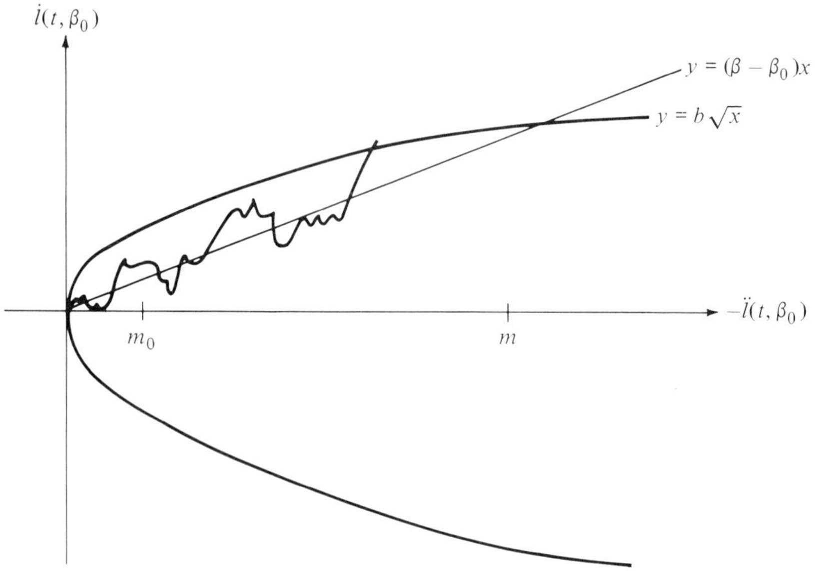

Figure 5.2.

parameter β satisfies $\beta = \beta_0 + \Delta/n^{1/2}$. Then for any $v > 0$ as $n \to \infty$

$$n^{-1/2}\dot{l}[\tau_n(\cdot, \beta_0), \beta_0] \to W_\Delta(\cdot)$$

in law on $[0, v]$, where W_Δ is Brownian motion with drift Δ.

The preceding analysis is predicated on the assumption that the triples (z_i, x_i, c_i) are independent and identically distributed, without which (5.46) and the approximating martingale $Q(t)$ make no sense. For purposes of experimental design this is not an unreasonable assumption. If the test is to have a bounded duration in real time, then any discussion of the power of the test requires some modeling assumptions about covariates, rate of censoring, baseline hazard rate, and arrival process. See Section 6 for an example. On the other hand, for purposes of data analysis it seems generally preferable to regard the covariates and censoring variables as ancillary and to condition on their observed values.

Sellke (1984) has recently proved (5.57) under the essentially minimal modeling assumption that the hazard function be of the form (5.34), with the z's and c's treated as constants. He also imposes conditions on the expected value of $-\ddot{l}$ to guarantee that eventually a sufficiently large amount of information will accumulate. These conditions are clearly implied by our stochastic model for the covariates and censoring variables. The additional observation which is the starting point for Sellke's analysis is that for $t_1 < t_2 < t_3 < t_4$

$$\dot{l}(t_2, s, \beta) - \dot{l}(t_1, s, \beta) \text{ and } \dot{l}(t_4, s, \beta) - \dot{l}(t_3, s, \beta)$$

are orthogonal martingales in s. This fact is easily proved using basic properties of stochastic integrals, but the essential ideas underlying Sellke's successful exploitation of the orthogonality are much too complicated to be discussed here.

In the case where β and z are vectors, $\beta = (\beta_1, \ldots, \beta_r)'$ and $z_i = (z_{i1}, \ldots, z_{ir})'$, of which one coordinate, say z_{i1} is a treatment indicator, one can presumably define a time scale and develop an appropriate large sample test of the hypothesis H_0: $\beta_1 = 0$ along the lines suggested in III.9. The situation is more complicated than in one dimension because the information used to define the time scale depends on estimators of the nuisance parameters $\beta_2, \ldots, \beta_r$ obtained under H_0. Some partial results are implicit in Sellke and Siegmund (1983) and Sellke (1982). See also Whitehead (1983), p. 184.

6. Survival Analysis—Examples and Applications

We begin by recalling the important results of Section 5. Let $\dot{l}(t, \beta)$ denote the derivative with respect to β of the log partial likelihood given in (5.38). Then for arbitrary β_0, when the true value of β is close to β_0, $\dot{l}(t, \beta)$ behaves approximately like Brownian motion with drift $(\beta - \beta_0)$ plotted in the time scale of Fisher information, $-\ddot{l}(t, \beta_0)$. (See Theorems 5.59 and 5.62; conceptually similar results are given in III.9.) This means that one can define a repeated significance test (for example) of H_0: $\beta = \beta_0$ against H_1: $\beta \neq \beta_0$ by means of the stopping rule

$$T = \inf\{t: -\ddot{l}(t, \beta_0) \geq m_0, |\dot{l}(t, \beta_0)| \geq b[-\ddot{l}(t, \beta_0)]^{1/2}\}$$

truncated as soon as $-\ddot{l}(t, \beta_0) \geq m$. Asymptotically this test has about the same power function as that defined by (4.11), and $E_\beta\{[-\ddot{l}(T, \beta_0)] \wedge m\}$ is given approximately by Theorem 4.27.

Of course, it may be inconvenient to make these calculations continuously, but possible to define a "group" sequentil test. Then group size k is defined by an increase of k units in $-\ddot{l}(t, \beta)$, and one can compute approximately the operating characteristics of the test by referring to the corresponding discrete time normal tests. See IV.3–5.

Tables 5.4, 5.5, and 5.6 report the results of a small Monte Carlo experiment to illustrate the insights and limitations of this approximation. For a much larger, somewhat different study, see Gail *et al.* (1982).

Suppose that we want to compare a new treatment with a standard or with placebo and that there are no other covariates than treatment assignment. Patients receive the new treatment or placebo according to independent tosses of a fair coin, so the z's are independent 0-1 variables indicating the treatment given. The null hypothesis of no effect is H_0: $\beta = 0$. Note that if z_i and c_i are independent and H_0 holds (so x_i and z_i are also independent), then by (5.51) $\tilde{\sigma}^2(t, s, 0) \simeq \frac{1}{4}$ in large samples and hence by (5.50) $-\ddot{l}(t, 0) \cong N(t, t)/4$. Consequently in this special case for β close to 0, measuring "time" by means of $-\ddot{l}(t,0)$ is essentially equivalent to measuring it by the number of deaths observed. In particular, a "fixed" sample size in the time scale of $-\ddot{l}(t,0)$ is essentially a sample having a fixed number of (observed) deaths.

Assume that for planning purposes one expects to recruit about 350 persons who arrive independently and uniformly over a three year interval, that lifetimes are exponentially distributed with mean of about three years under H_0, and that little loss to follow-up is expected. A simple calculation shows that conditional on the value of z, the probability that a randomly selected individual is alive at time t equals

$$\{\exp[-\lambda_z(t-t_I)^+] - \exp(-\lambda_z t)\}/\lambda_z \min(t, t_I), \tag{5.63}$$

where λ_z^{-1} is the mean lifetime of a person with covariate $z \in \{0, 1\}$, and $t_I = 3$ is the recruitment period. For exponential life distributions $\beta = \log(\lambda_0/\lambda_1)$.

From the viewpoint described above, a "fixed" sample test is one based on a fixed amount of Fisher information $-\ddot{l}(t,0)$ or what is almost the same thing, a fixed number of deaths $N(t,t)$. This is not at all the same as a test at a fixed *time* t. Nevertheless, for simplicity consider a test of H_0: $\beta = 0$ at $t = 5$ years. Under H_0 with $\lambda_0^{-1} = 3$, by (5.63) the expected number of deaths in five years is $(.675)(350) = 236$ and hence the expected amount of information is about $236/4 = 59$. If $\lambda_0^{-1} = 3$ and $\beta = .5$, so $\exp(\beta) = 1.65$ and $\lambda_1^{-1} \cong 5$ years, by (5.63) the expected number of deaths is 206, resulting in about 51.5 units of information. A normal approximation using these figures indicates that a .05 level two-sided test based on $\dot{l}(5,0)$ would have power about .95 if $\lambda_0^{-1} = 3$ and $\beta = .5$.

For a sequential test consider one which reviews the data after one year and every six months thereafter up to a maximum of $5\frac{1}{2}$ years. For $\lambda_0^{-1} = 3$ and $0 \le \beta \le .6$ there will be an accumulation of about 50–60 units of information at the end of five years or an average of roughly 10–12 units per year, although initially the accumulation is much slower. Define the following repeated significance test. If at some time $t = 1, 1.5, 2, 2.5, \ldots, 5$ $-\ddot{l}(t,0) \ge 11$, and $|\dot{l}(t,0)| \ge b[-\ddot{l}(t,0)]^{1/2}$, stop the experiment and reject H_0; otherwise continue until $-\ddot{l}(t,0) \ge 55$ or $t = 5\frac{1}{2}$ years, whichever comes first; if the test is truncated at t because $-\ddot{l}(t,0) \ge 55$ or $t = 5\frac{1}{2}$, reject H_0 if $|\dot{l}(t,0)| \ge c[-\ddot{l}(t,0]^{1/2}$.

This test is a compromise between the administratively convenient one, for which the data are reviewed at equally spaced points in time, and the one implicit in our preceding discussion, for which observations are to be at equal intervals of accumulated information. Theorems 5.63 and 5.66 lead one to hope that at least under the conditions and for the range of parameter values anticipated the test will behave approximately like a repeated significance test in discrete time with $m_0 = 2$, $m = 10$, and normally distributed observations having mean $\mu = 5.5$ and variance 5.5 (equivalently $\mu = (5.5)^{1/2}\beta$ and variance 1). An approximation to the expected information at termination of this test is 5.5 times the expected sample size of the approximating normal theory test.

Table 5.4. Fixed Time Test (Theoretical Calculations are in Parentheses.) $t = 5$ years; reject H_0: $\beta = 0$ if $|\dot{l}(5,0)| \geq 1.96[-\ddot{l}(5,0)]^{1/2}$

$\exp(\beta)$	Power	E(Time)	E(Info)	E(Deaths)
1.80	.98	5.0	49	200
	(.99)			(200)
1.65	.95	5.0	50	206
	(.95)			(205)
1.50	.81	5.0	52	211
	(.84)			(211)
1.40	.69	5.0	53	215
	(.69)			(215)
1.00	.045	5.0	59	237
	(.05)			(236)

Table 5.5. Repeated Significance Test ($b = 2.5$)

$\exp(\beta)$	Power	E(Time)	E(Info)	E(Deaths)
1.80	.97	2.9	25	101
	(.98)		(21)	
1.65	.91	3.3	30	122
	(.92)		(27)	
1.50	.77	3.9	38	154
	(.77)		(34)	
1.40	.58	4.3	44	177
	(.60)		(39)	
1.00	.053	4.8	56	226
	(.05)			

Tables 5.4, 5.5, and 5.6 report the results of a 400 repetition Monte Carlo experiment. Figures given parenthetically are theoretical calculations based on the normal approximations suggested above. (For the theoretical power calculations of the fixed time test of Table 5.4, (5.63) is used to obtain the expected number of deaths, which is divided by four to obtain approximately the expected information, which in turn is used as $(\text{variance})^{-1}$ of a normal distribution with mean β to obtain an approximation to the power.)

The theoretical approximations appear to be quite good even though those in Table 5.5 and 5.6 are being used in a way not entirely justified by Theorems 5.59 and 5.62. More interesting, however, is the comparison between the fixed time test of Table 5.4 and the sequential tests of Tables 5.5 and 5.6.

It is predicted by our theoretical results and readily observed in the tables that the expected information required to stop the test is much smaller for the sequential than the fixed time test when β deviates substantially from 0, and in the case of the modified repeated significance test this is accomplished with no

Table 5.6. Modified Repeated Significance Test ($b = 2.85, c = 2.05$)

$\exp(\beta)$	Power	E(Time)	E(Info)	E(Deaths)
1.80	.99	3.2	28	112
	(.99)		(26)	
1.65	.95	3.7	35	143
	(.95)		(33)	
1.50	.83	4.3	43	175
	(.84)		(40)	
1.40	.70	4.7	49	198
	(.69)		(47)	
1.00	.055	4.9	57	231
	(.05)			

loss of power. The expected number of deaths is about four times the expected information, so here too the sequential tests seem to offer a considerable improvement. Savings in the expected time required for the sequential tests to terminate can be substantial but are not as great percentage wise as the savings in expected information. In particular the expected time essentially always exceeds the three year intake period. This means that patients (potentially) in the trial do not benefit from its shorter duration unless they can be switched to the favored treatment after termination of the trial. On the other hand, patients not in the trial might benefit from a shorter trial which makes a new treatment accessible at an earlier date.

Balancing the relative merits of these considerations makes the situation quite complicated, and it becomes more so when one reinserts censoring, additional covariates, etc. into the model. The goal of this discussion cannot be to supply a definitive answer, but rather to provide basic tools for dealing with the problem.

It bears repeating that it is only because of our very simple model that the preceding direct comparison with a fixed *time* test is possible. In general, if one wants a test whose asymptotic power function depends approximately only on β and not on the baseline hazard function, rate of intake, amount of censoring, etc., then by Theorems 5.59 and 5.62 the duration of the test should be determined by the value of $-\ddot{l}(t, 0)$ and not by t itself. Presumably in practice one should compromise somewhat between what one anticipates in theoretical calculations prior to the start of a trial and what one actually sees during the trial, so that one does not have a test with power close to zero because of slower intake, more censoring, and/or longer baseline survival than anticipated.

To be specific, suppose one plans to use the test of Table 5.6, but intake is slower and survival longer than anticipated, so four years are required to recruit 350 patients, and $\lambda_0 = .25$. Although one may not have recognized this situation already at the end of one year when the first data review is scheduled, it will certainly be apparent before the end of the planned three year recruitment period. Suppose that the recruitment period is extended to four years and

Table 5.7. Modified Repeated Significance Test ($b = 2.85, c = 2.05$) Intake for 4 Years, $\lambda_0 = .25$, Maximum Follow Up for 7 Years

$\exp(\beta)$	Power	E(Time)	E(Info)	E(Deaths)
1.65	.94	4.8	34	138
1.50	.83	5.5	41	165
1.40	.71	6.0	46	187
1.00	.053	6.4	57	229

the follow up time to a maximum of seven years, but that otherwise the test remains the same as that described above (specifically the test does not terminate if at a scheduled review $-\ddot{l}(t,0) < 11$ and does not continue if $-\ddot{l}(t,0) > 55$). One expects that the power function and expected information will be essentially the same as before, but of course the expected time required for the test to terminate will increase. The outcome of a 400 repetition Monte Carlo experiment verifying these expectations is reported in Table 5.7.

A particularly important and interesting application is the randomized trial of propranolol conducted by the β-Blocker Heart Attack Trial Research Group (BHAT, 1982; see also DeMets *et al.* 1984). Over a period of about 27 months 3837 victims of acute myocardial infarction were randomized to a placebo group (1921) or a treatment group (1916), who then received a daily regimen of propranolol hydrochloride. The principal endpoint was survival time, which was assumed to follow a proportional hazards model.

Planned data reviews were at $t = 1, 1.5, 2, 2.5, 3, 3.5$, and 4 years. These were assumed to correspond to seven reviews at approximately equal increments of information time. The stopping rule was that of a truncated sequential probability ratio test (cf. IV.4), but for the following discussion we consider instead a repeated significance test. Let t_n denote the time of the nth planned inspection, $n = 1, 2, \ldots, 7$, and consider the stopping rule

$$T = \inf\{t_n\colon n \geq 2, |\dot{l}(t,0)| \geq b[-\ddot{l}(t,0)]^{1/2}\}. \tag{5.64}$$

Stop sampling at $\min(T, t_7)$ and reject $H_0\colon \beta = 0$ if either $T < t_7$ or $T > t_7$ and $|\dot{l}(t_7,0)| \geq c[-\ddot{l}(t_7,0)]^{1/2}$. Assuming that the normal approximation of Theorem 5.59 is applicable, one can easily use the approximations of IV.4 to see that the values $b = 2.65$ and $c = 2.05$ yield a .05 level test. Table 5.8 gives approximately the power and expected sample size functions for this test for the approximating normal model.

To relate Table 5.8 to the test defined by (5.64), it is necessary to know the rate of increase of $-\ddot{l}(t,0)$. Preliminary calculations described in BHAT (1981) indicate an expectation of about 600 deaths, hence the accumulation of about 150 units of information during the follow up period. Due to somewhat better baseline survival than expected, and perhaps to slightly slower intake, a more realistic figure would have been about 400 deaths for approximately 100 units of information. For these latter figures there would be on average about

Table. 5.8. Repeated Significance Test for Normal Data: $b = 2.65$; $c = 2.05$; $m_0 = 2$; $m = 7$

μ	Power	Expected Sample Size
1.50	.97	3.59
1.25	.90	4.47
1.00	.74	5.31
.75	.50	
.50	.27	
.00	.05	

Table 5.9. Propranolol Study

t	1	1.5	2	2.5	3	3.5
$\dot{l}/(-\ddot{l})^{1/2}$	1.68	2.24	2.37	2.30	2.34	2.82
Deaths/Total (TRT)	22/860	29/1080	50/1480	74/1846	106/1916	135/1916
Deaths/Total (CTRL)	34/848	48/1080	76/1486	103/1841	141/1921	183/1921
$-\ddot{l}(=$ Deaths/4)	14.0	19.3	31.5	44.3	61.8	79.5
$\dot{l}$	6.3	9.8	13.3	15.3	18.4	25.1
$\dot{l}/(-\ddot{l})$	.45	.51	.42	.35	.30	.32

$100/7 = 14.3$ units of information per inspection period, so the values of μ in Table 5.8 correspond to values of $\beta = \mu/(14.3)^{1/2} = \mu/3.78$. In particular the row for $\mu = 1.25$ in Table 5.8, where the power is .9, corresponds to $\beta = .33$.

Data for this experiment are given in Table 5.9 and graphed in Figure 5.3. The first three rows of Table 5.9 were supplied by D. DeMets in a private communication. Published sources (for example, DeMets *et al.*, 1984) give only the first row. The final three rows are actually fictitious reconstructions based on the assumption that $-\ddot{l}(t, 0)$ equals one-fourth the number of deaths up to time t. Because of the large number of persons at risk and the relatively small proportion of deaths, the approximation would appear to be an extremely good one. (The life table reported by BHAT, 1982, gives added support to the validity of this approximation.) For the stopping rule (5.64) with $b = 2.65$, the experiment would terminate at the sixth inspection—as did the actual experiment (see Problem 5.13).

A naive 80% approximate confidence interval for β is

$$\dot{l}(T, 0)/[-\ddot{l}(T, 0)] \pm 1.28/[-\ddot{l}(T, 0)]^{1/2} = .32 \pm .14 = [.18, .46].$$

By the method of IV.5 an 80% confidence interval for μ in the approximating normal test of Table 5.8 is [0.4, 1.54] with midpoint at .97. If we use the average amount of information actually accumulated during the first six inspection periods to convert this to an interval for β, the result is [.11, .42] with midpoint at .27.

Since the actual outcomes of the experiment of Freireich *et al.* described in Section 2 were survival times, which were subsequently reduced to Bernoulli

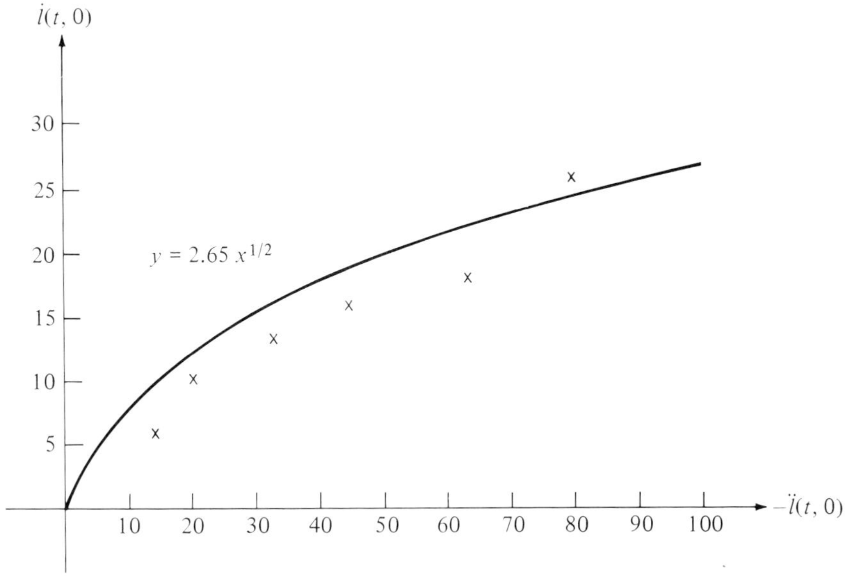

Figure 5.3.

data, it is interesting to consider an alternative analysis via the proportional hazards model. We first consider the experimental design.

Recall that in this study the "time of entry" was the time of a remission in leukemia, and "death" was the relapse of this remission. Hence "lifetime" denoted the duration of the remission.

For exponential lifetimes, or more generally for Weibull lifetimes with a common shape parameter for the treatment and control populations, the probability p (in the absence of censoring) that a patient in the treatment group outlives a patient on placebo is $\exp(\beta)/[1 + \exp(\beta)]$, where $\exp(\beta)$ is the ratio of hazard functions. Hence $p = .75$ corresponds to $\beta = \log 3 = 1.1$, and the design requirement for the derived Bernoulli data of power .95 at $p = .75$ might reasonably be replaced by requiring power .95 at $\beta = 1.1$ in the proportional hazards model.

The stopping rule actually used by Freireich *et al.* was a symmetric two-sided version of the linear stopping boundary (3.7) with the value of η appropriate for the simple alternative $p = .75$ (see Problem 5.6). If we assume that the Brownian approximation of the preceding section is applicable, the analogous stopping rule for the log rank statistic is of the form

$$T = \inf\{t\colon |\dot{l}(t,0)| \geq b + \eta[-\ddot{l}(t,0)]\} \tag{5.65}$$

truncated as soon as $-\ddot{l}(t,0) \geq m$. The hypothesis H_0: $\beta = 0$ is rejected if $-\ddot{l}(T,0) < m$. The value of η appropriate for the simple alternative $\beta = 1.1$ is $\eta = .55$; and for an approximate .05 level test having power .95 at $\beta = 1.1$, it is easy to see that $b = 3.3$ and $m = 14.6$. Freireich *et al.* actually reviewed their

Table 5.10. Trial of 6-Mercaptopurine and Placebo

Treatment	Survival in Weeks (+ Denotes Censoring)						
6-MP	6	6	6	6+	7	9+	10
	10	11+	13	16	17+	19+	20+
	22	23	25+	32+	32+	34+	35+
Placebo	1	1	2	2	3	4	4
	5	5	8	8	8	8	11
	11	12	12	15	17	22	23

Data from Freireich *et al.* (1963).

data every three months, so the true significance level would be smaller than the nominal .05.

The final survival record in weeks of 42 patients (21 pairs) after one year of experimentation is given in Table 5.10. Unfortunately, however, the published data do not contain times of entry into the study, so it is not possible to reconstruct the statistics $\dot{l}(t, 0)$ and $-\ddot{l}(t, 0)$ at different times during the experiment. The final value of the statistic $\dot{l}$ for these data (see, for example, Kalbfleisch and Prentice, 1980, p. 80 for the treatment of ties) is easily calculated to be 9.86 while $-\ddot{l} = 6.48$, so $\dot{l}/(-\ddot{l})^{1/2} = 3.9$. In the paired comparison analysis described in Section 2 there were 18 preferences for placebo in the 21 pairs, which yield a standardized value of S_n equal to 3.3. Hence the Bernoulli and proportional hazards models are reasonably consistent with each other. On the other hand $\dot{l}/(-\ddot{l})^{1/2}$ is enough larger that it seems possible that a test based on (5.65) might have indicated termination at the preceding inspection.

Several authors have analyzed the data in Table 5.10 without taking into account the use of a stopping rule in the experiment. (See McCullagh and Nelder, 1983, p. 186 ff. for references and a summary of several studies.) Their results are consistent with the simple minded (from (5.61)) estimate of β by $\dot{l}/(-\ddot{l}) = 9.86/6.48 = 1.52$ with standard error $1/(-\ddot{l})^{1/2} = .39$. Problem 5.10 indicates that these estimates may involve substantial bias.

Problems

5.1.* Let $W(t) = (W_1(t), \ldots, W_d(t))'$ be d-dimensional Brownian motion with drift $(\theta_1, \ldots, \theta_d)'$. Let $\theta = (\sum_1^d \theta_i^2)^{1/2}$. For $b > 0$, $m_0 > 0$ define $T = \inf\{t\colon t \geq m_0, \|W(t)\| \geq bt^{1/2}\}$. Suppose that $b \to \infty$, $m_0 \to \infty$, and $m \to \infty$ such that $bm^{-1/2} = \gamma_1 < \gamma_0 = bm_0^{-1/2}$.

(a) Generalize the argument of IV.8 to show that

$$P_0\{T \leq m\} \sim 2^{-(d+1)/2}[\Gamma(\tfrac{1}{2}d)]^{-1} \log(m/m_0) b^d e^{-b^2/2}$$

(b) Show that for $0 < \theta < \gamma_0$

$$E_\theta(T) = (m\gamma_1^2 - d)/\theta^2 + o(1).$$

(c) Use the argument suggested in Problem 4.7 to show that if $\theta = \gamma_1 + \Delta m^{-1/2}$,

$$E_\theta(T \wedge m) = m\gamma_1^2/\theta^2 - \{m^{1/2}(\theta - \tfrac{1}{2}\gamma_1)^{-1}\}[\phi(\Delta) - \Delta\Phi(-\Delta)] + o(m^{-1/2}).$$

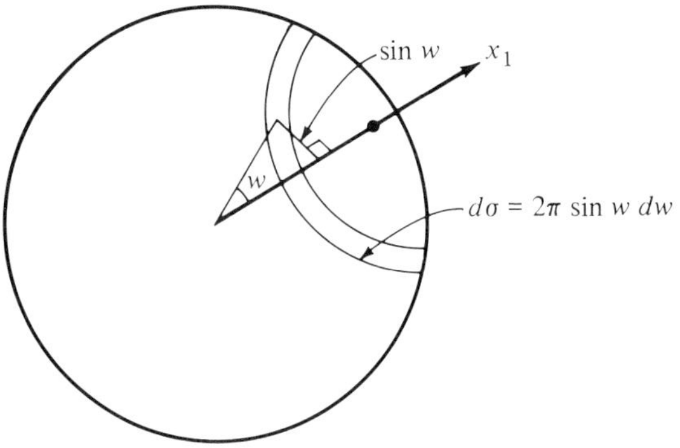

Figure 5.4.

(d) Show that for $\xi_0 \in (\gamma_1^2/\gamma_0, \gamma_1]$

(5.66) $$P_0\{T < m \mid \| W(m) \| = m\xi_0\} \sim (\gamma_1 \xi_0^{-1}) \exp\{-\tfrac{1}{2}m(\gamma_1^2 - \xi_0^2)\},$$

and use this result to obtain versions of (5.13) and (5.14) for Brownian motion. *Hint:* To uncondition (5.66) when $\theta > 0$ it is helpful to express the density of $W(m)$ in polar coordinates in the following form:

$$\begin{aligned} P_\mu\{\| W(m) \| \in d\rho, \omega_m \in dw\} &= \exp\{\theta\rho \cos w(s) - \tfrac{1}{2}\theta^2 m\} \rho^{d-1} \\ &\quad \cdot \exp(-\tfrac{1}{2}\rho^2/m) \, d\sigma(s) \, d\rho/(2\pi m)^{d/2} \\ &= C_{d-1}(2\pi m)^{-d/2} \exp\{\theta\rho \cos w - \tfrac{1}{2}\theta^2 m\} \rho^{d-1} \\ &\quad \cdot \exp(-\tfrac{1}{2}\rho^2/m)(\sin w)^{d-2} \, dw \, d\rho. \end{aligned}$$

Here s is a point on the unit sphere S_{d-1}, $w(s)$ is the angle made by the ray through s and the x_1 axis, σ denotes surface area on S_{d-1}, $C_d = 2\pi^{d/2}/\Gamma(\tfrac{1}{2}d)$ is the surface area of S_{d-1}, and the range of (w, ρ) is $(0, \pi) \times (0, \infty)$ (see Figure 5.4).

5.2. For the sequential t test of Section 4, use the idea suggested in Problem 4.7 to derive an approximation for $E_\theta(T \wedge m)$ of the form $K_1 m + K_2 m^{1/2} + o(m^{1/2})$, and show that it agrees with the Brownian approximation suggested in Section 4 up to the accuracy indicated.

5.3. Let $x_1, x_2, \ldots$ be independent and identically distributed with density function $f(x; \theta)$ for some unknown, real parameter θ. Let $l_n(\theta) = \sum_1^n \log f(x_k; \theta)$ be the log likelihood function and

$$\Lambda_n = \sup_\theta [l_n(\theta) - l_n(\theta_0)],$$

the likelihood ratio statistic for testing $H_0: \theta = \theta_0$ against $H_1: \theta \neq \theta_0$. Show heuristically that for large n $(j\Lambda_j)^{1/2}$, $nt_1 \le j \le nt_2$, behaves like (the absolute value of) a random walk with normal increments (a) for fixed $\theta \neq \theta_0$ and (b) for $\theta = \theta_0 + \Delta n^{-1/2}$.

5.4. For the proportional hazards model, suppose that $\beta = 0$, $y_1 = y_2 = \cdots = y_k = 0$, $c_1 = c_2 = \cdots = c_k = +\infty$, and that each z_i is either 0 or 1. Let $k^{(1)} = \sum_1^k z_i$ denote the number of persons having covariate equal to one. Consider the experiment of drawing j balls from an urn having k balls, $k^{(1)}$ of which are black and the rest white, and let S_j denote the number of black balls drawn. Show that the score process (5.39) observed at the instants when a death occurs has the same joint distributions as the process

$$S_j - \sum_{i=1}^{j} [(k^{(1)} - S_{i-1} + 1/(k - i + 1)].$$

In particular, this process is a martingale and can be used as the basis of an exact, nonparametric test of H_0: $\beta = 0$. Show that the score process observed at death times remains a martingale even if one no longer requires the c_i to be $+\infty$, but give an example to show that it does not if the y_i are no longer restricted to be equal.

5.5. Consider testing the null hypothesis of no treatment effect for two samples of censored survival data by a statistic of the form

$$\sum_i \int_{[0,t]} H(t,x)[z_i - \tilde{\mu}(t,x)] N_i(t,dx), \tag{5.67}$$

where N_i is defined by (5.35) and $\tilde{\mu}$ is given by (5.40) with $\beta = 0$. Show (a) if $H(t,x)$ is the (left continuous) Kaplan–Meier estimator of the survival function (based on the pooled data under H_0), then (5.67) is essentially the adaptation of the Wilcoxon statistic to censored data suggested by Peto and Peto (1972) and Prentice (1978); (b) if $H(t,x)$ is the size of the risk set $R(t,x)$, (5.67) is Gehan's (1965) adaptation of the Wilcoxon statistic. Argue heuristically that at least under H_0 results like Theorems 5.53 and 5.59 should hold in case (a). What can be used in place of $-\ddot{l}(t,0)$ to define an appropriate time scale? See Harrington *et al.* (1982) for additional information on statistics of the form (5.67) and Slud and Wei (1982) for a discussion of sequential use of Gehan's statistic.

5.6.* Let $x_1, x_2, \ldots$ be independent Bernoulli variables with $P\{x_i = 1\} = p = 1 - P\{x_i = 0\}$, and put $S_n = x_1 + \cdots + x_n$. Consider the test of H_0: $p = \frac{1}{2}$ against H_1: $p \neq \frac{1}{2}$, which stops sampling at $\min(T, m)$, where

$$T = \inf\{n: |2S_n - n| \geq b + \eta n\},$$

and rejects H_0 if and only if $T \leq m$. Using a Brownian approximation with mean $\mu = 2p - 1$ and variance $\sigma^2 = 4p(1 - p)$, show that for $b = 7$, $\eta = .26$, and $m = 66$ the test has significance level about .05 and power about .95 for $p = .75$. For the data obtained by Brown *et al.* (and reported in Section 2), this test would termiate with $T = 18$. What according to the Brownian approximation is the attained significance level of this result? In the study by Brown *et al.* two more observations, one a preference for antitoxin and one a preference for placebo, became available after the trial was terminated at $T = 18$. Use Problem 3.15 to show that the attained significance of the 16 preferences for antitoxin out of 20 preferences in the experiment is approximately .019. (For a better probability approximation, see Example 10.63.)

5.7. Consider again the study by Brown *et al.* but this time with the stopping rule (5.2) of a repeated likelihood ratio test. Use a Brownian approximation and argue along the lines of Problem 3.15 (in the notation developed there) that now the attained significance level is approximately

$$\tilde{G}[(b/U)^2 - \bar{m}] + \bar{m}U^{-2}\tilde{G}''[(b/U)^2 - \bar{m}], \tag{5.68}$$

where $\tilde{G}(m)$ denotes the right hand side of the expression in Theorem 4.21. Evaluate (5.68) for $b = 2.79$, $\bar{m} = 2$, $U = .60$ (corresponding to $S_{20} = 16$) and compare this with a Brownian approximation to the attained significance level of the datum $T = 18$ (ignoring the two additional observations).

5.8.* Consider the usual model for simple linear regression: for $i = 1, 2, \ldots$ the random variables y_i are independent and normally distributed with mean $\alpha + \beta x_i$ and variance 1. (a) Define and study the properties of repeated significance tests for $H_0\colon \alpha = 0$ against $H_1\colon \alpha \neq 0$ with β considered to be a nuisance parameter. (b) Do the same for $H_0\colon \beta = 0$ against $H_1\colon \beta \neq 0$ with α considered to be a nuisance parameter. (c) Generalize (a) to the case of two parallel regression lines (analysis of covariance).

Hint: For part (a) consider the invariance reduction defined by

$$u_i = y_i x_i^{-1} - y_1 x_1^{-1}, i = 2, 3, \ldots \qquad (u_i = y_i \text{ if } x_i = 0).$$

The likelihood ratio of $u_2, \ldots, u_n$ under an arbitrary value of α relative to $\alpha = 0$ is

$$\exp(\alpha S_n - \tfrac{1}{2}\alpha^2 t_n),$$

where

$$S_n = \sum_1^n y_i - \sum_1^n x_i \sum_1^n x_i y_i \Big/ \sum_1^n x_i^2 \text{ and } t_n = n\left[1 - \left(\sum_1^n x_i\right)^2 \Big/ n\sum_1^n x_i^2\right].$$

Remark: Presumably in most contexts one would consider the x's as constants or as ancillary statistics and condition on their values. It is also interesting to think about the case in which the x's are independent, identically distributed random variables with an approximately known distribution. This assumption would allow one to compute approximations to the error probabilities and expected sample size which might be particularly useful at the design stage of an experiment.

5.9. Use (5.63) and the fact that the expected number of deaths in Tables 5.5–5.7 is about four times the expected information to obtain a crude approximation to the expected time required for the test (under the assumptions of these tables). Compare your approximation numerically with the Monte Carlo estimates given in the tables.

5.10. Consider the sequential test defined in Problem 5.6 but assume that the data are reviewed only after every k observations. (Note that for the following theoretical calculation it is not necessary that k be an integer.) Let $\hat{p}$ denote the relative frequency of successes at the end of the test, and let $\hat{\mu} = 2\hat{p} - 1$ be the naive estimator of $\mu = 2p - 1$. Use (3.20) corrected for the excess over the boundary to obtain an approximation for the bias of $\hat{\mu}$ when p is sufficiently close to one that $P_p(T < m) \cong 1$. Evaluate the estimator $\hat{\mu}$ and the approximate bias for an experiment with $k = 5.25$, which terminates with 18 successes in 21 trials. (The expected excess over the boundary when $k = 5$ is about $.57k^{1/2}$; see Chapter X.)

5.11. Let $x_1, x_2, \ldots$ be independent and normally distributed with mean μ and variance σ^2, both unknown. Let $2 \le m_0 < m$ and let $\tilde{T}$ be any stopping rule such that $m_0 \le \tilde{T} \le m$. Use the notion of a confidence sequence as defined in Problem 4.11 together with Theorem 5.29 to obtain an approximate confidence interval for μ based on the data $x_1, \ldots, x_{\tilde{T}}$. Why does this confidence interval seem unsatisfactory as a solution to the problem of estimating μ by a confidence interval following the repeated likelihood ratio test defined in Section 4?

5.12.[†] Consider d-dimensional Brownian motion as in Problem 5.1 and a test of H_0: $\theta = 0$ defined in terms of the stopping rule $\tau \wedge m$, where $\tau = \inf\{t: \|W(t)\| \ge b + \eta t\}$. Find approximations for the significance level, power, and expected sample size. (The asymptotic normalization which yields simple approximations has $b/m \to \zeta$ and η fixed as $m \to \infty$.)

5.13. Use the approximations of III.5, 6 to show that a truncated sequential probability ratio test with $b = 5.48$ and $m = 7$ (cf. IV.4) has about the same power function and expected sample size as the repeated significance test of Table 5.8. Apply this test to the data of Table 5.9. Compute approximate 80% confidence intervals for β. Compute approximate attained significance levels for this test and for the one discussed in the text.

CHAPTER VI

Allocation of Treatments

In comparative studies generally and especially in clinical trials the allocation of treatments and the appropriate use of randomization are particularly important. The present chapter discusses three related aspects of this general subject, in the simple context of the comparison of two treatments. Section 1 is concerned with sequential two population randomization tests when allocation is at random. Section 2 gives a very brief introduction to the notion of biased randomization to force balancing in small experiments. Sections 3–5 involve the adaptive allocation of treatments during the course of experimentation in such a way that the less desirable treatment is used a minimum number of times. A central result is Theorem 6.20.

1. Randomization Tests

Consider a comparison of two treatments to which the response is success or failure. The paired comparison sequential test defined by (5.2) has the following interpretation as a randomization test. Suppose that no stochastic model is assumed for the responses, but that the null hypothesis of no treatment effect is interpreted to mean that the response of any particular individual is the same (success of failure) regardless of which treatment that individual receives. Suppose also that in each pair one person selected at random receives treatment A and the other treatment B. Then under the null hypothesis the joint distribution of the random variables $y_1, y_2, \ldots$ defined in V.2 is determined entirely by the randomization in the treatment allocation, so the y's are independent Bernoulli variables assuming the values zero or one with probability $\frac{1}{2}$ each. As a consequence the joint distribution of $S_n = y_1 + \cdots + y_n$,

$n = 1, 2, \ldots$ under the null hypothesis is the same as it was under the hypothesis $p = \frac{1}{2}$ in V.2. and the sequential test defined by (for example) (5.2) has the same significance level under the current randomization model as under the parametric assumptions of V.2.

Formulation of a two population sequential randomization test is slightly less direct. As a starting point we recall the randomization interpretation of the Fisher–Yates test for independence in a 2×2 table. Suppose again that the response of each individual is success or failure regardless of which treatment he receives. Suppose that in a population of N individuals, n are selected at random to receive treatment A with the remaining $N - n$ receiving treatment B. Here selection at random can mean that n are selected at random without replacement from the N, in which case n is a constant, or it can mean that a coin is tossed to determine the treatment assignment for each individual. In the latter case n is a binomial random variable, but we condition on its value and hence treat it as a constant. The total number of successes in the N trials is by hypothesis also a constant, v. The Fisher–Yates test is based on the number of successes s among the n assignments of treatment A. Under the hypothesis of no treatment effect the (conditional) distribution of s is determined entirely by the randomness in the treatment allocation scheme, and is easily seen to be the hypergeometric distribution

$$P_0\{s = i\} = \frac{\binom{n}{i}\binom{N-n}{v-i}}{\binom{N}{v}}. \tag{6.1}$$

The mean of s is nv/N, and an extreme departure in the observed value of s in either direction from its mean is interpreted as evidence against the hypothesis of no treatment effect.

To obtain a sequential randomization test, suppose that there are m groups of sizes $N_1, N_2, \ldots, N_m$, with each $N_i > 1$ and $N_1 + \cdots + N_m = N$. It may be useful to think about the particular case where the N_i are all about equal and large while m is a fairly small integer, say in the range from 2 to 10. However, these relations are not required in order to define the test.

Again suppose either that treatments are allocated by tossing a coin, in which case we condition on the number n_j of times that treatment A is assigned among the N_j persons in the jth group, or that n_j persons to receive treatment A are selected at random without replacement from the N_j in the jth group. In the latter case the n_j are actually constants which can be chosen by the experimenter. (This allocation scheme is discussed from a different point of view in Section 2.)

If the experiment is carried out sequentially in groups of size $N_j, j = 1, 2, \ldots, m$, it is possible to define a group sequential randomization test in terms of the s_j, the number of successes on treatment A out of the v_j successes in the jth group. For motivation it is helpful to recall that for m fixed and large values of the N_j, under the hypothesis of no treatment effect the s_j are independent and approximately normally distributed with means

$$\mu_j = n_j v_j / N_j \tag{6.2}$$

and variances

$$V_j = n_j(v_j/N_j)(1 - v_j/N_j)(N_j - n_j)/(N_j - 1). \tag{6.3}$$

Hence an approximate repeated significance test can be defined by the stopping rule

$$T = \inf\left\{k: k \geq m_0, \left|\sum_1^k (s_j - \mu_j)\right| \geq b\left(\sum_1^k V_j\right)^{1/2}\right\}, \tag{6.4}$$

truncated at m if stopping has not occurred earlier. Since under the null hypothesis of no treatment effect $P_0(T \leq j)$ depends only on the n_i, N_i, and v_i for $i \leq j$, the stopping rule (6.4) does indeed define a sequential randomization test. Also, to the extent that the V_j are all approximately the same, the distribution of T under H_0 can be approximated by the null hypothesis distribution of a repeated significance test for a normal mean.

Remark. It is by no means necessary that the N_j be large in order to define a sequential randomization test. In fact one obtains a paired comparison randomization test as a special case of the sampling scheme described above if $N_1 = N_2 = \cdots = 2$, one person is selected at random without replacement from each group to receive treatment A, and groups with two successes or two failures are discarded. Note, however, that it is *not* possible to take all the $N_j = 1$ and hence have a "continuously" monitored sequential randomization test.

Remark 6.5. It is natural to ask if the experimenter can force the V_j to be approximately equal in order to make a simple normal approximation to the significance level valid. He can obviously make the N_j the same; and if he allocates treatments by selecting at random without replacement from the N_j persons the n_j to receive treatment A, he can make these equal also. The v_j are beyond his control, although they will be about equal if the n_j/N_j are approximately equal, and if for each treatment the responses are actually independent Bernoulli trials with constant probability of success. An alternative would be to allocate the treatments by tossing a coin and define the N_j sequentially to be as large as necessary so that the V_j are about equal to some constant; but this is probably more complicated than necessary.

It is of interest to know whether the procedure described above is reasonably efficient. To discuss this question it is helpful to reintroduce a probability model for the responses. We assume that treatment A (B) leads to success with probability p_1 (p_2), and that the responses of different persons are stochastically independent. Then the null hypothesis distribution (6.1) in general becomes

$$P_\theta\{s = i\} = \frac{\binom{n}{i}\binom{N-n}{\nu-i}e^{\theta i}}{\sum_j \binom{n}{j}\binom{N-n}{\nu-j}e^{\theta j}}, \tag{6.6}$$

where $\theta = \log(p_1 q_2/p_2 q_1)$, and the hypothesis of no treatment effect is H_0: $p_1 = p_2$, or equivalently $\theta = 0$.

Suppose that m is fixed and the N_j tend to infinity in such a way that each N_j is proportional to N (with a constant of proportionality which may depend on j). Assume also that $\theta \sim \Delta/N^{1/2}$ for some constant Δ. Finally assume that the proportion of times treatment A is assigned in the jth group, n_j/N_j, converges to a constant in (0, 1) not depending on j. (This will be the case, for example, if treatment assignment is made by tossing a coin.) Then a tedious but straightforward computation shows that the joint asymptotic distribution of

$$\left|\sum_{j=1}^{k}(s_j - \mu_j)\right| \Big/ \left(\sum_1^k V_j\right)^{1/2}, \qquad k = 1, \dots, m \tag{6.7}$$

is the same as that of the square root of the likelihood ratio statistic (5.7) observed at the times $N_1 + \cdots + N_k$, $k = 1, 2, \dots, m$. Hence a sequential test defined in terms of the random variables in (6.7), which can be interpreted as a randomization test without any parametric model, has at local alternatives the same asymptotic properties as a test defined similarly in terms of the likelihood ratio statistic inside the parametric model.

2. Forcing Balanced Allocation

A characteristic of clinical trials is that although the treatment structure may be simple, the number of covariates and/or strata may be very large and complex. It is often desirable that the treatment assignment be approximately balanced within the individual strata, which can be a problem in small strata if treatments are allocated by tossing a coin. The problem is exacerbated if one uses a sequential design. Even in a two population problem without strata, treatment assignment by tossing a coin can lead to severe imbalance of allocation in the initial stages of an experiment. If the data lead one to consider stopping after a small number of observations, lack of balance can be a serious problem. The sequential clinical trial of Brown *et al.* (1960) described in V.2 seems to have suffered from such a defect.

To deal with the problem of imbalance, a number of authors beginning with Efron (1971) have suggested various schemes for favoring the treatment which has been used less frequently in the past when making the current allocation.

The subject is a complicated one containing a large number of *ad hoc* schemes. The purpose of this section is to call attention to some of the relevant literature by means of a brief discussion of one aspect of forced balancing: control of selection bias.

Selection bias refers to the effect that an experimenter can have if he knows (or guesses) the next treatment assignment and admits the next patient in a way which favors a particular treatment. The simplest example would be refusing to admit a very ill patient if the treatment favored by the experimenter is the next to be allocated. If there are two treatments which are allocated by tossing a fair coin, an experimenter can expect to guess the next treatment assignment fifty percent of the time. Any excess over this level is called selection bias, a term introduced by Blackwell and Hodges (1957).

More formally, let $\delta_n = +1$ or -1 according as treatment A or treatment B is assigned for the nth patient. Let $D_n = \delta_1 + \cdots + \delta_n$ ($D_0 = 0$) denote the excess number of times treatment A has been allocated in the first n allocations. Complete balance is indicated by $D_n = 0$, and to achieve something approximating this it seems reasonable to use an allocation scheme satisfying

$$(6.8)\qquad P\{\delta_{n+1} = 1 | \delta_1, \ldots, \delta_n\} \begin{cases} \geq \frac{1}{2} & \text{if } D_n < 0 \\ = \frac{1}{2} & \text{if } D_n = 0 \\ \leq \frac{1}{2} & \text{if } D_n > 0. \end{cases}$$

If (6.8) holds and one wants to guess the next allocation, the obvious strategy is to guess treatment A or B for the $n + 1$st patient according as $D_n <$ or > 0. If $D_n = 0$ the guess can be arbitrary. Under these assumptions the selection bias of the first n allocations is

$$(6.9)\qquad \sum_{i=0}^{n-1} [P\{D_i \delta_{i+1} < 0\} + \tfrac{1}{2} P\{D_i = 0\}] - \tfrac{1}{2} n.$$

It is easy to see that (6.9) equals 0 if allocation is made by tossing a fair coin.

As an alternative to randomization by coin tossing, consider a permuted block allocation. One selects an even number $N = 2k$; and in each block of $2k$ patients, k selected at random without replacement receive treatment A while the other k receive treatment B. This scheme was seen in the preceding section to have the advantage that randomization tests based on it are particularly convenient. If k is small the method leads to serious imbalances in allocation with very small probability, and in the limit as k becomes infinite it approaches allocation by fair coin tossing. We now compute the selection bias at time $n = 2k$ of a permuted block design.

For the permuted block design of size $2k$, the sample path D_i, $i = 0, 1, \ldots, 2k$ takes exactly k steps away from 0 and exactly k steps toward zero. For the guessing scheme given above, among those steps that do not begin from the 0 state there are correct guesses on the steps toward 0, but not otherwise. Hence the first sum in (6.9) is exactly $k = n/2$, and the selection bias of the permuted block design at time $2k$ is

$$\frac{1}{2}\sum_{i=0}^{2k-1} P\{D_i = 0\}.$$

From the hypergeometric distribution one sees that

$$P\{D_{2i} = 0\} = \frac{\binom{2i}{i}\binom{2k-2i}{k-i}}{\binom{2k}{k}},$$

and from combinatorial identity $\sum_{i=0}^{k}\binom{2i}{i}\binom{2k-2i}{k-i} = 2^{2k}$ (cf. Problem 6.2) it follows that the selection bias of one complete block of a permuted block allocation rule is

$$2^{2k-1}/\binom{2k}{k} - \tfrac{1}{2}. \tag{6.10}$$

For example, if $k = 8$, in the 16 allocations which constitute a block one can guess the next allocation about two times more than with allocation by fair coin tossing or ten times in all. In return for this excess in selection bias one has a considerably greater tendency towards balance. After 16 allocations there is exact balance, and the probability of finding an imbalance at least as large as five at some time during the 16 allocations that constitute a block is about 1/11. For allocation by fair coin tossing, the corresponding probability is roughly 0.6 (see Problem 3.3).

Of course, if one uses a permuted block design with a fixed block size $2k$, the total selection bias at time n is proportional to n as n becomes large. It seems reasonable to use blocks with increasing block sizes as an experiment proceeds and the effect or relatively small imbalances becomes less significant. For the sake of variety we consider next a different possibility: allocation by an adaptive biased coin design (Wei, 1978).

Suppose that the probabilities used to assign treatments are of the form

$$P\{\delta_{n+1} = 1 \mid \delta_1, \ldots, \delta_n\} = p(n^{-1}D_n), \tag{6.11}$$

where $p(x) = 1 - p(-x)$ and to be consistent with (6.8)

$$p(x) \begin{cases} \geq \tfrac{1}{2} & \text{for } x < 0 \\ = \tfrac{1}{2} & \text{for } x = 0 \\ \leq \tfrac{1}{2} & \text{for } x > 0. \end{cases} \tag{6.12}$$

A particularly simple special case is

$$p(x) = (1 - x)/2. \tag{6.13}$$

By symmetry $E(D_n) = 0$. It is easy to see that

$$E(D_{n+1}^2 \mid \delta_1, \ldots, \delta_n) = D_n^2 + 2D_n[2p(n^{-1}D_n) - 1] + 1. \tag{6.14}$$

It follows from (6.12) and (6.14) that $E(D_{n+1}^2) \leq E(D_n^2) + 1$, hence

$$E(D_n^2) \le n$$

and consequently

$$n^{-1} D_n \to 0 \tag{6.15}$$

in probability. (A slightly stronger argument shows that (6.15) actually holds with probability one. See Wei, 1978, or apply Theorem 1 of Robbins and Siegmund, 1971, along the lines of their Application 1.)

For $x > 0$ define $q(x) = 1 - p(x)$. From (6.9) it follows that the selection bias at time n of an allocation rule defined by (6.11) and (6.12) equals

$$\begin{aligned} &\sum_{i=0}^{n-1} [P\{D_i \delta_{i+1} < 0\} + \tfrac{1}{2} P\{D_i = 0\}] - \tfrac{1}{2} n \\ &= \sum_{i=0}^{n-1} [E\{q(i^{-1}|D_i|); D_i \ne 0\} - \tfrac{1}{2} P\{D_i \ne 0\}] \\ &= \sum_{i=1}^{n-1} E\{q(i^{-1}|D_i|) - q(0)\}. \end{aligned} \tag{6.16}$$

In view of (6.15) it seems clear intuitively that the selection bias can be kept bounded as $n \to \infty$ by making sufficiently many derivatives of $q(x)$ vanish at $x = 0$. (A closer analysis shows that it suffices to have $q'(0) = q''(0) = 0$ and q''' bounded.) The possibility of an asymptotically small selection bias makes the adaptive biased coin seem attractive; but since our main interest in schemes for forced balancing is their effect on very short trials, it seems unlikely that an asymptotic analysis would provide convincing evidence in favor of a particular allocation rule.

Consider now the special case (6.13). Taking expectations in (6.14) yields the recursive relation

$$E(D_{n+1}^2) = E(D_n^2)(1 - 2/n) + 1,$$

which after some algebraic manipulation yields

$$E(D_n^2) = (n - 2)/3 + 2/3. \tag{6.17}$$

Substitution of (6.13) into (6.16) and use of (6.17) together with the Schwarz inequality provide an upper bound for the selection bias in this special case, to wit

$$.5\left(1 + 3^{-1/2} \sum_{i=3}^{n-1} i^{-1/2}\right). \tag{6.18}$$

One can compute numerically the exact selection bias of (6.13)—at least for small n–but in view of its simplicity the upper bound in (6.18) may be useful. For $n = 16$ it provides the bound of 1.86 compared to an exact value of 1.6 given by Wei (1977). For this value of n the selection bias of the biased coin (6.13) is slightly smaller than that found earlier for block design of block length n.

3. Data Dependent Allocation Rules

An important reason for introducing stopping rules into clinical trials is to reduce the number of patients receiving an inferior treatment by stopping a trial as soon as the relative merits of the treatments are reasonably well established. In the rest of this chapter we consider the possibility of reducing the number of assignments of an inferior treatment by adaptively allocating an apparently inferior treatment less frequently as the experiment proceeds. As we shall see, these allocation schemes are completely different than those discussed in Sections 1 and 2.

Consider the following simplified model of a clinical trial. The immediate response of the ith patient to receive treatment A (B) is x_i (y_i). At any stage of the trial, after observing $x_1, \ldots, x_k, y_i, \ldots, y_n$ the experimenter can stop the experiment and declare (i) A is the better treatment, (ii) B is better, or (iii) there is essentially no difference between A and B; or he can continue the trial, in which case the next patient is assigned to treatment A or B according to some allocation rule which may depend on the data already observed. We shall be interested in allocation rules which permit valid influences at the termination of the experiment and which minimize in some sense the number of times the inferior treatment is used during the trial.

Assume that $x_1, x_2, \ldots (y_1, y_2, \ldots)$ are independent and normally distributed with mean μ_1 (μ_2) and variance 1. Let $\delta = (\mu_1 - \mu_2)/2^{1/2}$ and for definiteness suppose that a large response is desirable. Then treatment A is superior to treatment B if $\delta > 0$ while the converse holds if $\delta < 0$. Let $\bar{x}_k = k^{-1} \sum_1^k x_i$, $\bar{y}_n = n^{-1} \sum_1^n y_j$, and $Z_{k,n} = 2^{1/2} kn(\bar{x}_k - \bar{y}_n)/(k + n)$. Since problems of inference about δ are invariant under a common change of location for both the x and y populations, it is reasonable to restrict consideration to invariant procedures, i.e. those depending on the x's and y's only through the values of $u_i = x_i - x_1$, $i = 1, 2, \ldots$ and $v_j = y_j - x_1, j = 1, 2, \ldots$ or equivalently only on $Z_{k,n}$, $k, n \geq 1$. (See Lehmann, 1959, Chapter 6 or Cox and Hinkley, 1974, Chapter 5 for a discussion of the concept of invariance.)

The following fundamental result shows that if one considers only procedures defined in terms of $Z_{k,n}$ and $kn/(k + n)$, then the problem of inference and the selection of an allocation rule can be treated separately. One can use the results of Chapter IV to choose, say, a test of $H_0\colon \delta = 0$ against $H_1\colon \delta \neq 0$ under the simplifying assumption of pairwise allocation. Then one can select an allocation rule.

Let $\mathscr{E}_{k,n}$ denote the class of events which can be defined in terms of $Z_{i,j}$ or equivalently u_i, v_j for $i = 1, 2, \ldots, k, j = 1, 2, \ldots, n$. Let $W(t)$ be Brownian motion with drift δ and $\mathscr{E}(t)$ the class of events defined by $W(s)$, $s \leq t$. Put $t_{k,n} = 2kn/(k + n)$. Let $\mathscr{L}(X)$ denote the probability law of the random variable X.

Theorem 6.19. *For arbitrary* $k, n = 1, 2, \ldots$ *and* $i, j = 0, 1, \ldots$

$$\mathscr{L}(Z_{k+i,n+j} - Z_{k,n} | \mathscr{E}_{k,n}) = \mathscr{L}[W(t_{k+i,n+j}) - W(t_{k,n}) | \mathscr{E}(t_{k,n})].$$

A proof of Theorem 6.19 is given at the end of this section.

An important consequence of Theorem 6.19 is that if one uses an allocation rule which on the basis of the observations $Z_{i,j}$, $i + j = 2, 3, \ldots, k + n$ (or equivalently u_i, $i = 1, \ldots, k$, v_j, $j = 1, 2, \ldots, n$) decides whether the next observation is to be $Z_{k+1,n}$ or $Z_{k,n+1}$, there exists an "isomorphic" rule which on the basis of the values of $W(t_{i,j})$, $i + j = 2, 3, \ldots, k + n$ selects the next time, $t_{k+1,n}$ or $t_{k,n+1}$ at which $W(t)$ is to be observed, in such a way that the processes $\{Z_{k,n}, 2kn/(k + n), k + n = 2, 3, \ldots\}$ and $\{W(t_{k,n}), t_{k,n}, k + n = 2, 3, \ldots\}$ have the same joint distribution. In other words, an invariant allocation rule can be thought of as a random sequence $\tau_1 < \tau_2 < \cdots$ of times at which Brownian motion is to be observed with τ_{j+1} depending on the values of $W(\tau_1), \ldots, W(\tau_j)$.

Now consider a fixed sequential test for the drift of a Brownian motion process $W(t)$, e.g. a repeated significance test. Suppose that the stopping rule of the test is defined by the first time the process $(t, W(t))$ leaves some region in $[0, \infty) \times (-\infty, \infty)$ with a continuous boundary. Since $W(\cdot)$ is continuous and $(t_{k+1,n} - t_{k,n}) \vee (t_{k,n+1} - t_{k,n}) \leq 2$, with high probability the "observed" process $(t_{k,n}, W(t_{k,n}))$ must leave the region at about the same time and the same place. This means that the characteristics of the statistical procedure and the distribution of $t_{k,n}$ upon termination are to a large extent independent of the allocation rule used and can be computed relative to any convenient rule, e.g. pairwise allocation. (An obvious exception occurs if $t_{k,n}$ can remain bounded, i.e. if one treatment is allocated a bounded number of times in an infinitely long trial. Then it may happen that $t_{k,n}$ never reaches the value t at which the test stops. To rule out this possibility we assume that $k \wedge n \to \infty$ as $k + n \to \infty$.)

To be more specific, consider pairwise allocation and a repeated significance test defined by the stopping rule

$$T = \inf\{n: n \geq m_0, |Z_{n,n}| \geq bn^{1/2}\}.$$

Stop sampling at $\min(T, m)$, and reject H_0: $\delta = 0$ if $T \leq m$ or $T > m$ and $|Z_{m,m}| \geq cm^{1/2}$. For an invariant allocation rule there exists the analogous procedure: stop sampling at

$$(K, N) = \inf\{(k, n): |Z_{k,n}| \geq bt_{k,n}^{1/2}\} \tag{6.20}$$

or when $t_{k,n}$ first exceeds m, whichever occurs first, and reject H_0 if $t_{K,N} \leq m$ or $t_{K,N} > m$ and the terminal value of $|Z_{k,n}|$ exceeds $cm^{1/2}$. These two tests have roughly the same power function, and

$$E_\delta\{t_{K,N} \wedge m\} \cong E_\delta(T \wedge m). \tag{6.21}$$

In this way we obtain a family of tests whose power functions are essentially independent of the allocation rule used, so we can consider various allocation rules in an attempt to minimize the expected number of assignments of the inferior treatment.

The approximate equality (6.21) has two important consequences. Let $\tilde{K}(\tilde{N})$ denote the total number of allocations of treatment A (B), so $t_{\tilde{K},\tilde{N}} \cong t_{K,N} \wedge m$. (i) Since $\tilde{K}\tilde{N}/(\tilde{K} + \tilde{N}) \leq \min(\tilde{K}, \tilde{N})$ with equality only if $\max(\tilde{K}, \tilde{N}) = \infty$, it follows from (6.21) that

(6.22) $$\min(E_\delta(\tilde{K}), E_\delta(\tilde{N})) \geq \tfrac{1}{2}E_\delta(T \wedge m),$$

and there can be approximate equality in (6.22) only if $\max(E_\delta(\tilde{K}), E_\delta(\tilde{N}))$ is extremely large. (ii) Since $x(1-x) \leq \frac{1}{4}$ for $0 \leq x \leq 1$ with equality only at $x = \frac{1}{2}$, (6.21) implies that

(6.23) $$E_\delta(\tilde{K} + \tilde{N}) \geq 4E_\delta[\tilde{K}\tilde{N}/(\tilde{K} + \tilde{N})] \cong 2E_\delta(T \wedge m),$$

and the inequality is an approximate equality if and only if $\tilde{K} \cong \tilde{N}$. To summarize (i) and (ii), any deviation from pairwise allocation increases the total expected sample size, and no allocation rule can reduce the expected number of assignments of the inferior treatment below one half the expected number using pairwise allocation. This latter bound could in principle be approximately achieved only by making the expected number of times the superior treatment is used infinitely large.

Proof of Theorem 6.19. It is readily verified that

$$Z_{k+i,n+j} - Z_{k,n} = \frac{(n+j)\sum_{k+1}^{k+i} x_r - (k+i)\sum_{n+1}^{n+j} y_r}{k+n+i+j} + \frac{(kj-ni)\left(\sum_1^k x_r + \sum_1^n y_r\right)}{(k+n)(k+n+i+j)}.$$

Each of the two terms on the right hand side of this expression is uncorrelated with u_r, $r \leq k$, and v_s, $s \leq n$. Since all these random variables are jointly normally distributed, it follows that the conditional distribution of $Z_{k+i,n+j} - Z_{k,n}$ given $\mathscr{E}_{k,n}$ is normal with mean

$$\frac{i(n+j)\mu_1 - j(k+i)\mu_2}{k+n+i+j} + \frac{(kj-ni)(k\mu_1 + n\mu_2)}{(k+n)(k+n+i+j)}$$
$$= \frac{[in(n+j) + kj(k+i)](\mu_1 - \mu_2)}{(k+n)(k+n+i+j)} = (t_{k+i,n+j} - t_{k,n})\delta$$

and variance

$$\frac{(n+j)^2 i + (k+i)^2 j}{(k+n+i+j)^2} + \frac{(kj-ni)^2(k+n)}{[(k+n)(k+n+i+j)]^2} = t_{k+i,n+j} - t_{k,n}.$$

This is also the conditional distribution of $W(t_{k+i,n+j}) - W(t_{k,n})$ given $\mathscr{E}(t_{k,n})$, as claimed. □

4. Loss Function and Allocation

To evaluate and compare allocation rules it is helpful to specify a loss function which measures the relative costs of using treatment A or B as a function of $\delta = (\mu_1 - \mu_2)/2^{1/2}$. Given δ, assume that the cost of allocating treatment A is $g(\delta)$ and the cost of allocating treatment B is $h(\delta)$. A specific example considered in more detail below is

(6.24) $$g(\delta) = h(-\delta) = \begin{cases} 1 & \text{if } \delta > 0 \\ 1 + d|\delta| & \text{if } \delta < 0 \end{cases}.$$

This cost function involves a unit experimental cost per observation, whichever treatment is used, and an ethical cost proportional to the difference in treatment effects if the inferior treatment is used. The expected total cost of sampling is

(6.25) $$g(\delta)E_\delta(\tilde{K}) + h(\delta)E_\delta(\tilde{N}),$$

where $\tilde{K}(\tilde{N})$ denotes the number of times treatment A (B) is allocated during the experiment.

The overall risk function is the sum of (6.25) and the risk of an incorrect decision when the trial is finished. A test defined as in the preceding section has a power function which is approximately the same for essentially all invariant allocation rules. Hence in choosing an allocation rule one can ignore the component of the risk function associated with the terminal decision and attempt to minimize (6.25).

Simple algebra shows that (6.25) equals

(6.26) $$(h + g)E_\delta[\tilde{K}\tilde{N}/(\tilde{K} + \tilde{N})] + E_\delta\{[\tilde{K}\tilde{N}/(\tilde{K} + \tilde{N})][g\tilde{K}/\tilde{N} + h\tilde{N}/\tilde{K}]\}.$$

Moreover, by the results of the preceding section the first expectation in (6.26) is essentially independent of the allocation rule, so it suffices to minimize the second. By calculus $gx + hx^{-1} \geq 2(gh)^{1/2}$ with equality if and only if $x = (h/g)^{1/2}$. Hence (6.26) always exceeds

(6.27) $$(h^{1/2} + g^{1/2})^2 E_\delta[\tilde{K}\tilde{N}/(\tilde{K} + \tilde{N})],$$

and there would be equality if and only if

(6.28) $$P_\delta\{\tilde{K}/\tilde{N} = [h(\delta)/g(\delta)]^{1/2}\} = 1.$$

Since δ is unknown, it is impossible to achieve (6.29), but one can hope to estimate δ sequentially by $Z_{k,n}/t_{k,n}$ and obtain something approximating (6.28). An obvious possibility after observing $x_1, \ldots, x_k$ and $y_1, \ldots, y_n$, is to allocate the next observation to treatment A (i.e. observe x_{k+1}) if and only if

(6.29) $$k/n < [h(Z_{k,n}/t_{k,n})/g(Z_{k,n}/t_{k,n})]^{1/2}.$$

The law of large numbers implies that (6.28) will hold approximately in large samples, and to the extent it does one has the approximations

$$E_\delta(\tilde{K}) = E_\delta\{[\tilde{K}\tilde{N}/(\tilde{K} + \tilde{N})](1 + \tilde{K}/\tilde{N})\} \cong [1 + (h/g)^{1/2}]E_\delta\{\tilde{K}\tilde{N}/(\tilde{K} + \tilde{N})\},$$
(6.30)

(6.31) $$E_\delta(\tilde{N}) \cong [1 + (g/h)^{1/2}]E_\delta\{\tilde{K}\tilde{N}/(\tilde{K} + \tilde{N})\};$$

and by (6.26) the risk (6.25) is approximately the lower bound (6.27).

Table 6.1 gives a numerical example of a repeated significance test with pairwise allocation and with the allocation rule (6.29), where g and h are defined by (6.24) with $d = 20$. Figures given in parentheses are the results of a

Table 6.1. Modified Repeated Significance Test with Adaptive Allocation. Test Parameters: $b = 2.95$; $c = 2.07$; $m_0 = 7$; $m = 49$

δ	Power		$E_\delta(\tilde{K})$		$E_\delta(\tilde{N})$		$2E_\delta[\tilde{K}\tilde{N}/(\tilde{K}+\tilde{N})]$		Risk	
.6	.98		22.9		22.9		22.9		321	
	.98	(.98)	53	(50)	15	(15)	22.9	(23)	243	(248)
.4	.78		37.2		37.2		37.2		372	
	.78	(.80)	74	(69)	25	(25)	37.2	(36)	298	(290)
.2	.28		—		—		—	(46)	—	(278)
	.28	(.28)	—	(71)	—	(36)	—	(46)	—	(254)
.0	.05		—		—		—		—	
	.05	(.06)	—	(53)	—	(53)	—	(49)		

First row in each cell is for pairwise allocation; second row is for allocation by (6.29), where g and h are given by (6.24) with $d = 20$. Monte Carlo estimates are in parentheses.

2500 repetition Monte Carlo experiment. Other results are based on the approximations of Chapter IV in the case of pairwise allocation and on those suggested in (6.21), (6.27), (6.30), and (6.31) for the allocation rule (6.29).

The Monte Carlo estimates of the power function and of $E_\delta[\tilde{K}\tilde{N}/(\tilde{K}+\tilde{N})]$ are consistent with the results of Section 3 which imply that these quantities are essentially the same for both allocation schemes. The approximations (6.27), (6.30), and (6.31) are crude, but provide a useful picture of the effect of the allocation rule (6.29).

One should not put too much emphasis on the results of Table 6.1, which depend heavily on the choice of d. Nevertheless, the decrease in risk effected by the allocation rule (6.29), while not as large as the decrease in going from a fixed sample to a sequential test, is sufficient to make the procedure seem attractive. Since the model under discussion is artificially simple, the following comments are designed to indicate some disadvantages of adaptive allocation rules in practical situations.

These disadvantages include (i) the fact that the allocation rule (6.29) is non-randomized, (ii) questionable performance for group sequential tests, where one cannot continuously change the allocation ratio, or for survival data, where many treatment assignments must be made before responses start becoming available, and (iii) questionable performance if the population contains a large number of small strata and some minimum amount of balance seems advisable in each stratum. The allocation rule (6.29) can be modified to take these factors into account. For example, one can define a randomized version of (6.29) by making $[h(Z_{k,n}/t_{k,n})/g(Z_{k,n}/t_{k,n})]^{1/2}$ the odds in favor of allocating treatment A to the next patient. (This modification would control selection bias to some extent, but it would not provide for a randomization test.) Individually each modification results in some deterioration in the theoretical performance of the allocation rule. If all the modifications are used at the same time (as probably would be appropriate in an actual clinical trial), the deterioration can be quite substantial—perhaps to the point of making the adaptive allocation rules more trouble than they are worth. See Siegmund (1983) for a more complete discussion and a Monte Carlo experiment.

An additional consideration is the increase in expected total sample size which the sampling rule (6.29) requires in order to reduce the number of allocations of the inferior treatment. If the comparison is between a new treatment and a standard, and the new treatment is better, then an increase in the total sample size resulting in a longer trial is contrary to the interests of patients not in the trial, who presumably continue to receive the standard treatment. A possible modification of (6.24) to incorporate this consideration into the cost structure is the following. Suppose that treatment A is experimental and treatment B is standard. Let

$$g(\delta) = \begin{cases} 1 + d'\delta, & \delta \geq 0 \\ 1 + d|\delta|, & \delta < 0 \end{cases}$$

and

$$h(\delta) = \begin{cases} 1 + (d + d')\delta, & \delta \geq 0 \\ 1, & \delta < 0. \end{cases}$$

The meaning of the constant term 1 and the constant d are the same as before. The term $d'\delta$ measures the cost to patients not in the trial of prolonging the trial by taking another observation when the experimental drug is superior by an amount δ. It seems fairly clear from (6.29) that unless d' is substantially smaller than d, when $\delta > 0$ the optimal allocation is not sufficiently different from pairwise to justify the nuisance of adaptive allocation. (The situation is different when $\delta < 0$, but with this asymmetric cost structure it may be reasonable to use a stopping rule which terminates whenever H_0: $\delta = 0$ appears to be true, and then the allocation rule for $\delta < 0$ plays a minor role. See III.8 and Problem 4.1.)

An overall conclusion seems to be that the potential advantages of adaptive allocation are probably less than indicated by our theoretical discussion, and hence the requirements of a particular trial should be thoroughly considered before adopting the approach considered here.

Even if we do not seriously consider adaptive allocation, Theorem 6.19 is still an interesting result, because it indicates that the results of Chapter IV, where we considered only pairwise allocation, are approximately valid in a two population context.

Problems

6.1. Adapt the discussion of Section 1 to continuous response variables.

6.2. Let $x_1, x_2, \ldots$ be independent random variables with $P\{x_i = +1\} = P\{x_i = -1\} = \frac{1}{2}$, and put $S_n = x_1 + \cdots + x_n$, $\tau_0 = \inf\{n\colon n \geq 1, S_n = 0\}$. Let $u_j = P\{S_j = 0\}$.
 (a) Use the results of Problem 3.3(b) to show that $P\{\tau_0 > 2k\} = u_{2k}$ $(k = 1, 2, \ldots)$.
 (b) Let $L_{2k} = \sup\{j\colon j \leq 2k, S_j = 0\}$. Use (a) to show that $P\{L_{2k} = 2j\} = u_{2j}u_{2k-2j}$ and conclude that $\sum_{j=0}^{k} \binom{2j}{j}\binom{2k-2j}{k-j} = 2^{2k}$.

6.3. Consider the likelihood ratio statistic (5.7) for testing H_0: $p_1 = p_2$. Assume that $p_1 - p_2 = \Delta m^{-1/2}$. Suppose $m_0, m, n_1, n_2 \to \infty$ with m_0/m and $n_1/(n_1 + n_2)$ bounded away from 0 and 1. Show that the signed square root of the likelihood ratio statistic behaves like Brownian motion in the time scale of $n_1 n_2/(n_1 + n_2)$ for $m_0 \le n_1 n_2/(n_1 + n_2) \le m$. Hence the results of Section 3 are asymptotically applicable to the repeated likelihood ratio test of V.2.

6.4. Let $u_i = x_i - x_1$ and $v_j = y_j - x_1$ as in Section 3. Let Q_δ be the joint distribution of $u_1, u_2, \ldots, v_1, v_2, \ldots$, and $Q = \int_{-\infty}^{\infty} Q_\delta d\delta/(2\pi)^{1/2}$. Find the likelihood ratio of $u_1, \ldots, u_k, v_1, \ldots, v_n$ under Q_δ relative to Q_0 and under Q relative to Q_0. Let (K, N) be defined by (6.21). Use the argument of IV.8 to show (without reference to Brownian motion) that if one neglects excess over the boundary,

$$P_0\{t_{K,N} \le m\} \cong 2[1 - \Phi(b)] + b\phi(b)\log(m/m_0).$$

CHAPTER VII

Interval Estimation of Prescribed Accuracy

1. Introduction and Heuristic Stopping Rule

For a fixed sample problem the accuracy of an estimator of a parameter θ typically depends on the unknown value of θ and perhaps also on the value of an unknown nuisance parameter λ. To achieve estimators of prescribed accuracy, it seems natural to proceed sequentially, to get some preliminary idea of the value of (θ, λ) and use the prelininiminary information to determine the final sample size.

Procedures of this general character have been studied since Stein's (1945) paper on a two stage confidence interval of prescribed width for a normal mean. In this chapter we describe the results of a somewhat different approach initiated by Anscombe (1953) and extended by Chow and Robbins (1965). In spite of the naturalness of the problem for theoretical inquiry, it seems to have had little effect on statistical practice. An exception may be simulation studies, where the use of sequential sampling designs usually poses no problems of implementation. In this case, however, prescribed proportional accuracy may be more sensible than prescribed absolute accuracy.

Although the problem considered in this chapter is different in spirit from the rest of the book, the mathematical methods of solution have a great deal in common. There is in addition a connection with the time change argument of III.9. We begin with a general heuristic argument and then discuss two examples which exhibit particularly interesting special properties.

We shall use the notation and assumptions of III.9: $x_1, x_2, \ldots$ are independent with probability density function $f(x; \theta, \lambda)$; $l(n; \theta, \lambda)$ is the log likelihood function of $x_1, \ldots, x_n$; and $i(n; \theta, \lambda)$ is the matrix of second partial derivatives of $-l(n; \theta, \lambda)$. If $(\hat{\theta}_n, \hat{\lambda}_n)$ denotes the maximum likelihood estimator of (θ, λ) based on $x_1, \ldots, x_n$, then under general regularity conditions for large n

$$\{i^*(n;\theta,\lambda)\}^{1/2}(\hat{\theta}_n - \theta)$$

is approximately normally distributed with mean 0 and variance 1 (Cox and Hinkley, 1974, p. 296). Here

$$i^*(n;\theta,\lambda) = i_{\theta\theta}(n;\theta,\lambda) - i_{\theta\lambda}^2(n;\theta,\lambda)/i_{\lambda\lambda}(n;\theta,\lambda).$$

Hence an approximate $(1-2\gamma)$ 100% confidence interval for θ is

$$\hat{\theta}_n \pm z_\gamma/[i^*(n;\hat{\theta}_n,\hat{\lambda}_n)]^{1/2}, \tag{7.1}$$

where $\gamma = 1 - \Phi(z_\gamma)$. If we seek a confidence interval of length smaller than $2d$, then (7.1) suggests sampling sequentially and using the stopping rule

$$T = T(d,\gamma) = \inf\{n: n \geq m_0, i^*(n;\hat{\theta}_n,\hat{\lambda}_n) \geq (z_\gamma d^{-1})^2\}. \tag{7.2}$$

Moreover, since $i(n;\theta,\lambda)$ is a matrix of sums of independent, identically distributed random variables, it follows from the law of large numbers that

$$n^{-1}i^*(n;\theta,\lambda) \to I^*(\theta,\lambda) = I_{\theta\theta} - I_{\theta\lambda}^2/I_{\lambda\lambda} \tag{7.3}$$

with probability one (see III.9 for notation). Hence it seems plausible from (7.2) that as $d \to 0$

$$P_{\theta,\lambda}\{d^2T(d,\lambda) \to z_\gamma^2/I^*(\theta,\lambda)\} = 1, \; E_{\theta,\lambda}\{d^2T(d,\lambda)\} \to z_\gamma^2/I^*(\theta,\lambda). \tag{7.4}$$

Also by a Taylor series expansion similar to (3.59), but with the left hand side evaluated at $(\hat{\theta}_n, \hat{\lambda}_n)$, and an application of Theorem 2.40

$$P_{\theta,\lambda}\{|\hat{\theta}_T - \theta| \leq d\} \to 1 - 2\gamma. \tag{7.5}$$

Hence asymptotically for small d, the interval $\hat{\theta}_T \pm d$ has approximately the desired coverage probability $1 - 2\gamma$.

By (7.4) this is accomplished with an asymptotically minimal expected sample size in the sense that by (7.3), (7.1) can define an interval of length $\leq 2d$ only if $n \geq z_\gamma^2/d^2 I^*(\theta,\lambda)$ up to terms of smaller order than d^{-2}.

2. Example—The Normal Mean

The heuristic discussion of Section 1 is quite general, but the asymptotic theory indicated by (7.4) and (7.5) is rather crude. In the special case of a normal mean (with unknown variance as nuisance parameter) sharper results can be obtained, and these in turn suggest a slight modification of the definition (7.2).

Suppose $x_1, x_2, \ldots, x_n, \ldots$ are independent and normally distributed with unknown mean μ and variance σ^2, and we seek an approximate confidence interval for μ of length at most $2d$, regardless of the value of σ^2. Then $\hat{\mu}_n = \bar{x}_n = n^{-1}\sum_{i=1}^n x_i$, $\hat{\sigma}_n^2 = n^{-1}\sum_{i=1}^n (x_i - \bar{x}_n)^2$, and $i^*(n;\mu,\sigma^2) = n\sigma^{-2}$, so a slight generalization of (7.2) is

$$T = \inf\{n: n \geq m_0, \sum_{i=1}^n (x_i - \bar{x}_n)^2 \leq a_n d^2 n^2/z_\gamma^2\}, \tag{7.6}$$

where $\{a_n\}$ is a sequence of positive constants converging to 1 as $n \to \infty$. From the orthogonal (Helmert) transformation

$$\begin{pmatrix} 2^{-1/2} & -2^{-1/2} & 0 & 0 & \cdots & 0 & 0 \\ (2\cdot 3)^{-1/2} & (2\cdot 3)^{-1/2} & -2(2\cdot 3)^{-1/2} & 0 & \cdots & 0 & 0 \\ \vdots & & & & & & \\ [(n-1)n]^{-1/2} & \cdots & \cdots & \cdots & \cdots & [(n-1)n]^{-1/2} & -(n-1)[(n-1)n]^{-1/2} \\ n^{-1/2} & n^{-1/2} & \cdots & \cdots & \cdots & \cdots & n^{-1/2} \end{pmatrix}$$

applied to $x_1, \ldots, x_n$ (regarded as a column vector) or by direct calculation, one may easily verify that $y_{i-1} = [(i-1)/i]^{1/2}(x_i - \bar{x}_{i-1})$, $i = 2, 3, \ldots, n$ are independent and normally distributed with mean 0 and variance σ^2, that $y_1, \ldots, y_{n-1}$ are independent of $\bar{x}_n$, and that $\sum_{i=1}^{n}(x_i - \bar{x}_n)^2 = \sum_1^{n-1} y_i^2$. Hence T defined by (7.6) can be re-expressed as

$$T = \inf\{n : n \geq m_0, \sigma^{-2} \sum_1^{n-1} y_i^2 \leq a_n n^2/n_0\}, \tag{7.7}$$

where $n_0 = (\sigma z_\gamma/d)^2$. (Note that n_0 is the fixed sample size required to obtain a confidence interval of width $2d$ if σ^2 were known.) Moreover, since $\{T = n\}$ and $\bar{x}_n$ are independent

$$P_{\mu,\sigma^2}\{|\hat{\mu}_T - \mu| \leq d \,|\, T\} = 2\Phi[z_\gamma(T/n_0)^{1/2}] - 1.$$

It follows that the coverage probability of $\hat{\mu}_T \pm d$ is

$$P_{\mu,\sigma^2}\{|\hat{\mu}_T - \mu| \leq d\} = E_{\mu,\sigma^2}\{\Psi(z_\gamma^2 T/n_0)\}, \tag{7.8}$$

where $\Psi(x) = 2\Phi(x^{1/2}) - 1$, so in the normal case the rather complicated probability on the left hand side of (7.5) involves only the marginal distribution of T.

For this problem it is fairly straightforward to give a completely rigorous justification of (7.4) and (7.5), which depends on the fact that $n_0^{-1} T \to 1$ with probability one as $d \to 0$ (see Problem 7.1). Here we obtain a second order correction for these approximations, which requires that we also know the asymptotic behavior of $E_{\mu,\sigma^2}(T - n_0)^2/n_0$ as $d \to 0$.

Assume that for some $a_0 > 0$,

$$a_n = 1 - n^{-1}a_0 + o(n^{-1}) \qquad (n \to \infty). \tag{7.9}$$

Theorem 7.10. *Assume condition* (7.9). *If* $m_0 \geq 4$, *then*

$$E_{\mu,\sigma^2}(T) = n_0 + \rho + a_0 - 3 + o(1) \qquad \text{as } n_0 \to \infty, \tag{7.11}$$

and for fixed $0 < \gamma < \frac{1}{2}$

$$\begin{aligned} &P_{\mu,\sigma^2}\{|\hat{\mu}_T - \mu| \leq d\} \\ &= 1 - 2\gamma + n_0^{-1}[z_\gamma^2 \Psi'(z_\gamma^2)(\rho + a_0 - 3) + z_\gamma^4 \Psi''(z_\gamma^2)] + o(n_0^{-1}) \end{aligned} \tag{7.12}$$

as $\sigma/d \to \infty$. *Here* ρ *can be computed numerically and equals approximately* .82.

Remarks. The constant ρ arises because of the excess of the random walk $\sigma^{-2}\sum_1^{n-1} y_i^2$ under the boundary at the time T of first passage. Its existence and calculation are discussed in Chapter IX, where it is shown to be related to $\rho(\mu)$ defined in (4.36). The following informal argument in support of Theorem 7.10 fails to explain the restriction placed on m_0, which unfortunately is a rather technical matter. To see what can go wrong if $m_0 = 3$, note that since $\sigma^{-2}(y_1^2 + y_2^2)$ has a negative exponential distribution with expectation 2,

$$P_{\mu,\sigma^2}\{T = 3\} = 1 - \exp(-9a_3/2n_0) \sim 9a_3/2n_0,$$

so $E_{\mu,\sigma^2}(T - n_0; T = 3)$ does not converge to 0. This means that although T is close to n_0 with probability close to one, small values of T still make a nonvanishing contribution to $E_{\mu,\sigma^2}(T)$. See IX.4 for a more detailed discussion.

Since the a_0 appearing in (7.9) is at our disposal, it seems reasonable to choose it to make the coefficient of n_0^{-1} in (7.12) vanish. It is easily verified that with this choice the right hand side of (7.11) becomes $n_0 + \frac{1}{2}(z_\gamma^2 + 1) + o(1)$. It would be interesting to know if this is asymptotically a minimum up to the term $o(1)$, among those procedures whose coverage probability is $1 - 2\gamma + o(d^2)$.

Proof of Theorem 7.10. Neglecting the excess under the boundary, we obtain from (7.7) and Wald's identity (2.18) the asymptotic equality

$$E_{\mu,\sigma^2}(T - 1) = \sigma^{-2}E_{\mu,\sigma^2}\left(\sum_1^{T-1} y_i^2\right) \cong n_0^{-1}E_{\mu,\sigma^2}(a_T T^2). \tag{7.13}$$

In the range $n \geq \frac{1}{2}n_0$ it is easy to see from (7.9) that

$$\begin{aligned} n_0^{-1}a_n n^2 = {} & n_0 + 2n + n_0^{-1}(n - n_0)^2 - a_0 + o(1) \\ & + a_0(n^{-1}n_0 - 1) + o\{n^{-1}n_0^{-1}(n - n_0)^2\}. \end{aligned}$$

Hence by (7.13)

$$\begin{aligned} E_{\mu,\sigma^2}(T) \cong {} & n_0 + a_0 - 1 - n_0^{-1}E_{\mu,\delta^2}\{(T - n_0)^2; T \geq \tfrac{1}{2}n_0\} \\ & - n_0^{-1}E_{\mu,\sigma^2}\{a_T T^2; T < \tfrac{1}{2}n_0\} + 2E_{\mu,\sigma^2}(T; T < \tfrac{1}{2}n_0) \\ & - (n_0 + a_0)P_{\mu,\sigma^2}\{T < \tfrac{1}{2}n_0\} + o(1) \\ & - E_{\mu,\sigma^2}\{a_0(T^{-1}n_0 - 1) + T^{-1}n_0^{-1}o[(T - n_0)^2]; T \geq \tfrac{1}{2}n_0\}. \end{aligned} \tag{7.14}$$

Since y_i has mean 1 and variance 2, an argument along the lines of Problem 3.11 shows that as $d \to 0$, T is asymptotically normally distributed with mean n_0 and variance $2n_0$. Hence the first expectation on the right hand side of (7.14) should converge to 2 and the last five terms should be negligible. This yields (7.11) except for the constant ρ, which comes from analyzing the excess under the boundary.

To see (7.12), observe that by (7.8) and the boundedness of Ψ

$$P_{\mu,\sigma^2}\{|\hat{\mu}_T - \mu| \leq d\} = E_{\mu,\sigma^2}\{\Psi(z_\gamma^2 T/n_0); T \geq \tfrac{1}{2}n_0\} + o(n_0^{-1}). \tag{7.15}$$

For $n \geq \frac{1}{2}n_0$,

$$\Psi(z_\gamma^2 n/n_0) = \Psi(z_\gamma^2) + n_0^{-1}(n - n_0)z_\gamma^2\Psi'(z_\gamma^2) + \tfrac{1}{2}n_0^{-2}(n - n_0)^2 z_\gamma^4\Psi''(\zeta z_\gamma^2),$$

where $|\zeta - 1| \le |n_0^{-1}n - 1|$. Since $\Psi''(\zeta z_\gamma^2)$ is bounded provided $n \ge \frac{1}{2}n_0$, substituting this Taylor expansion into (7.15) and arguing as above yields (7.12). See IX.4 for a more precise argument. □

3. Example—The Log Odds Ratio

Suppose that for $i = 1, 2, x_{i1}, x_{i2}, \ldots$ are independent Bernoulli variables with success probability p_i (and put $q_i = 1 - p_i$). Assume also that $\{x_{1j}\}$ and $\{x_{2k}\}$ are independent. Let $\theta = \log(p_1 q_2/p_2 q_1)$ be the log odds ratio, which after n observations from each population is estimated by $\hat{\theta}_n = \log(s_{1n} f_{2n}/s_{2n} f_{1n})$, where $s_{in} = x_{i1} + \cdots + x_{in}$ and $f_{in} = n - s_{in}$ are, respectively, the number of successes and the number of failures in the first n trials on the ith population. For the nuisance parameter λ taken to be $\lambda = \log(p_2/q_2)$, say, a calculation gives $i^*(n; \theta, \lambda) = n/\{(p_1 q_1)^{-1} + (p_2 q_2)^{-1}\}$, so an approximate $(1 - 2\gamma)$ 100% large sample confidence interval for θ is

$$\hat{\theta}_n \pm z_\gamma (n/s_{1n} f_{1n} + n/s_{2n} f_{2n})^{-1/2}, \tag{7.16}$$

where as usual $1 - \Phi(z_\gamma) = \gamma$.

The parameter θ has several attractive features. One is that it provides a meaningful comparison of p_1 and p_2 even when both are near 0 or 1. However, in these boundary cases the convergence to normality required to justify use of (7.16) is very slow.

If observations are taken in pairs, one from each population, it is easy to see that (7.2) becomes

$$T = \inf n\colon \left\{n \ge m_0, n\left(\frac{1}{s_{1n} f_{1n}} + \frac{1}{s_{1n} f_{2n}}\right) \le (z_\gamma^{-1} d)^2\right\} \tag{7.17}$$

and to verify (7.4) and (7.5). However, unlike the fixed sample case sketched above the convergence now is *uniform* in p_i for $0 < p_1, p_2 < 1$.

Theorem 7.18. *Let T be defined by* (7.17) *and put $c = (z_\gamma d^{-1})^2$. As $d \to 0$, uniformly in $0 < p_1, p_2 < 1$,*

$$E_{p_1, p_2}(T)/\{c[(p_1 q_1)^{-1} + (p_2 q_2)^{-1}]\} \to 1$$

and

$$P_{p_1, p_2}\{c^{1/2}(\hat{\theta}_T - \theta) \le x\} \to \Phi(x).$$

There is also a version of Theorem 7.18 for a single Bernoulli population. Although this case is not as interesting statistically, except perhaps when it arises from the paired comparison model (5.1), it provides the basic tools and illustrates the basic features of the problem. Hence the detailed discussion given below is restricted to this somewhat simpler case.

Let $x_1, x_2, \ldots$ be independent Bernoulli variables equal to 1 or 0 with probabilities p and $q = 1 - p$ respectively. Let $s_n = x_1 + \cdots + x_n$ and $f_n = n - s_n$. For $c > 0$ define

$$T = \inf\{n: n^{-1}s_n f_n \geq c\}. \tag{7.19}$$

Theorem 7.20. *For T defined by* (7.19), *as $c \to \infty$, uniformly in $0 < p < 1$,*

$$pqE_p(T)/c \to 1 \tag{7.21}$$

and

$$P_p\{c^{1/2}[\log(s_T/f_T) - \log(p/q)] \leq x\} \to \Phi(x). \tag{7.22}$$

The proof of Theorem 7.20 is given following three lemmas, the first of which implies (7.21). Repeated use is made of the algebraic identity

$$n^{-1}s_n f_n = (q - p)(s_n - np) + npq - (s_n - np)^2/n. \tag{7.23}$$

Lemma 7.24. *For T defined in* (7.19)

$$c \leq pqE_p(T) < (c + 1)/[1 - (4c)^{-1}].$$

PROOF. By (7.19), (7.23), and the obvious lower bound $T \geq 4c$,

$$\begin{aligned} c &\leq (q - p)(s_T - pT) + qpT < c + 1 + (s_T - pT)^2/T \\ &\leq c + 1 + (s_T - pT)^2/(4c). \end{aligned}$$

By Wald's identities (2.18) and Problem 2.11

$$E_p(s_T - pT) = 0 \text{ and } E_p(s_T - pT)^2 = pqE_p(T).$$

Hence

$$c \leq pqE_p(T) < (c + 1) + pqE_p(T)/(4c),$$

which can be solved to yield the lemma. □

Lemma 7.25. *For T defined in* (7.19) *there exists a c_0 such that for all $c \geq c_0$*

$$(pq)^2E_p(T - c/pq)^2 \leq 7c.$$

PROOF. Squaring (7.23) yields

$$\begin{aligned} (n^{-1}s_n f_n - c)^2 = {} & (q - p)^2(s_n - pn)^2 + (pq)^2(n - c/pq)^2 + (s_n - np)^4/n^2 \\ & + 2\{(q - p)(s_n - pn)(pqn - c) - (q - p)(s_n - np)^3/n \\ & - (pqn - c)(s_n - np)^2/n\}. \end{aligned}$$

By the Schwarz inequality and Problem 2.11

$$|E_p\{(s_T - pT)(T - c/pq)\}| \leq \{pqE_p(T)E_p(T - c/pq)^2\}^{1/2}.$$

Hence since $(T^{-1}s_T f_T - c)^2 < 1$, another application of Problem 2.11 yields

$$\begin{aligned} 1 > &(q-p)^2 pqE_p(T) + (pq)^2 E_p(T - c/pq)^2 \\ &- 2pq|p-q|\{pqE_p(T)E_p(T-c/pq)^2\}^{1/2} \\ &- 2|q-p|pqE_p(T) - 2(pq)^2 E_p(T), \end{aligned}$$

which may be rearranged to give

$$\begin{aligned} &(pq)^2 E_p(T-c/pq)^2 - 2pq|p-q|\{pqE_p(T)E_p(T-c/pq)^2\}^{1/2} \\ &\quad + pq(q-p)^2 E_p(T) < 1 + 2pqE_p(T), \end{aligned}$$

or

$$|pq\{E_p(T-c/pq)^2\}^{1/2} - |p-q|[pqE_p(T)]^{1/2}| \leq \{1 + 2pqE_p(T)\}^{1/2}.$$

Taking account of the possibilities that the quantity inside the absolute value sign may be positive or negative, rearranging, squaring, using $(a+b)^2 \leq 2(a^2+b^2)$, and using Lemma 7.24 yield

$$\begin{aligned} (pq)^2 E_p(T-c/pq)^2 &\leq \{[pqE_p(T)]^{1/2} + [1 + 2pqE_p(T)]^{1/2}\}^2 \\ &\leq 2 + 6pqE_p(T) \leq 2 + 6(c+1)/[1-(4c)^{-1}]. \end{aligned}$$

This completes the proof. □

Lemma 7.26. *There exists a constant $k > 0$ such that for each $0 < \varepsilon < 1$ and $c \geq c_0$, where c_0 is defined as in Lemma 7.25,*

$$P_p\{|S_T - pT| \geq \varepsilon pqT\} \leq k/(\varepsilon^2 c).$$

Proof. Let $0 < \delta < 1$ and $n_0 = c/(pq)$. By Lemma 7.25 for all $c \geq c_0$

$$P_p\{|T - n_0| > \delta c/pq\} \leq (\delta c)^{-2}(pq)^2 E_p(T - c/pq)^2 \leq 7/\delta^2 c.$$

Hence by Problem 2.11 once again and Lemma 7.24

$$\begin{aligned} P_p\{|s_T - pT| > \varepsilon pqT\} &\leq 7/(\delta^2 c) + P_p\{|s_T - pT| > \varepsilon pqT, |T - n_0| \leq \delta c/pq\} \\ &\leq 7/(\delta^2 c) + P_p\{|s_T - pT| > \varepsilon(1-\delta)c\} \\ &\leq 7/\delta^2 c + E_p(s_T - pT)^2/[\varepsilon(1-\delta)c]^2 \\ &\leq 7/\delta^2 c + 2/[\varepsilon^2(1-\delta)^2 c]. \end{aligned}$$

□

Proof of Theorem 7.20. The convergence in (7.21) is an immediate consequence of Lemma 7.24. To prove (7.22) observe that by the mean value theorem

$$\begin{aligned} c^{1/2}[\log(s_T/f_T) - \log(p/q)] &= c^{1/2}[(s_T - pT)/pqT][pq/\zeta_T(1-\zeta_T)] \\ &= n_0^{1/2}[(s_T - pT)/(pq)^{1/2}T][pq/\zeta_T(1-\zeta_T)], \end{aligned}$$

where $|\zeta_T - p| \leq |T^{-1}s_T - p|$, and $n_0 = c/(pq)$. Hence it suffices to show that uniformly in $0 < p < 1$

$$P_p\{n_0^{1/2}(s_T - pT)/(pq)^{1/2}T \le x\} \to \Phi(x)$$

and

$$pq/\zeta_T(1 - \zeta_T) \to 1 \text{ in probability.}$$

The second statement follows easily from Lemma 7.26. With the help of Lemmas 7.25 and 7.26 the first can be obtained by careful scrutiny of the proof of Theorem 2.40 and the fact that $(s_{n_0} - pn_0)/(n_0 pq)^{1/2}$ is asymptotically normal uniformly in p as $c \to \infty$. □

Remarks. It is interesting to consider Theorem 7.18 as defining a random time change in the sense of III.9, particularly in comparison with the logistic example given there, which is a two population test in the special case that the z_i are indicators of population membership. One interpretation of Theorem 7.18 (more precisely for an appropriately strengthened Theorem 7.18) is that for large c, $c\hat{\theta}_T$ behaves like Brownian motion with drift θ at time c; and the approximation is uniform in θ. Hence $c\hat{\theta}_T$ could be used as a statistic for testing H_0: $\theta = 0$ by a repeated significance test in the time scale c. However, if $|\theta|$ is very large because one of the p_i is near 0 or 1, the time scale determined via (7.17) by the rate of increase of

$$\left[n\left(\frac{1}{s_{1n}f_{1n}} + \frac{1}{s_{2n}f_{2n}}\right)\right]^{-1}$$

grows slowly. The effect is that the data can indicate an obviously false null hypothesis, but the test does not terminate because the time scale has not advanced. The test of III.7 does not have the same obvious defect because the time change there is defined by the observed Fisher information under the null hypothesis. For the special case of two Bernoulli populations $\dot{l}(n; 0, \hat{\lambda}_{0,n}) = \frac{1}{2}(s_{1n} - s_{2n})$ and $i^*(n; 0, \hat{\lambda}_{0,n}) = (s_{1n} + s_{2n})(f_{1n} + f_{2n})/4n$. The time scale grows slowly only if *both* p_1 and p_2 are near 0 or near 1.

A second order approximation for $P_p\{|\hat{\theta}_T - \theta| \le d\}$ along the lines of (7.12) seems rather difficult to obtain—to some extent because of the discreteness of s_n. For T defined by (7.19) a second order approximation for $E_p(T)$ follows easily from Theorem 9.28, but it is nonuniform in p in the most dramatic way possible. As $c \to \infty$, for any p such that $(pq^{-1})^2$ is irrational

$$pqE_p(T) = c + \tfrac{1}{2}(1 - pq) + o(1). \tag{7.27}$$

Surprisingly, if $(pq^{-1})^2$ is rational, the corresponding result is different and much more complicated (see Lalley, 1981). Since the left hand side of (7.27) is a continuous function of p, the indicated asymptotic relation cannot be uniform in any p interval, no matter how small.

Problems

7.1. Suppose $x_1, x_2, \ldots$ are independent and identically distributed with $\mu = Ex_1$, $\sigma^2 = Ex_1^2 - \mu^2$, and assume also that the distribution function of x_1 is continuous. Let T be defined by

$$T_d = \inf\{n\colon n \geq m_0, \sum_{i=1}^{n} (x_i - \bar{x}_n)^2 \leq d^2n^2/z^2\},$$

and let $n_0 = (\sigma z/d)^2$. Show that as $d \to 0$, $n_0^{-1} T_d \to 1$ with probability one, $n_0^{-1} ET_d \to 1$, and $P\{|\bar{x}_T - \mu| \leq d\} \to 2\Phi(z) - 1$. *Hint*: The hard part is that concerning ET_d. That $\liminf n_0^{-1} ET_d \geq 1$ follows easily from the convergence with probability one. To show $\limsup n_0^{-1} ET_d \leq 1$, replace T_d by

$$\tau_d = \inf\left\{n\colon n \geq m_0, \sum_{i=1}^{n} (x_i - \mu)^2 \leq L_d(n)\right\},$$

where $L_d(n)$ is the line tangent to the curve d^2n^2/z^2 at $(n_0, d^2n_0^2/z^2)$. Now use the fact that for d small enough

$$\sum_{1}^{\tau} (x_i - \mu)^2 \geq \sum_{1}^{\tau-1} (x_i - \mu)^2 > L_d(\tau - 1)$$

and Wald's identity (2.18).

7.2. Under the assumptions of Section 2, discuss the problem of estimating $\mu \neq 0$ by a confidence interval having prescribed *proportional* accuracy, i.e. such that given $\delta > 0$, $P_{\mu,\sigma^2}\{|\hat{\mu} - \mu| \leq \delta|\mu|\} \cong 1 - 2\gamma$. Note that when σ is known this problem has been briefly discussed in IV.5. For a more difficult version of the problem try to achieve the proportional accuracy requirement not only in the limit as $\delta \to 0$ but *uniformly* for all $\mu \neq 0$, at least when σ is known. *Hint*: Use a test of power one.

7.3. In the context of Section 2, discuss the problem of achieving prescribed proportional *or* prescribed absolute accuracy, whichever is "easier."

7.4. Discuss the proportional accuracy requirement of Problem 7.2 in the following context, which arises naturally in regenerative simulation (cf. Crane and Lemoine, 1977). The data are independent, identically distributed pairs (y_n, τ_n), $n = 1, 2. \ldots,$ where the τ_i are strictly positive random variables and the parameter of interest is $\theta = E(y_1)/E(\tau_1)$.

7.5. For the following variant of the procedure of Section 2, Starr (1966) has calculated numerically the expected sample size and coverage probability. His results provide a convenient way of checking the asymptotic theory. Let

$$\tilde{T} = \inf\{n\colon n \geq m_0, n \text{ odd}, \sum_{i=1}^{n} (x_i - \bar{x}_n)^2 \leq n(n-1)a_n d^2/z_\gamma^2\},$$

where $a_n = 1 - \frac{1}{2}n^{-1}(1 + z_\gamma^2) + o(n^{-1})$. This value of a_n comes from using percentiles of the t distribution with $(n - 1)$ degrees of freedom in place of the normal percentile z_γ. Stopping only at odd n means that the random walk $\sigma^{-2}(y_1^2 + y_2^2)$, $\sigma^{-2}(y_1^2 + y_2^2 + y_3^2 + y_4^2), \ldots, \sigma^{-2}(y_1^2 + y_2^2 + \cdots + y_{2k}^2)$ is a random walk of negative exponential random variables with mean 2, which leads to several nice computational properties. Show that neglecting excess under the stopping boundary leads to the approximations

(7.28) $$E_{\mu,\sigma^2}(\tilde{T}) \cong n_0 - \tfrac{1}{2}(3 - z_\gamma^2) + o(1)$$

and

(7.29) $$P_{\mu,\sigma^2}\{|\bar{x}_{\tilde{T}} - \mu| \leq d\} \cong 1 - 2\gamma - 2n_0^{-1} z_\gamma \phi(z_\gamma) + o(n_0^{-1}).$$

To correct these approximations for excess under the boundary add a constant $\tilde{\rho}$ to (7.28) and $n_0^{-1}\tilde{\rho} z_\gamma \phi(z_\gamma)$ to the right hand side of (7.29). An approximate evaluation of $\tilde{\rho}$ shows that $\tilde{\rho} \cong 3/2$ (see Problem 9.16). A numerical comparison with the computations of Starr shows that the corrected (7.28) gives excellent approximations, whereas the corrected (7.29) is disappointingly inaccurate.

7.6. Stein's two-stage procedure. Let t_n denote the $1 - \gamma$ percentile of Student's t-distribution with n degrees of freedom. Under the assumptions of Section 2, let $m_0 \geq 2$ be arbitrary and define

$$N = \max(m_0, [t_{m_0 - 1}^2 m_0 \hat{\sigma}_{m_0}^2/(m_0 - 1)d^2] + 1).$$

(Here $[x]$ denotes the largest integer $\leq x$.)

(a) Show that $P_{\mu,\sigma^2}\{|\hat{\mu}_N - \mu| \leq d\} \geq 1 - 2\gamma$ for all μ, σ^2, i.e. $\hat{\mu}_N \pm d$ is a $(1 - 2\gamma)$ 100% confidence interval for μ of width $2d$.

(b) Prove that if $d \to 0$ and $m_0 d^2 \to 0$, then $E_{\mu,\sigma^2}(N) - (\sigma z_\gamma/d)^2 \to \infty$. Hence this procedure is less efficient than the fully sequential one of Section 2.

7.7. Let $x_1, x_2, \ldots$ be independent and normally distributed with mean μ and variance σ^2. Consider some specific test of H_0: $\sigma^2 = \sigma_0^2$ against H_1: $\sigma^2 > \sigma_0^2$ derived under the assumption that μ is known (for example, a sequential probability ratio test). Show how when μ is unknown the orthogonal transformation of Section 2 can be used to adapt this test without changing its power function. What is the relation between the distributions of the sample sizes of the two tests?

CHAPTER VIII

Random Walk and Renewal Theory

1. The Problem of Excess over a Boundary

Wald's approximations to the power function and expected sample size of a sequential probability ratio test are based on ignoring the discrepancy between the (log) likelihood ratio and the stopping boundary, thus replacing a random variable by a constant. In what follows we develop methods to approximate this discrepancy and hence to obtain more accurate results. Some of the approximations have already been stated and used in III.5 and IV.3. The present chapter is concerned with linear stopping boundaries; and the more difficult non-linear case is discussed in Chapter IX. An alternative method for linear problems is given in Chapter X.

To begin with the simplest possible case, consider a "one-sided" sequential probability ratio test with $a = \log A = -\infty$ (cf. II.4). This problem is of limited statistical interest and serves here primarily as an introduction to more difficult problems. However, it also occurs in Insurance Risk Theory and in Queueing Theory. See Appendix 2 for a brief discussion of the connection.

To develop the desired results in a suitably general framework, let $z_1, z_2, \ldots$ be independent, identically distributed random variables, and let $S_n = z_1 + \cdots + z_n$. For $b > 0$ define

$$\tau = \tau(b) = \inf\{n\colon n \geq 1, S_n \geq b\}, \tag{8.1}$$

where it is understood that $\inf \phi = +\infty$. Also let

$$\tau_+ = \inf\{n\colon n \geq 1, S_n > 0\}. \tag{8.2}$$

For example, under the assumptions of II.4 let $z_n = \log[f_1(x_n)/f_0(x_n)]$, where $x_1, x_2, \ldots$ are independent and identically distributed, and f_i $(i = 0, 1)$ are probability density functions. Then $E_0(z_1) < 0$ while $E_1(z_1) > 0$, and the objects of primary interest are $P_0\{\tau < \infty\}$ and $E_1\tau$. Moreover, Wald's identities (2.18) and (2.24) yield the representations.

(8.3) $$P_0\{\tau < \infty\} = e^{-b}E_1\{\exp[-(S_\tau - b)]\}$$

and

(8.4) $$E_1\tau = (E_1 z_1)^{-1}[b + E_1(S_\tau - b)]$$

(cf. (2.36) and (2.37)). (For a proof that $E_1\tau < \infty$ and hence (2.18) is applicable, see Proposition 8.21 below.) From (8.3) and (8.4) it is apparent that the key to good approximations for $P_0\{\tau < \infty\}$ and $E_1\tau$ is a reasonable approximation to the P_1-distribution of $S_\tau - b$, which in Chapter II was replaced by 0.

For general random walks $\{S_n\}$, not necessarily of the form of a log likelihood ratio, if $\mu = Ez_1 > 0$ it is trivial to generalize (8.4) to obtain

(8.5) $$E\tau = \mu^{-1}[b + E(S_\tau - b)].$$

To find an appropriate generalization of (8.3), assume that the distribution function F of z_1 can be imbedded in an exponential family as in II.3. More precisely assume that

$$e^{\psi(\theta)} = \int e^{\theta x} dF(x)$$

is convergent in some non-degenerate (real) interval containing 0, so that

(8.6) $$dF_\theta(x) = e^{\theta x - \psi(\theta)} dF(x)$$

defines a family of distribution functions indexed by θ. Let P_θ denote the probability under which $z_1, z_2, \ldots$ are independent and identically distributed with distribution function F_θ. By differentiating the identity $\int e^{\theta x - \psi(\theta)} dF(x) = 1$ under the integral one sees that

(8.7) $$\psi'(\theta) = E_\theta z_1 \text{ and } \psi''(\theta) = \operatorname{var}_\theta z_1 \geq 0.$$

In particular ψ is strictly convex (unless F is degenerate at a point). From (8.6) it follows that the likelihood ratio of $z_1, \ldots, z_n$ under P_{θ_0} relative to P_{θ_1} is

$$\exp\{(\theta_0 - \theta_1)S_n - n[\psi(\theta_0) - \psi(\theta_1)]\}.$$

Hence by Proposition 2.24, for any stopping time T and event A prior to T

(8.8) $$\begin{aligned} &P_{\theta_0}(A \cap \{T < \infty\}) \\ &\quad = E_{\theta_1}[\exp\{(\theta_0 - \theta_1)S_T - T[\psi(\theta_0) - \psi(\theta_1)]; A\{T < \infty\}]. \end{aligned}$$

A particularly interesting special case occurs when $\theta_0 = 0$, $\mu = E_0(z_1) = \psi'(0) < 0$, and there exists a $\theta_1 \neq 0$ such that $\psi(\theta_1) = 0 (= \psi(0))$. Necessarily $\theta_1 > 0$ and by (8.7) $E_{\theta_1}(z_1) > 0$—see the first graph in Figure 2.1. Then (8.8) with $T = \tau$ becomes

(8.9) $$P_0\{\tau < \infty\} = e^{-b\theta_1}E_{\theta_1}\{\exp[-\theta_1(S_\tau - b)]\},$$

which generalizes (8.3).

2. Reduction to a Problem of Renewal Theory and Ladder Variables

Consider the problem posed in Section 1: $S_n = z_1 + \cdots + z_n$ is a random walk with $\mu = E(z_1) > 0$, τ is defined by (8.1), and we seek an approximation to the distribution of $S_\tau - b$.

In the special case that

(8.10) $$P\{z_1 > 0\} = 1,$$

it is easy to conjecture an answer. Let $F(x) = P\{z_1 \le x\}$. Then if (8.10) holds,

(8.11) $$\begin{aligned} P\{S_\tau - b > y\} &= \sum_{n=0}^{\infty} P\{S_n < b, S_{n+1} > b + y\} \\ &= \sum_{n=0}^{\infty} \int_{[0,b)} P\{S_n \in dx\}[1 - F(b + y - x)] \\ &= \int_{[0,b)} U(dx)[1 - F(b + y - x)], \end{aligned}$$

where

(8.12) $$U(x) = \sum_{n=0}^{\infty} P\{S_n \le x\}.$$

Since S_n, $n = 0, 1, \ldots$ increases at rate μ, and $U(x + h) - U(x)$ is the expected length of time the random walk spends in the interval $(x, x + h]$, a reasonable conjecture for large x is that

(8.13) $$U(x + h) - U(x) \cong h/\mu.$$

Obviously some care must be taken in the interpretation of (8.13). For example, if the z_n are integer-valued, the increases in U occur only at integers, and only integer combinations of x and h would be reasonable. But if we temporarily restrict consideration to continuous random variables and suppose that for large x $U(dx)$ in (8.11) can be replaced by dx/μ, the change of variable $u = b - x + y$ leads to the conjecture

(8.14) $$P\{S_\tau - b > y\} \to \mu^{-1} \int_{(y,\infty)} [1 - F(u)]\, du$$

as $b \to \infty$.

The simplest possible example occurs when

(8.15) $$1 - F(x) = e^{-\lambda x} \qquad (x \ge 0).$$

It is readily verified from (8.12) that $U(dx) = \lambda dx$. Hence by (8.11) $P\{S_\tau - b > y\} = e^{-\lambda y}$, and (8.14) holds with equality for all b.

Of course, the assumption (8.10) is usually too restrictive, but it is possible to reduce the general case to that of positive z_1. Suppose that $P\{z_1 < 0\} > 0$ but $\mu > 0$, so by the strong law of large numbers $P\{\tau(b) < \infty\} = 1$. Recall the definition (8.2). Let $\tau_+^{(0)} = 0$, $\tau_+^{(1)} = \tau_+$, and for arbitrary k

$$\tau_+^{(k)} = \inf\{n: n > \tau_+^{(k-1)}, S_n > S_{\tau_+^{(k-1)}}\}. \tag{8.16}$$

The random variables $\tau_+^{(k)}$ and $S_{\tau_+^{(k)}}$ are called the kth (strict) ladder epoch and kth (strict) ladder height, respectively. It is easy to see that $\{(\tau_+^{(k)} - \tau_+^{(k-1)}, S_{\tau_+^{(k)}} - S_{\tau_+^{(k-1)}}), k = 1, 2, \ldots\}$ are independent and identically distributed. Moreover, a picture shows that τ must equal $\tau_+^{(k)}$ for some k, so if

$$\sigma = \inf\{k: S_{\tau_+^{(k)}} > b\},$$

then $\tau_+^{(\sigma)} = \tau$ and $S_{\tau_+^{(\sigma)}} - b = S_\tau - b$. Since $P\{S_{\tau_+} > 0\} = 1$ by definition, (8.14) may be applied to the distribution of S_{τ_+} to obtain the conjecture

$$\begin{aligned}\lim_{b\to\infty} P\{S_\tau - b > y\} &= \lim_{b\to\infty} P\{S_{\tau_+^{(\sigma)}} - b > y\} \\ &= (ES_{\tau_+})^{-1} \int_{(y,\infty)} P\{S_{\tau_+} > u\}\, du.\end{aligned} \tag{8.17}$$

Since $E(S_\tau - b) = \int_{(0,\infty)} P\{S_\tau - b > y\}\, dy$, (8.17) leads easily to the conjecture

$$E(S_\tau - b) \to ES_{\tau_+}^2 / 2ES_{\tau_+} \qquad (b \to \infty), \tag{8.18}$$

and hence by (8.5)

$$E\tau(b) = \mu^{-1}[b + ES_{\tau_+}^2/2ES_{\tau_+} + o(1)] \qquad (b \to \infty). \tag{8.19}$$

Under the exponential family imbedding assumption described at the end of Section 1, (8.9) and (8.17) suggest

$$P_0\{\tau < \infty\} \sim e^{-b\theta_1}(\theta_1 E_{\theta_1} S_{\tau_+})^{-1}(1 - E_{\theta_1} e^{-\theta_1 S_{\tau_+}}) \qquad (b \to \infty). \tag{8.20}$$

Making (8.19) and (8.20) into usable approximations still requires substantial work. In addition to investigating precise conditions under which (8.14) and hence (8.17)–(8.20) are true, it is also necessary to compute the distribution of S_{τ_+}, which enters into the approximations. These problems are discussed in Sections 3 and 4.

3. Renewal Theory

Let $z_1, z_2, \ldots$ be indepenent, identically distributed random variables with positive mean μ, and set $S_n = z_1 + \cdots + z_n$. Let τ be defined by (8.1) and U by (8.12). Observe that if $P\{z_1 > 0\} = 1$, then $E\tau(b) = U(b-)$. The following result shows that in this case (8.13) is true "on the average."

Proposition 8.21. *$E\tau(b) < \infty$ for all b and $\lim_{b\to\infty} b^{-1} E\tau(b) = \mu^{-1}$.*

PROOF. Suppose initially that for some c, $P\{z_1 \le c\} = 1$. By Wald's identity and (8.1)

$$\mu E(\tau \wedge n) = ES_{\tau \wedge n} \le b + c.$$

Letting $n \to \infty$ shows that

$$\mu E\tau \le b + c < \infty,$$

and letting $b \to \infty$ shows that

(8.22) $$\limsup_b b^{-1} E\tau \le \mu^{-1}.$$

The reverse inequality

(8.23) $$b^{-1} E\tau = b^{-1}\mu^{-1} ES_\tau \ge \mu^{-1}$$

follows immediately from Wald's identity. This proves the proposition for z's which are bounded above.

In general let $c > 0$ and define $z_n^{(c)} = z_n I_{\{z_n \le c\}} + c I_{\{z_n > c\}}$, where c is so large that $\mu^{(c)} = Ez_1^{(c)} > 0$. Define $\tau^{(c)}$ by analogy with (8.1). Since $\tau \le \tau^{(c)}$, the preceding argument shows that $E\tau < \infty$, and by (8.22)

$$\limsup_b b^{-1} E\tau \le (\mu^{(c)})^{-1}.$$

Since $\mu^{(c)} \uparrow \mu$ as $c \uparrow \infty$, this proves (8.22) in general; and the reverse inequality (8.23) is again immediate from Wald's identity. □

In order to give a precise statement of (8.13) it is necessary to distinguish arithmetic and non-arithmetic random variables.

Definition. A random variable Y is called arithmetic if for some constant $d > 0$, $P\{Y \in \{\dots. -2d, -d, 0, d, 2d, \dots\}\} = 1$. The largest such constant d is called the span of Y.

Obviously if z_1 is arithmetic with span d, then S_n is a multiple of d with probability one, so U (defined in (8.12)) increases only by jumps at $\dots -2d$, $-d$, 0, d, $2d$, $\dots$.

Theorem 8.24. *Suppose $P\{z_1 > 0\} = 1$. If z_1 is non-arithmetic, then for any $h > 0$*

$$U(x + h) - U(x) \to h/\mu \qquad (x \to \infty).$$

If z_1 is arithmetic with span d, then

$$U[(k + 1)d] - U(kd) \to d/\mu \qquad (k \to \infty).$$

The actual content of Theorem 8.24 is that the indicated limits exist, since by Proposition 8.21 and the fact that $U(b-) = E\tau(b)$ the only possible values for these limits are the given ones. Since the proof of Theorem 8.24 is long and does not seem to provide important insights into the processes of primary interest in this book, it is discussed in Appendix 4.

Theorem 8.25. *Assume $P\{z_1 > 0\} = 1$ and let $\mu = Ez_1$. Let g be a positive, non-increasing function on $[0, \infty)$ with $\int_{[0,\infty)} g(x)\,dx < \infty$.*

(i) *If* z_1 is non-arithmetic, *then*

$$\lim_{b\to\infty} \int_{[0,b)} g(b-x)U(dx) = \mu^{-1} \int_{[0,\infty)} g(x)\,dx; \tag{8.26}$$

in particular

$$\lim_{b\to\infty} P\{S_\tau - b > y\} = \mu^{-1} \int_{(y,\infty)} P\{z_1 > x\}\,dx, \tag{8.27}$$

and if $Ez_1^2 < \infty$, then

$$\lim_{b\to\infty} E(S_\tau - b) = Ez_1^2/2\mu. \tag{8.28}$$

(ii) *If z_1 is arithmetic with span d and $b \to \infty$ through multiples of d, the corresponding results are*

$$\lim_{b\to\infty} \sum_{j \le b/d} g(b - jd)\{U(jd) - U[(j-1)d]\} = d\mu^{-1} \sum_0^\infty g(j), \tag{8.29}$$

$$\lim_{b\to\infty} P\{S_\tau - b = jd\} = d\mu^{-1} P\{z_1 \ge jd\}, \tag{8.30}$$

and if $Ez_1^2 < \infty$,

$$\lim_{b\to\infty} E(S_\tau - b) = Ez_1^2/2\mu + d/2. \tag{8.31}$$

PROOF. To prove (8.26), let $m = 1, 2, \ldots$ and $h > 0$ be fixed. Observe that

$$U(x+h) - U(x) \le U(h) < \infty \tag{8.32}$$

because the number of visits to $(x, x+h]$ equals at most one (the first visit if there is a visit) plus a random variable whose distribution is stochastically smaller than the number of visits to $(0, h]$. Since g is non-increasing.

$$\int_{[0,b]} g(b-x)U(dx) \le \sum_{i=1}^{m} g((i-1)h)\{U[b-(i-1)h] - U(b-ih)\}$$
$$+ U(h) \sum_{i=m}^{\infty} g(ih).$$

Hence by Theorem 8.24

$$\limsup_{b\to\infty} \int_{[0,b]} g(b-x)U(dx) \le \mu^{-1} \sum_{i=0}^{m-1} g(ih)h + U(h) \sum_{i=m}^{\infty} g(ih).$$

Putting $m = c/h$ and letting $c \to \infty$, then $h \to 0$ yields

$$\limsup_{b} \int_{[0,b]} g(b-x)U(dx) \le \mu^{-1} \int_{(0,\infty)} g(x)\,dx.$$

A similar but easier lower bound completes the proof of (8.26).

The conclusion (8.27) follows from (8.26) and the representation (8.11). Integrating (8.11) gives

$$E(S_\tau - b) = \int_{[0,b)} \left(\int_{(b-x,\infty)} [1 - F(y)]\, dy \right) U(dx),$$

which with (8.26) and the condition $Ez_1^2 = 2\int_{(0,\infty)} \int_{(x,\infty)} [1 - F(y)]\, dy\, dx < \infty$ yields (8.28).

The proof of (ii) is similar and is omitted. □

For applications we are primarily interested in the case $P\{z_1 < 0\} > 0$ and a re-interpretation of Theorem 8.25 in terms of S_{τ_+}, as discussed in Section 2. The following result makes (8.17) and (8.18) precise.

Corollary 8.33. *Assume* $0 < Ez_1 < \infty$. *If* z_1 *is non-arithmetic then*

$$\lim_{b\to\infty} P\{S_\tau - b > y\} = (ES_{\tau_+})^{-1} \int_{(y,\infty)} P\{S_{\tau_+} > x\}\, dx.$$

If in addition $Ez_1^2 < \infty$, *then*

$$\lim_{b\to\infty} E(S_\tau - b) = ES_{\tau_+}^2 / 2ES_{\tau_+}.$$

If z_1 *is arithmetic with span d, then as* $b \to \infty$ *through multiples of d*

$$\lim_b P\{S_\tau - b = jd\} = d(ES_{\tau_+})^{-1} P\{S_{\tau_+} \geq jd\};$$

and if also $Ez_1^2 < \infty$, *then*

$$\lim_b E(S_\tau - b) = ES_{\tau_+}^2 / 2ES_{\tau_+} + d/2.$$

Proof. Because of the argument in Section 2 leading to (8.17), it suffices to show that z_1 non-arithmetic (arithmetic with span d) implies S_{τ_+} non-arithmetic (arithmetic with span d), and $Ez_1^2 < \infty$, implies $ES_{\tau_+}^2 < \infty$. That S_{τ_+} inherits the non-arithmetic (arithmetic) property of z_1 is fairly obvious intuitively, but for a simple proof see Problem 8.1. That $Ez_1^2 < \infty$ implies $ES_{\tau_+}^2 < \infty$ follows from Problem 8.6. □

Let H denote the limiting distribution of $S_\tau - b$. Recall from Problem 3.11 that $(\tau - b/\mu)/(b\sigma^2/\mu^3)^{1/2}$ is asymptotically normal as $b \to \infty$. The following result shows that these two variables are asymptotically independent.

Theorem 8.34. *As* $b \to \infty$ *(through multiples of d in the arithmetic case), for all* $-\infty < x < \infty$, $y > 0$

$$P\{\tau \leq b\mu^{-1} + x(b\sigma^2/\mu^3)^{1/2}, S_\tau - b \leq y\} \to \Phi(x)H(y).$$

Proof. Let $m = m(b, x) = [b\mu^{-1} + x(b\sigma^2/\mu^3)^{1/2}]$. (Here $[u]$ denotes the integer part of u.) Then

$$(8.35)\qquad P\{\tau > m, S_\tau - b \le y\} = E(P\{S_\tau - b \le y|\tau > m, S_m\}; \tau > m).$$

Let $\tilde{b} = b - b^{1/4}$, and observe that

$$(8.36)\qquad \begin{aligned} &E(P\{S_\tau - b \le y|\tau > m, S_m\}; \tau > m, \tilde{b} < S_m < b) \\ &\le P\{\tilde{b} < S_m < b\} \to 0 \end{aligned}$$

by the central limit theorem. Moreover, uniformly on $\{\tau > m, S_m \le \tilde{b}\}$

$$P\{S_{\tau(b)} - b \le y|\tau > m, S_m = v\} = P\{S_{\tau(b-v)} - (b - v) \le y\} \to H(y)$$

by Corollary 8.33. Hence by (8.35), (8.36), and the asymptotic normality of $(\tau - b\mu^{-1})/(b\sigma^2/\mu^3)^{1/2}$

$$\begin{aligned} P\{\tau > m, S_\tau - b \le y\} &= E(P\{S_\tau - b \le y|\tau > m, S_m\}; \tau > m, S_m \le \tilde{b}) + o(1) \\ &= H(y)P\{\tau > m, S_m \le \tilde{b}\} + o(1) \to H(y)\{1 - \Phi(x)\}. \end{aligned}$$

□

Corollary 8.37. *As $b \to \infty$ $S_\tau - b$ and $(S_\tau - \mu\tau)/\tau^{1/2}$ are asymptotically independent.*

4. Ladder Variables

The main result of this section is Theorem 8.45, which evaluates $E(s^{\tau_+}e^{i\lambda S_{\tau_+}})$. It gives a method for computing numerically the right hand sides of (8.19) and (8.20) in some simple cases and provides the basis for a more sophisticated algorithm for a large class of random walks.

Let $z_1, z_2, \ldots$ be independent and identically distributed, and let $S_n = z_1 + \cdots + z_n$. Define τ_+ by (8.2) and let

$$\tau_- = \inf\{n: n \ge 1, S_n \le 0\}.$$

In analogy with (8.16) let $\tau_-^{(0)} = 0$, $\tau_-^{(1)} = \tau_-$, and

$$\tau_-^{(j)} = \inf\{n: n > \tau_-^{(j-1)}, S_n \le S_{\tau_-^{(j-1)}}\} \qquad (j = 1, 2, \ldots).$$

Proposition 8.38. *For $B \subset (-\infty, 0]$, $n = 0, 1, \ldots$*

$$P\{\tau_+ > n, S_n \in B\} = \sum_{j=0}^{\infty} P\{\tau_-^{(j)} = n, S_{\tau_-^{(j)}} \in B\}.$$

For $B \subset (0, \infty)$, $n = 0, 1, \ldots$

$$P\{\tau_- > n, S_n \in B\} = \sum_{j=0}^{\infty} P\{\tau_+^{(j)} = n, S_{\tau_+^{(j)}} \in B\}.$$

PROOF. Let $B \subset (-\infty, 0]$, $n = 0, 1, \ldots$. Then since $(z_1, z_2, \ldots, z_n)$ and $(z_n, z_{n-1}, \ldots, z_1)$ have the same distribution,

$$\begin{aligned}P\{\tau_+ > n, S_n \in B\} &= P\{S_1 \le 0, S_2 \le 0, \ldots, S_n \le 0, S_n \in B\}\\ &= P\{S_n - S_{n-1} \le 0, S_n - S_{n-2} \le 0, \ldots, S_n - S_0 \le 0, S_n \in B\}\\ &= P\{S_n = \min_{0 \le k \le n} S_k, S_n \in B\} = \sum_{j=0}^{\infty} P\{\tau_-^{(j)} = n, S_{\tau_-^{(j)}} \in B\}.\end{aligned}$$

The proof of the second assertion is similar. □

Corollary 8.39. $E\tau_+ = 1/P\{\tau_- = \infty\}$; $E\tau_- = 1/P\{\tau_+ = \infty\}$.

Proof. Putting $B = (-\infty, 0]$ in (8.38) and summing over n yields

$$\begin{aligned}E\tau_+ &= \sum_{n=0}^{\infty} P\{\tau_+ > n\} = \sum_{j=0}^{\infty} P\{\tau_-^{(j)} < \infty\}\\ &= \sum_{j=0}^{\infty} (P\{\tau_- < \infty\})^j = 1/P\{\tau_- = \infty\}.\end{aligned}$$

Again the second assertion follows similarly. □

Essentially the same computation yields

Corollary 8.40. *For* $|s| < 1$ *and* $-\infty < \lambda < \infty$

$$\sum_{n=0}^{\infty} s^n E(\exp(i\lambda S_n); \tau_+ > n) = [1 - E(s^{\tau_-} \exp(i\lambda S_{\tau_+}); \tau_- < \infty)]^{-1}$$

and

$$\sum_{n=0}^{\infty} s^n E(\exp(i\lambda S_n); \tau_- > n) = [1 - E(s^{\tau_+} \exp(i\lambda S_{\tau_+}); \tau_+ < \infty)]^{-1}.$$

Proof. Since the distribution of $(\tau_-^{(j)}, S_{\tau_-^{(j)}})$ is the j-fold convolution of the distribution of (τ_-, S_{τ_-}), by (8.38)

$$\begin{aligned}\sum_{n=0}^{\infty} s^n E\{\exp(i\lambda S_n); \tau_+ > n\} &= \sum_{j=0}^{\infty}\sum_{n=0}^{\infty} s^n E\{\exp(i\lambda S_{\tau_-^{(j)}}); \tau_-^{(j)} = n\}\\ &= \sum_{j=0}^{\infty} E\{s^{\tau_-^{(j)}} \exp(i\lambda S_{\tau_-^{(j)}}); \tau_-^{(j)} < \infty\}\\ &= \sum_{j=0}^{\infty} [E\{s^{\tau_-} \exp(i\lambda S_{\tau_-}); \tau_- < \infty\}]^j\\ &= [1 - E\{s^{\tau_-} \exp(i\lambda S_{\tau_-}); \tau_- < \infty\}]^{-1}. \quad \square\end{aligned}$$

Let $g(\lambda) = Ee^{i\lambda z_1}$ and $G_\pm(s, \lambda) = E(s^{\tau_\pm} \exp(i\lambda S_{\tau_\pm}); \tau_\pm < \infty)$.

Theorem 8.41. *For* $|s| \le 1$ and $-\infty < \lambda < \infty$

(8.42) $$[1 - G_+(s, \lambda)][1 - G_-(s, \lambda)] = 1 - sg(\lambda).$$

PROOF. Let $|s| < 1$. Consider $E(e^{i\lambda S_n}; \tau_+ \geq n)$ $(n = 1, 2, \ldots)$. It equals $E(\exp(i\lambda S_{\tau_+}); \tau_+ = n) + E(\exp(i\lambda S_n); \tau_+ > n)$ and also equals $E(e^{i\lambda S_{n-1}} e^{i\lambda z_n};$ $\tau_+ > n - 1) = g(\lambda) E(e^{i\lambda S_{n-1}}; \tau_+ > n - 1)$. Equating these two expressions, multiplying by s^n, summing from $n = 1$ to $n = \infty$, and taking account of (8.40) yields

$$G_+(s, \lambda) + [1 - G_-(s, \lambda)]^{-1} - 1 = sg(\lambda)[1 - G_-(s, \lambda)]^{-1},$$

which is equivalent to (8.42) when $|s| < 1$. That equality continues to hold when $|s| = 1$ follows by continuity. □

Theorem 8.43. *For* $0 \leq s < 1$, $-\infty < \lambda < \infty$

$$1 - G_+(s, \lambda) = \exp\left[-\sum_1^\infty n^{-1} s^n E(e^{i\lambda S_n}; S_n > 0)\right]$$

and

$$1 - G_-(s, \lambda) = \exp\left[-\sum_1^\infty n^{-1} s^n E(e^{i\lambda S_n}; S_n \leq 0)\right].$$

PROOF. For $0 \leq s < 1$, $|G_\pm(s, \lambda)| < 1$, so it is possible to take logarithms in (8.42) and expand in the series $-\log(1 - x) = \sum_1^\infty x^n/n$. The result of doing this with $1 - sg(\lambda)$ is

$$\begin{aligned} -\log(1 - sg(\lambda)) &= \sum_1^\infty n^{-1} s^n \int_{(-\infty, \infty)} e^{i\lambda x} P\{S_n \in dx\} \\ &= \int_{(-\infty, \infty)} e^{i\lambda x} V_1(dx), \end{aligned}$$

say, where $V_1(dx) = \sum_1^\infty n^{-1} s^n P\{S_n \in dx\}$. Since $(\tau_\pm^{(j)}, S_{\tau_\pm^{(j)}})$ is the j-fold convolution of $(\tau_\pm, S_{\tau_\pm})$, a similar calculation yields

$$-\log[1 - G_+(s, \lambda)] - \log[1 - G_-(s, \lambda)] = \int_{(-\infty, \infty)} e^{i\lambda x} V_2(dx),$$

where

$$V_2(dx) = \begin{cases} \sum_1^\infty n^{-1} E(s^{\tau_+^{(n)}}; \tau_+^{(n)} < \infty, S_{\tau_+^{(n)}} \in dx) & (x > 0) \\ \sum_1^\infty n^{-1} E(s^{\tau_-^{(n)}}; \tau_-^{(n)} < \infty, S_{\tau_-^{(n)}} \in dx) & (x \leq 0). \end{cases}$$

Hence by (8.42) $\int_{(-\infty, \infty)} e^{i\lambda x} V_1(dx) = \int_{(-\infty, \infty)} e^{i\lambda x} V_2(dx)$, and since V_1 and V_2 are both finite measures, they are equal by the uniqueness theorem for characteristic functions. In particular

$$\sum_1^\infty n^{-1} E(s^{\tau_+^{(n)}}; \tau_+^{(n)} < \infty, S_{\tau_+^{(n)}} \in dx) = \sum_1^\infty n^{-1} s^n P\{S_n \in dx\}, \qquad x > 0$$

and

$$\sum_1^\infty n^{-1}E(s^{\tau_-^{(n)}}; \tau_-^{(n)} < \infty, S_{\tau_-^{(n)}} \in dx) = \sum_1^\infty n^{-1}s^n P\{S_n \in dx\}, \qquad x \le 0.$$

Taking Fourier transforms of these equalities (the first over $(0, \infty)$, the second over $(-\infty, 0])$ and then exponentiating yields the theorem. □

Corollary 8.44. $P\{\tau_+ = \infty\} = \exp[-\sum_1^\infty n^{-1}P\{S_n > 0\}]$ and $P\{\tau_- = \infty\} = \exp[-\sum_1^\infty n^{-1}P\{S_n \le 0\}]$.

PROOF. These results follow immediately by setting $\lambda = 0$ and letting $s \uparrow 1$ in the expressions for $1 - G_\pm(s, \lambda)$ in Theorem 8.43. □

Corollary 8.45. *Assume that $\mu = E(z_1) > 0$ and that z_1 is nonarithmetic. Let $\tau = \inf\{n: S_n \ge b\}$. Then for each $\alpha > 0$*

$$\lim_{b\to\infty} E\{\exp[-\alpha(S_\tau - b)]\} = (\alpha\mu)^{-1}\exp\left[-\sum_1^\infty n^{-1}Ee^{-\alpha S_n^+}\right].$$

PROOF. By Corollary 8.33

$$\begin{aligned}\lim_{b\to\infty} E\{\exp[-\alpha(S_\tau - b)]\} &= [ES_{\tau_+}]^{-1}\int_{(0,\infty)} e^{-\alpha x}P\{S_{\tau_+} \ge x\}\,dx \\ &= [\alpha ES_{\tau_+}]^{-1}\{1 - E\exp(-\alpha S_{\tau_+})\}.\end{aligned} \tag{8.46}$$

By Wald's identity and Corollary 8.39 $E(S_{\tau_+}) = \mu E(\tau_+) = \mu/P(\tau_- = \infty)$. Corollary 8.45 follows by substituting this expression for $E(S_{\tau_+})$ into (8.46) and by using Theorem 8.43 and Corollary 8.44 to evaluate $1 - E\exp(-\alpha S_{\tau_+})$ and $P\{\tau_- = \infty\}$, respectively.

Remark 8.47. Corollary 8.45 together with (8.46) provides a precise road map from (8.9) to (8.20) and a useful algorithm to evaluate the right hand side of (8.20). Under the exponential imbedding assumption described in Section 1, one can find alternative expressions for the right hand side of (8.20). By Wald's likelihood ratio identity $P_0\{\tau_+ < \infty\} = E_{\theta_1}[\exp(-\theta_1 S_{\tau_+})]$. As above, $E_{\theta_1}(S_{\tau_+}) = \psi'(\theta_1)E_{\theta_1}\tau_+ = \psi'(\theta_1)/P_{\theta_1}\{\tau_- = \infty\}$. Hence (8.20) can also be expressed as

$$P_0\{\tau < \infty\} \sim \exp(-b\theta_1)P_0\{\tau_+ = \infty\}P_{\theta_1}\{\tau_- = \infty\}/\theta_1\psi'(\theta_1), \tag{8.48}$$

where $P_0\{\tau_+ = \infty\}$ and $P_{\theta_1}\{\tau_- = \infty\}$ are given by Corollary 8.44.

In the special case that z_1 is normal with mean $-\mu < 0$, it is easy to see that $\theta_1 = 2\mu$ and the P_{θ_1} distribution of z_1 is normal with mean μ, so (8.48) and (8.44) yield

$$P_0\{\tau < \infty\} \sim e^{-2b\mu}(2\mu^2)^{-1}\exp\left[-2\sum_1^\infty n^{-1}\Phi(-\mu n^{1/2})\right]. \tag{8.49}$$

The factor multiplying $e^{-2b\mu}$ on the right hand side of (8.49) is precisely $v(2\mu)$, where $v(\cdot)$ is the function defined in (4.37) and evaluated approximately in (4.38).

A similar but rather tedious calculation gives an expression for the right hand side of (8.18). If $Ez_1 > 0$ and $Ez_1^2 < \infty$, one may differentiate the expression for $1 - G_+(s, \lambda)$ twice with respect to λ, set $\lambda = 0$, and let $s \to 1$ to obtain

$$E(S_{\tau_+}^2)/2ES_{\tau_+} = Ez_1^2/2Ez_1 - \sum_1^\infty n^{-1}ES_n^-, \tag{8.50}$$

where $x^- = -\min(x, 0)$.

Corollary 8.45 is directly applicable when explicit expressions for the distribution of S_n, $n = 1, 2, \ldots$ are available. A computational procedure which is valid under more general conditions is given next.

Theorem 8.51. *Let $g(\lambda) = Ee^{i\lambda z_1}$. Assume that z_1 is strongly nonarithmetic in the sense that* $\limsup_{|\lambda|\to\infty} |g(\lambda)| < 1$. *Suppose also that $\mu = Ez_1 > 0$ and $Ez_1^2 < \infty$. Then for $\alpha > 0$*

$$(\alpha\mu)^{-1}\exp\left(-\sum_1^\infty n^{-1}Ee^{-\alpha S_n^+}\right)$$

$$= \exp\left\{-(2\pi)^{-1}\int_{-\infty}^{\infty}\left(\frac{1}{\alpha + i\lambda} - \frac{1}{i\lambda}\right)\left[\log\left(\frac{1}{1 - g(\lambda)}\right) + \log(-i\mu\lambda)\right] d\lambda\right\}$$

$$= (1 + \mu\alpha)^{-1}\exp\left\{-(2\pi)^{-1}\int_{-\infty}^{\infty}\left(\frac{1}{\alpha + i\lambda} - \frac{1}{i\lambda}\right)\left[\log\left(\frac{1}{1 - g(\lambda)}\right)\right.\right.$$

$$\left.\left. + \log\left(\frac{-i\mu\lambda}{1 - i\mu\lambda}\right)\right] d\lambda\right\}.$$

Remarks. The first of these expressions was obtained by Woodroofe (1979). See Spitzer (1960) for a similar result. The second is slightly less convenient, but requires less analysis for its proof (see below). For purposes of numerical computation the integrands can be expressed as real-valued functions, since the integral of the imaginary parts must vanish. For example, the real part of the first integrand is

$$\lambda^{-1}\left(\frac{\alpha^2}{\alpha^2 + \lambda^2}\right)\left[\operatorname{Im}\log\left(\frac{1}{1 - g(\lambda)}\right) - \pi/2\right]$$

$$- \left(\frac{\alpha}{\alpha^2 + \lambda^2}\right)\left[\operatorname{Re}\log\left(\frac{1}{1 - g(\lambda)}\right) + \log\mu\lambda\right].$$

Woodroofe (1979) has also obtained an expression for the series in (8.50). Alternatively, an approximation to $E(S_{\tau_+}^2)/2ES_{\tau_+}$ which seems adequate for most purposes is given in Problem 10.2.

PROOF OF THE SECOND EXPRESSION IN THEOREM 8.51. Let

$$u(x) = e^{-\alpha x}, \qquad x \geq 0$$
$$= e^{\beta x}, \qquad x \leq 0.$$

Let z_0 be normal with mean 0 and variance δ and independent of S_n, $n = 1, 2, \ldots$. Then $z_0 + S_n$ has characteristic function $e^{-\delta\lambda^2/2} g^n(\lambda)$ and probability density function

$$f_n(x) = (2\pi)^{-1} \int_{-\infty}^{\infty} e^{-i\lambda x - \delta\lambda^2/2} g^n(\lambda)\, d\lambda.$$

Hence

$$\begin{aligned}
\Sigma n^{-1} Eu(z_0 + S_n) &= \Sigma n^{-1} \int_{-\infty}^{\infty} u(x) f_n(x) dx \\
&= \Sigma n^{-1} \left\{ \int_0^{\infty} e^{-\alpha x} (2\pi)^{-1} \int_{-\infty}^{\infty} e^{-i\lambda x - \delta\lambda^2/2} g^n(\lambda)\, d\lambda\, dx \right. \\
&\quad \left. + \int_{-\infty}^{0} e^{\beta x} (2\pi)^{-1} \int_{-\infty}^{\infty} e^{-i\lambda x - \delta\lambda^2/2} g^n(\lambda)\, d\lambda\, dx \right\} \\
&= \Sigma n^{-1} (2\pi)^{-1} \int_{-\infty}^{\infty} \left[\frac{1}{\alpha + i\lambda} + \frac{1}{\beta - i\lambda} \right] e^{-\delta\lambda^2/2} g^n(\lambda)\, d\lambda \\
&= (2\pi)^{-1} \int_{-\infty}^{\infty} \left[\frac{1}{\alpha + i\lambda} + \frac{1}{\beta - i\lambda} \right] e^{-\delta\lambda^2/2} \log\left(\frac{1}{1 - g(\lambda)} \right) d\lambda.
\end{aligned}$$

To see that the interchange of summation and integration in the last equality above is justified, first note that for $\lambda \neq 0$

$$\left| \sum_{n=1}^{k} n^{-1} g^n(\lambda) \right| \leq \sum_{1}^{\infty} n^{-1} |g(\lambda)|^n = -\log(1 - |g(\lambda)|).$$

Since by assumption $|g(\lambda)|$ is bounded away from 1 for λ bounded away from 0, the integrands are dominated at ∞; and some simple Taylor series expansions show that they are also dominated at 0.

Now letting $\delta \to 0$ and noting that $\left| \dfrac{1}{\alpha + i\lambda} + \dfrac{1}{\beta - i\lambda} \right| = 0(\lambda^{-2})$ is integrable at ∞, we obtain

$$(8.52) \quad \Sigma n^{-1} Eu(S_n) = (2\pi)^{-1} \int_{-\infty}^{\infty} \left[\frac{1}{\alpha + i\lambda} + \frac{1}{\beta - i\lambda} \right] \log\left(\frac{1}{1 - g(\lambda)} \right) d\lambda.$$

Consider the special case $P\{z_1 \in dx\} = \mu^{-1} e^{-\mu^{-1} x} dx$, $x \geq 0$. Direct calculation shows that the left hand side of (8.52) is $\log[\mu\alpha/(1 + \mu\alpha)]$, which according to (8.52) equals

$$(2\pi)^{-1} \int_{-\infty}^{\infty} \left[\frac{1}{\alpha + i\lambda} + \frac{1}{\beta - i\lambda} \right] \log\left(\frac{-i\mu\lambda}{1 - i\mu\lambda} \right) d\lambda.$$

Adding and subtracting this special case in (8.52) yields

$$\begin{aligned}\Sigma n^{-1} Eu(S_n) &= \log[(1+\mu\alpha)/\mu\alpha] \\ &\quad + (2\pi)^{-1}\int_{-\infty}^{\infty}\left[\frac{1}{\alpha+i\lambda}+\frac{1}{\beta-i\lambda}\right]\left[\log\left(\frac{1}{1-g(\lambda)}\right)\right. \\ &\quad \left. + \log\left(\frac{-i\mu\lambda}{1-i\mu\lambda}\right)\right] d\lambda.\end{aligned}$$

Letting $\beta \to 0$ gives the desired conclusion. □

Remark 8.53. A more elaborate smoothing argument shows that the condition $\limsup_{|\lambda|\to\infty} |g(\lambda)| < 1$ imposed in Theorem 8.51 is unnecessary. For an application of this more general result see Theorem 9.69.

Proof of Remark 8.53. Let $y_1, y_2, \ldots$ be independent standard normal random variables and put $z_n^{(\sigma)} = z_n + \sigma y_n$, $0 < \sigma < 1$, $n = 1, 2, \ldots$. Let $S_n^{(\sigma)} = \sum_1^n z_k^{(\sigma)}$. The characteristic function of $z_1^{(\sigma)}$ is $g^{(\sigma)}(\lambda) = g(\lambda)e^{-\lambda^2\sigma^2/2}$, which satisfies the conditions of Theorem 8.51. Hence the conclusion is valid for the random walk $S_n^{(\sigma)}$, $n = 1, 2, \ldots$, so it suffices to show

$$\lim_{\sigma\to 0} \Sigma n^{-1} E[\exp(-\alpha S_n^{(\sigma)+})] = \Sigma n^{-1} E[\exp(-\alpha S_n^{+})] \tag{8.54}$$

and

$$\begin{aligned}&\lim_{\sigma\to 0}\int_{-\infty}^{\infty} \operatorname{Re}\left\{\left(\frac{1}{\alpha+i\lambda}-\frac{1}{i\lambda}\right)\left[\log\left(\frac{1}{1-g^{(\sigma)}(\lambda)}\right)+\log(-i\mu\lambda)\right]\right\} d\lambda \\ &\quad = \int_{-\infty}^{\infty} \operatorname{Re}\left\{\left(\frac{1}{\alpha+i\lambda}-\frac{1}{i\lambda}\right)\left[\log\left(\frac{1}{1-g(\lambda)}\right)+\log(-i\mu\lambda)\right]\right\} d\lambda.\end{aligned} \tag{8.55}$$

Since $S_n^{(\sigma)} \to S_n$ pointwise as $\sigma \to 0$, to prove (8.54) it suffices to show that the tail of the series is small uniformly in σ. This follows easily from Chebyshev's inequality, which yields

$$\begin{aligned}E[\exp(-\alpha S_n^{(\sigma)+})] &\le \exp(-n\alpha\mu/2) + P\{S_n^{(\sigma)} \le n\mu/2\} \\ &\le \exp(-n\alpha\mu/2) + 4(\operatorname{var}(z_1) + \sigma^2)/n\mu^2.\end{aligned}$$

Note that this also establishes the convergence of the right hand side of (8.54).

To prove (8.55) we show that the integrand is dominated as $\sigma \to 0$ by considering separately the product of the real parts and the product of the imaginary parts of the two factors in the integrand. That the product the imaginary parts is dominated away from $\lambda = 0$ is obvious and near 0 follows easily from the expansion

$$\operatorname{Im}\log\left(\frac{1}{1-g^{(\sigma)}(\lambda)}\right) - \pi/2 = 0(\lambda) \qquad (\lambda \to 0),$$

which holds uniformly in σ. The product of real parts equals

$$[\alpha/(\alpha^2 + \lambda^2)][-\log|1 - g^{(\sigma)}(\lambda)| + \log(\mu\lambda)],$$

of which the first factor is integrable. Since

$$-\log|1 - g^{(\sigma)}(\lambda)| \geq -\log 2,$$

it follows from Fatou's lemma and the convergence of (8.54) that

$$\int_{-\infty}^{\infty} \left(\frac{\alpha}{\alpha^2 + \lambda^2}\right) |-\log|1 - g(\lambda)|| \, d\lambda < \infty.$$

For real $0 < z < 1$ and complex w with $|w| \leq 1$,

$$\begin{aligned} |1 - zw|^2 &= |1 - z + z(1 - w)|^2 \\ &= (1 - z)^2 + z^2|1 - w|^2 + 2z(1 - z)\operatorname{Re}(1 - w) \\ &\geq (1 - z)^2 + z^2|1 - w|^2 \geq \tfrac{1}{4}\min(1, |1 - w|^2). \end{aligned}$$

Hence

$$-\log|1 - g^{(\sigma)}(\lambda)| \leq \log[4/\min(1, |1 - g(\lambda)|],$$

which gives the required dominating, integrable function. □

5. Applications to Sequential Probability Ratio Tests and Cusum Tests

In order to obtain statistically interesting approximations from the results of Sections 1–4, one must occasionally resort to heuristic reasoning which seems difficult to justify theoretically. For a related mathematically rigorous approach, see Chapter X.

Assume that $z_1, z_2, \ldots$ are independent with distribution function of the form (8.6). By (8.7) $\mu = \psi'(\theta)$ is a continuous strictly increasing function of θ, and for present purposes it is convenient to regard the family of distributions as indexed by μ. Hence we shall write

(8.56) $$P_\mu\{z_1 \in dx\} = dF_\mu(x) = e^{\theta x - \psi(\theta)} dF(x).$$

To emphasize that θ is now regarded as a function of μ the notation $\theta(\mu)$ is occasionally used.

Let $a \leq 0 < b$, $S_n = z_1 + \cdots + z_n$, and consider the stopping rules

(8.57) $$N = \inf\{n: S_n \notin (a, b)\}$$

and

$$\tau = \inf\{n: S_n \geq b\}.$$

Let $\mu_0 < 0$ and assume there exists $\mu_1 > 0$ such that $\psi(\theta_0) = \psi(\theta_1)$, where $\theta_i = \theta(\mu_i)$. For a nonarithmetic distribution function F, (8.48) now takes the form

$$P_{\mu_0}\{\tau < \infty\} \sim \nu_+ e^{-b\Delta} \qquad (b \to \infty), \tag{8.58}$$

where $\Delta = \theta_1 - \theta_0$ and

$$\nu_+ = \nu_+(\mu_1) = P_{\mu_0}\{\tau_+ = \infty\} P_{\mu_1}\{\tau_- = \infty\}/\Delta\psi'(\theta_1). \tag{8.59}$$

It is tempting to apply this to the stopping rule (8.57) by writing

$$\begin{aligned} P_{\mu_0}\{S_N \geq b\} &= P_{\mu_0}\{\tau < \infty\} - P_{\mu_0}\{S_N \leq a, \tau < \infty\} \\ &= P_{\mu_0}\{\tau < \infty\} - E_{\mu_0}[P_{\mu_0}\{\tau < \infty | S_N\}; S_N \leq a] \end{aligned} \tag{8.60}$$

and applying (8.58) to each term separately. Treating (8.58) as an equality for the moment, but using the notation $\cong$ as a reminder that an approximation is involved, we obtain

$$\begin{aligned} P_{\mu_0}\{S_N \geq b\} &\cong \nu_+ e^{-\Delta b} - \nu_+ E_{\mu_0}[e^{-\Delta(b - S_N)}; S_N \leq a] \\ &= \nu_+ e^{-\Delta b}[1 - E_{\mu_0}(e^{\Delta S_N}; S_N \leq a)] \\ &= \nu_+ e^{-\Delta b}(1 - P_{\mu_1}\{S_N \leq a\}) \\ &= \nu_+ e^{-\Delta b} P_{\mu_1}\{S_N \geq b\}. \end{aligned} \tag{8.61}$$

A similar analysis with the roles of a and b, μ_0 and μ_1 interchanged leads to

$$P_{\mu_1}\{S_N \leq a\} \cong \nu_- e^{\Delta a} P_{\mu_0}\{S_N \leq a\}, \tag{8.62}$$

where in analogy with (8.59)

$$\nu_- = P_{\mu_1}\{\tau_- = \infty\} P_{\mu_0}\{\tau_+ = \infty\}/\Delta|\psi'(\theta_0)|. \tag{8.63}$$

Simultaneous solution yields (cf. (2.34))

$$P_{\mu_0}\{S_N \geq b\} \cong \frac{\nu_+ e^{-\Delta b} - \nu_+ \nu_- e^{-\Delta(b-a)}}{1 - \nu_+ \nu_- e^{-\Delta(b-a)}}. \tag{8.64}$$

EXAMPLE 8.65. Suppose that F is the standard normal distribution function, so $\psi(\theta) = \theta^2/2$, $\mu = \theta$, and F_μ is a normal distribution function with mean μ and variance 1. By symmetry and (8.48) (cf. also (8.49)), $\nu_+ = \nu_- = \nu(\Delta)$, where $\nu(\cdot)$ is defined in (4.37). Using the approximation (4.38) in (8.64) yields

$$P_{\mu_0}\{S_N \geq b\} \cong \frac{e^{-\Delta(b+\rho)} - e^{-\Delta(b-a+2\rho)}}{1 - e^{-\Delta(b-a+2\rho)}}, \tag{8.66}$$

which is just the Wald approximation of II.3 modified by using $b + \rho$ and $a - \rho$ in place of b and a. The numerical accuracy of this and similar approximations has been documented in Problem 2.2, III.5, and III.6.

Although (8.64) is invariably a considerable improvement over (2.34) for numerical purposes, it is difficult in general to provide a precise mathematical justification. If we assume that $|a|$ and $b \to \infty$ at the same rate, so $b|a|^{-1}$ is bounded away from 0 and ∞, the error implicit in the approximation of the first term on the right hand side of (8.60) via (8.58) can be larger than the second term, so there is no simple mathematically justifiable reason for including the second term in the final approximation. Of course, omitting the second term leads to the (asymptotically correct) approximation

$$P_{\mu_0}\{S_N \geq b\} \sim P_{\mu_0}\{\tau < \infty\},$$

which cannot be expected to be good when μ_0 is close to 0.

An alternative to the preceding analysis is to let $\mu_0 \to 0$ as b and $|a| \to \infty$. If $\mu_0 a$ remains bounded, then $P_{\mu_0}\{S_N \geq b\}$ is bounded away from 0 and 1; and an analysis of the excess over the boundary provides theoretical justification for an approximation similar to (8.66) (see Chapter X).

For the special case $\mu_0 = 0$, an analysis based on Wald's identity (2.18) yields (cf. Problem 2.10, 8.3)

$$P_0\{S_N \geq b\} \cong \frac{|a + \rho_-|}{b + \rho_+ + |a + \rho_-|}, \tag{8.67}$$

where

$$\rho_\pm = E_0(S_{\tau_\pm}^2)/2E_0(S_{\tau_\pm}) \tag{8.68}$$

is the expectation of the limiting distribution appearing in (8.33). In this case, Stone's (1965) refinement of the renewal theorem allows one to prove that the error made in the approximation (8.67) is exponentially small as $b \to \infty$, $|a| \to \infty$, and $b|a|^{-1}$ is bounded away from 0 and ∞. It is also reassuring to note that the right hand side of (8.64) converges to the right hand side of (8.67) as $\mu_0 \to 0$ (see Problem 8.3). Similar calculations yield approximations for $E_\mu(T)$ (Problem 8.4).

Suppose now that $a = 0$, which occurs for the cusum tests described in II.6. Then (8.61) remains valid, but (8.62) should be replaced by

$$\begin{aligned} P_{\mu_1}\{S_N \leq 0\} &= P_{\mu_1}\{\tau_- < \infty\} - E_{\mu_1}[P_{\mu_1}\{\tau_- < \infty | S_N\}; S_N \geq b] \\ &\cong P_{\mu_1}\{\tau_- < \infty\} - \nu_- E_{\mu_1}(e^{-\Delta S_N}; S_N \geq b) \qquad (8.69) \\ &= P_{\mu_1}\{\tau_- < \infty\} - \nu_- P_{\mu_0}\{S_N \geq b\}. \end{aligned}$$

One can easily obtain approximations for $E_\mu(T)$ by a similar analysis. A somewhat more detailed discussion is given in Siegmund (1975a), who indicates that at least in the case of normal random variables the approximations are very precise. For a slightly different approach which requires less numerical computation and is about as accurate, see X.2.

6. Conditioned Random Walks

This section contains an asymptotic version of (3.13) for random walks which can be imbedded in an exponential family. The main result is fairly complicated, and except for a few special cases one would presumably want to make additional approximations before using it for numerical calculations. On the other hand, it provides heuristic motivation for (3.28), and after some refinement for (4.41) and (4.49) as well (see IX.2).

Assume that $z_1, z_2, \ldots$ are independent with probability distribution of the form (8.56), and put $S_n = z_1 + \cdots + z_n$. Let $\tau = \tau(b) = \inf\{n: S_n \geq b\}$. Let $\mathscr{E}_n$ denote the class of events determined by $z_1, \ldots, z_n$, and

$$P_\xi^{(m)}(A) = P_\mu(A \mid S_m = \xi) \qquad (A \in \mathscr{E}_m).$$

(By sufficiency this conditional probability does not depend on μ.)

An additional technical condition is required to permit conditional probabilities to be expressed in terms of ratios of probability density functions and to insure that local limit theorems apply. The assumption can be either that z_1 is arithmetic or that z_1 has a well behaved continuous distribution. We shall consider in detail only the latter case, for which a convenient assumption is that for all μ there exists an n_0 such that

$$\int_{-\infty}^{\infty} |E_\mu \exp(i\lambda z_1)|^{n_0}\, d\lambda < \infty.$$

This implies that for all $n \geq n_0$ the P_μ distribution of S_n has a bounded density function $f_{\mu,n}$ and as $n \to \infty$

$$f_{\mu,n}(\sigma n^{1/2} y + n\mu)\sigma n^{1/2} \to \phi(y) \tag{8.70}$$

uniformly in y, where $\sigma^2 = \psi''(\theta)$ is the P_μ-variance of z_1 (Feller, 1971, p. 516). Assume also that z_1 has a bounded probability density function.

Theorem 8.72 below gives an asymptotic evaluation of $P_\xi^{(m)}\{\tau < m\}$ when $b = m\zeta$ and $\xi = m\xi_0$ for some fixed $\zeta > 0$ and $\xi_0 < \zeta$. Primary applications are to calculation of the unconditional probability

$$P_\mu\{\tau < m, S_m \leq c\} = \int_{(-\infty, c)} P_\xi^{(m)}\{\tau < m\} P_\mu\{S_m \in d\xi\} \tag{8.71}$$

for $c \leq b$. The situation is complicated by the fact that when $c = b$, for a wide range of values of μ the important values of ξ in (8.71) are $\xi = m\zeta - x$ with x bounded. For such ξ a slightly different approximation is appropriate, but it seems possible to combine the two cases heuristically to give a single formula for numerical computations.

We begin with a statement of Theorem 8.72 and then return to a discussion of applications. The proof is deferred to the end of the section.

Theorem 8.72. *Let $b = m\zeta$ and $\xi = m\xi_0$ for arbitrary, fixed $\zeta > 0$ and $\xi_0 < \zeta$. Assume that there exist $\mu_2 < 0 < \mu_1$ (necessarily unique) such that*

$$\psi(\theta(\mu_2)) = \psi(\theta(\mu_1)) \tag{8.73}$$

and

$$1 = \mu_1^{-1}\zeta + |\mu_2|^{-1}(\zeta - \xi_0). \tag{8.74}$$

Let $\theta_i = \theta(\mu_i)$ $(i = 1, 2)$, $\theta_0 = \theta(\xi_0)$, and $\sigma_i^2 = \psi''(\theta_i)$ $(i = 0, 1, 2)$. *Then as* $m \to \infty$

$$P_\xi^{(m)}\{\tau < m\} \sim K(\zeta, \xi_0)\exp\{-m[(\theta_1 - \theta_2)\zeta + (\theta_2 - \theta_0)\xi_0 - \psi(\theta_2) + \psi(\theta_0)]\}, \tag{8.75}$$

where

$$K(\zeta, \xi_0) = \frac{\sigma_0|\mu_2|^{1/2}P_{\mu_2}\{\tau_+ = \infty\}P_{\mu_1}\{\tau_- = \infty\}}{\sigma_2(\zeta - \xi_0)^{1/2}(\theta_1 - \theta_2)\mu_1}\left\{1 + \frac{\sigma_1^2|\mu_2|^3\zeta}{\sigma_2^2\mu_1^3(\zeta - \xi_0)}\right\}^{-1/2}. \tag{8.76}$$

Remark. We have already encountered the condition (8.73) on several occasions. The role of (8.74) will become clear during the proof of the theorem. It is usually routine to verify the existence of μ_1 and μ_2 and to compute them. For example, if $\psi(\theta)$ and $\psi'(\theta)$ both diverge (continuously) to $+\infty$ as θ approaches the endpoints of its interval of definition, then for $\mu_2 < 0 < \mu_1$ satisfying (8.73), as μ_1 increases from 0 to ∞, $\mu_1^{-1}\zeta + |\mu_2|^{-1}(\zeta - \xi_0)$ decreases continuously from ∞ to 0, and hence satisfies (8.74) for exactly one value μ_1.

Example 8.77. Suppose F in (8.56) is the standard normal distribution, so $\psi(\theta) = \theta^2/2$, $\mu = \theta$, and the P_μ distribution of z_1 is normal with mean μ and variance 1. By symmetry $\mu_1 = -\mu_2$, and it is easily verified that $\mu_1 = 2\zeta - \xi_0$. By (8.48) the right hand side of (8.75) becomes

$$v[2(2\zeta - \xi_0)]\exp[-2m\zeta(\zeta - \xi_0)], \tag{8.78}$$

where v is defined in (4.37) (see also (8.49)). Note that the exponential factor in (8.78) is the exact probability given in (3.13) for the corresponding Brownian motion problem.

This asymptotic result provides heuristic motivation for the approximation (3.28) as follows. If for the function v in (8.78) we use the approximation suggested in (4.38), then up to the asymptotically negligible factor $e^{2\rho^2/m}$, (8.78) can be rewritten

$$P_\xi^{(m)}\{\tau < m\} \cong \exp[-2m^{-1}(b + \rho)(b + \rho - \xi)]. \tag{8.79}$$

Substituting this into

$$P_\mu\{\tau < m, S_m \le c\} = \int_{-\infty}^{c} P_\xi^{(m)}\{\tau < m\}\phi[m^{-1/2}(\xi - m\mu)]m^{-1/2}\,d\xi$$

and integrating yields (3.28).

An argument like that given in Example 8.77 for normal random variables can be generalized to the exponential family assumption of Theorem 8.72. However, in general the unconditioning integral can only be evaluated asymptotically, and hence it seems doubtful that the approximation will be as good as (3.28) over as wide a range of the parameters μ, ζ, and m.

When $b = c$ these approximations are questionable because the primary contribution to the integral in (8.71) often comes from values of ξ close to b, for which Theorem 8.72 does not apply. However, this case is easy to treat directly, and the result is about the same as using (8.75) for ξ close to b. See IX.3—especially (9.64)–(9.66)—for detailed calculations in a related context.

Proof of Theorem 8.72. Let f_n denote the probability density function of S_n. Recall that $f_{\mu,n}$ denotes the P_μ density of S_n, so $f_n = f_{\tilde{\mu},n}$ for some value $\tilde{\mu}$ which plays no role in what follows. For a yet to be specified $\mu_1 > 0$ and all $n < m$ the likelihood ratio of $z_1, \ldots, z_n$ under $P_\xi^{(m)}$ relative to P_{μ_1} is

(8.80) $$f_{m-n}(m\xi_0 - S_n)\exp[-\theta_1 S_n + n\psi(\theta_1)]/f_m(m\xi_0).$$

Also

(8.81) $$f_m(m\xi_0) = f_{\xi_0,m}(m\xi_0)\exp[-m\theta_0\xi_0 + m\psi(\theta_0)],$$

and for a yet to be specified μ_2

$$f_{m-n}(m\xi_0 - S_n) = f_{\mu_2,m-n}(m\xi_0 - S_n)\exp[-\theta_2(m\xi_0 - S_n) + (m-n)\psi(\theta_2)].$$
(8.82)

Choosing μ_1 and μ_2 to satisfy (8.73), and substituting (8.81) and (8.82) into (8.80) yields a basic identity:

$$P_\xi^{(m)}\{\tau < m\} = \exp\{-m[(\theta_2 - \theta_0)\xi_0 + \psi(\theta_0) - \psi(\theta_2)]\}$$
$$\times E_{\mu_1}\{f_{\mu_2,m-\tau}(m\xi_0 - S_\tau)\exp[-(\theta_1 - \theta_2)S_\tau]/f_{\xi_0,m}(m\xi_0); \tau < m\},$$

which in terms of $R_m = S_\tau - \zeta m$ can be rewritten

$$P_\xi^{(m)}\{\tau < m\}\exp\{m[(\theta_1 - \theta_2)\zeta + (\theta_2 - \theta_0)\xi_0 + \psi(\theta_0) - \psi(\theta_2)]\}$$
$$= E_{\mu_1}\{\exp[-(\theta_1 - \theta_2)R_m]f_{\mu_2,m-\tau}[m(\xi_0 - \zeta) - R_m]/f_{\xi_0,m}(m\xi_0); \tau < m\}.$$
(8.83)

By (8.70)
$$f_{\xi_0,m}(m\xi_0) \sim (2\pi m\sigma_0^2)^{-1/2} \qquad (m \to \infty).$$

Easy approximations show that for $\varepsilon > 0$ chosen so that $\zeta\mu_1^{-1}(1+\varepsilon) < 1$

$$P_{\mu_1}\{\tau > m\zeta\mu_1^{-1}(1+\varepsilon)\} + P_{\mu_1}\{\tau < m, R_m > (\log m)^2\} = o(m^{-1/2});$$

and since $f_{\mu_2,m-\tau}$ is bounded, uniformly in τ, the expectation in (8.83) equals that taken over the smaller event

$$\{\tau \le m\zeta\mu_1^{-1}(1+\varepsilon), R_m \le (\log m)^2\}$$

plus a term tending to 0 as $m \to \infty$. By (8.70) and (8.74), uniformly on this smaller event

$$f_{\mu_2,m-\tau}[m(\xi_0 - \zeta) - R_m]$$
$$= \sigma_2^{-1}(m-\tau)^{-1/2}\phi\{[\mu_2(\tau - m\zeta/\mu_1) - R_m]/\sigma_2(m-\tau)^{1/2}\} + o(m^{-1/2}),$$

so the expectation in (8.83) has the same limit as

$$(2\pi)^{1/2}\sigma_0\sigma_2^{-1}E_{\mu_1}\{\exp[-(\theta_1-\theta_2)R_m](1-\tau/m)^{-1/2}$$
$$\times\,\phi\{\mu_2(\tau-m\zeta/\mu_1)/\sigma_2(m-\tau)^{1/2}\};\tau<m\zeta\mu_1^{-1}(1+\varepsilon),R_m<(\log m)^2\}.$$

Since $P_{\mu_1}\{m^{-1}\tau\to\zeta\mu_1^{-1}\}=1$, this last expression can be evaluated by Theorem 8.34 to yield (8.75) □

Remark. We are now in a position to understand the role of (8.74). Roughly speaking ζ/μ_1 is the proportion of the m steps which the process requires to ascend to $m\zeta$ at rate μ_1, and $(\zeta-\xi_0)/|\mu_2|$ is the proportion required to descend from $m\zeta$ to $m\xi_0$ at (negative) rate μ_2. These must add up to one. More precisely, in order to analyze $f_{\mu_2,m-\tau}(m\xi_0-S_\tau)$ by means of (8.70) it is necessary that $m\xi_0-S_\tau\cong m(\xi_0-\zeta)$ be roughly at the center of the distribution $f_{\mu_2,m-\tau}$. Under P_{μ_1}, τ is close to $m\zeta/\mu_1$ with probability close to one, and hence the desired centering occurs if μ_2 is chosen to satisfy $m(1-\zeta/\mu_1)\mu_2=m(\xi_0-\zeta)$, which is precisely (8.74).

Problems

8.1. Let $g(\lambda)=E(e^{i\lambda z_1})$. Show that z_1 is arithmetic if and only if there exists a $\lambda\neq 0$ such that $g(\lambda)=1$. If z_1 is arithmetic with span d and λ_0 denotes the smallest positive λ such that $g(\lambda)=1$, then $d=\lambda_0/2\pi$. Use (8.42) to show that if z_1 is nonarithmetic (arithmetic with span d), then S_{τ_+} and S_{τ_-} are nonarithmetic (arithmetic with span d).

8.2. Suppose $E(z_1)=\mu>0$. Find the limiting behavior as $b\to\infty$ of

$$\sum_n P\{\tau(b)>n, b-x-h<S_n\leq b-x\}.$$

8.3. Under the conditions of Section 5, suppose that $b\to\infty$, $a\to-\infty$ and $b/|a|$ is constant. Show that

$$P_0\{S_N\geq b\}=\frac{|a+\rho_-|}{b-a+\rho_+-\rho_-}+o(b^{-1}). \tag{8.84}$$

Obtain a similar approximation when $a=0$. Show that the right hand side of (8.64) converges to the right hand side of (8.67) as $\mu_0\to 0$.

Remark. A stronger version of the renewal theorem shows that the remainder in (8.84) is actually exponentially small (see Appendix 4).

8.4. Find approximations for $E_\mu(N)$ similar in spirit to those given for $P_\mu\{S_N\geq b\}$ in Section 5 and Problem 8.3.

8.5. Suppose $\mu>0$ and $\sigma^2<\infty$. Differentiate $(1-G_+(s,0))/(1-s)=\exp(\sum_1^\infty n^{-1}s^nP\{S_n\leq 0\})$ and let $s\uparrow 1$ to show $E(\tau_+^2)/E(\tau_+)=1+2\sum_1^\infty P\{S_n\leq 0\}\leq\infty$. Argue using Wald's second moment identity that $E(\tau_+^2)<\infty$ and hence $\sum P\{S_n\leq 0\}<\infty$.

8.6. Show that if $\mu>0$ and $E[(z_1^+)^r]<\infty$ for some $r>1$, then $E(S_{\tau_+}^r)<\infty$. *Hint*: Use Proposition 8.38 and the identity

$$P\{S_{\tau_+} > x\} = \sum_{n=0}^{\infty} \int_{(-\infty,0]} P\{\tau_+ > n, S_n \in dy\} P\{z_1 > x - y\}. \tag{8.85}$$

8.7. Show that $\mu > 0$ and $E[(z_1^+)^{r+1}] < \infty$ for some $r > 0$ imply that

$$E[(S_{\tau(b)} - b)^r] \to E(S_{\tau_+}^{r+1})/(r+1)E(S_{\tau_+}).$$

8.8. Assume $P\{z_1 > 0\} = 1$ and $E[(z_1^+)^{1+r}] < \infty$ for some $r > 0$. Use (8.11) and (8.32) to prove that $P\{S_{\tau(b)} - b > x\} \le U(1) \int_{x-2}^{\infty} P\{z_1 > y\}\, dy$ and hence $\int^{\infty} x^{r-1} \sup_b P\{S_{\tau(b)} - b > x\}\, dx < \infty$. Use Problem 8.6 to show that this latter conclusion holds without requiring $P\{z_1 > 0\} = 1$ (provided, of course, that $\mu > 0$). Show that this result gives another solution to Problem 8.7.

8.9. Show that if $E(z_1) = 0$, then $P\{\tau_+ < \infty\} = 1 = P\{\tau_- < \infty\}$. (*Hint*: Use Corollary 8.41 and Wald's identity.) Assume that $E(z_1) = 0$ and $E(S_{\tau_-}) < \infty$ (for which it is sufficient that $E(z_1^2) < \infty$—see Feller, 1971, p. 612). Show that $E[(z_1^+)^{r+1}] < \infty$ for some $r > 0$ implies $E(S_{\tau_+}^r) < \infty$ and $E[\exp(\lambda z_1)] < \infty$ for some $\lambda > 0$ implies that $E[\exp(\lambda S_{\tau_+})] < \infty$. *Hint*: Use (8.85), (8.38), and (8.32).

8.10. Suppose $z_n = 1 - x_n$, where $P\{x_n \in dx\} = e^{-x}\, dx$ for $x > 0$ and 0 otherwise. Show that S_{τ_-} has an exponential distribution on $(-\infty, 0)$ and S_{τ_+} is distributed uniformly on $(0, 1)$. Hence $E(S_{\tau_+}^2)/2E(S_{\tau_+}) = \frac{1}{3}$ and $E(S_{\tau_-}^2)/2E(S_{\tau_-}) = -1$.

8.11. Let $F_n(x)$ denote the empirical distribution of n independent, uniform on $[0, 1]$ random variables. Show that

$$P\left\{\sup_x (x - F_n(x)) > \zeta\right\} = P\left\{\max_{1 \le j \le n} (W_j - j) \ge n\zeta - 1 \,|\, W_{n+1} - (n+1) = -1\right\}, \tag{8.86}$$

where W_j is a sum of j independent standard exponential random variables. Use this representation and the method of proof of Theorem 8.72 to obtain an asymptotic expression for the left hand side of (8.86) as $n \to \infty$.

8.12. Apply Theorem 8.72 to obtain approximations in the spirit of (3.36) for the error probabilities of a truncated sequential probability ratio test for the mean of a negative exponential distribution. Compare this approximation numerically with that given in X.3.

8.13. Assume that $\mu = E(z_1) > 0$. Let $M = \min(S_1, S_2, \ldots)$. Prove that $[E(S_{\tau_+})]^{-1} \times P\{S_{\tau_+} > x\} = \mu^{-1} P\{M > x\}$ for all $x > 0$. For an application, see IX.5.
Hint: Let $\sigma = \sup\{n\colon S_n = M\}$. Start from $P\{\sigma = n, M > x\}$ and use time reversal as in Proposition 8.38.

8.14. Suppose that $Ez_1 > 0$ and $Ez_1^2 < \infty$.
(a) Differentiate (8.42) and use the identity

$$\min_{0 \le n < \infty} S_n = \sum_{k=0}^{\infty} S_{\tau_-^{(k)}} I_{\{\tau_-^{(k)} < \infty,\, \tau_-^{(k+1)} = \infty\}}$$

to show that

$$E(S_{\tau_+}^2)/2ES_{\tau_+} = Ez_1^2/2Ez_1 + E\left\{\min_{0 \le n < \infty} S_n\right\}.$$

(b) Show by a direct argument that

$$E\left\{\min_{0\le k\le n} S_k\right\} = -\sum_{k=1}^{n} E(S_k^-)/k.$$

Compare (a) and (b) with (8.50).

Hint: For (b) decompose $E\{\min_{0\le k\le n} S_k\}$ according as $S_n > 0$ or $S_n \le 0$, on $\{S_n \le 0\}$ write $\min_{0\le k\le n} S_k = z_1 + \min_{1\le k\le n}(S_k - S_1)$, and use the equality $E(z_1 | S_n) = n^{-1} S_n$.

CHAPTER IX

Nonlinear Renewal Theory

1. Introduction and Examples

This chapter is concerned with first passages of random walks to nonlinear boundaries. Suitable generalizations of the renewal theory of Chapter VIII are developed in order to justify and generalize the approximations suggested in IV.3.

The following class of examples contains a number of important special cases and provides motivation for the general theoretical approach which follows. Let $x_1, x_2, \ldots$ be independent and identically distributed random variables with mean μ and variance σ^2. Let g be a function which is positive and twice continuously differentiable in a neighborhood of μ, and let

$$T = T_a = \inf\left\{n: n \geq m_0, ng\left(n^{-1}\sum_1^n x_i\right) \geq a\right\}. \tag{9.1}$$

As a special case consider a first passage time of the form

$$T = \inf\left\{n: \sum_1^n x_i \geq cn^\gamma\right\}, \tag{9.2}$$

which can be put in the form (9.1) with $g(x) = (x^+)^{(1-\gamma)^{-1}}$ and $a = c^{(1-\gamma)^{-1}}$, provided $0 \leq \gamma < 1$ and $\mu > 0$. A second example of (9.1) is a repeated likelihood ratio test of H_0: $\theta = \theta_0$ against H_1: $\theta \neq \theta_0$ in the exponential family (8.6). The likelihood ratio statistic is of the form $ng(n^{-1}\sum_1^n x_i)$, where $g(x) = \hat{\theta}(x)x - \psi[\hat{\theta}(x)] - [\theta_0 x - \psi(\theta_0)]$ and $\hat{\theta}(x)$ is the maximum likelihood estimator of θ when $n^{-1}\sum_1^n x_i = x$. Since $\psi'[\hat{\theta}(x)] = x$ and ψ is convex, it is easy to see that $g(\mu) > 0$ provided $\theta \neq \theta_0$.

Expanding g in a Taylor series about μ, we have for $n^{-1}\sum_1^n x_i$ in a neighborhood U of μ

$$(9.3)\quad ng(n^{-1}\Sigma x_i) = ng(\mu) + g'(\mu)\sum_1^n (x_i - \mu) + \frac{1}{2}n^{-1}\left[\sum_1^n (x_i - \mu)\right]^2 g''(\zeta_n),$$

where $|\zeta_n - \mu| \le |n^{-1}\sum_1^n x_i - \mu|$. This can be written more abstractly in the form

$$(9.4)\qquad Z_n = \tilde{S}_n + \eta_n,$$

where $\tilde{S}_n = ng(\mu) + g'(\mu)\sum_1^n (x_i - \mu)$ is a random walk with positive drift $\tilde{\mu} = g(\mu)$ and $\eta_n = Z_n - \tilde{S}_n$. Moreover, by the law of large numbers with probability converging to one, $n^{-1}\sum_{i=1}^n x_i \in U$, so

$$\eta_n = \frac{1}{2}n^{-1}\left[\sum_{i=1}^n (x_i - \mu)\right]^2 g''(\zeta_n)$$

converges in law to $(\sigma^2/2)g''(\mu)\chi_1^2$, where χ_1^2 denotes a chi-squared random variable with one degree of freedom. It seems clear that to a considerable extent the behavior of Z_n in (9.4) is determined by the behavior of $\tilde{S}_n$. This is made more precise in what follows.

There are two different and complementary viewpoints behind (9.1) and (9.2). The first involves a nonlinear function of a random walk crossing a fixed boundary, the second an ordinary random walk but a nonlinear boundary. With (9.1) goes the Taylor expansion (9.3) which approximates the nonlinear function by a linear one plus a comparatively insignificant quadratic term. The analogous linearization of (9.2) is to approximate the nonlinear boundary by its tangent at the point where with high probability the first passage takes place. In spite of the conceptual similarity of these approaches, there are important technical differences. Note, for example, that the viewpoint embodied by (9.1) and (9.3) generalizes immediately to vector valued random variables $x_1, x_2, \ldots$.

Section 2 contains some general results based on the representation (9.4). Applications are given in Sections 3 and 4. A different approach based directly on (9.2) is described briefly in Section 5.

2. General Theorems

Let $z_1, z_2, \ldots$ be independent and identically distributed with $E(z_1) = \tilde{\mu} > 0$. Let $\tilde{S}_n = \sum_{i=1}^n z_i$ and let $\mathscr{E}_n$ denote the σ-algebra of events determined by $z_1, \ldots, z_n$ $(n = 1, 2, \ldots)$. We shall be interested in developing renewal theory for processes of the form

$$Z_n = \tilde{S}_n + \eta_n,$$

where η_n is adapted to $\mathscr{E}_n$ and satisfies certain additional conditions.

Definition. The random variables η_n, $n = 1, 2, \ldots$ are called slowly changing if

$$n^{-1}\max[|\eta_1|, \ldots, |\eta_n|] \to 0 \tag{9.5}$$

in probability and for every $\varepsilon > 0$ there exist n^* and $\delta > 0$ such that for all $n \geq n^*$

$$P\left\{\max_{1 \leq k \leq n\delta} |\eta_{n+k} - \eta_n| > \varepsilon\right\} < \varepsilon. \tag{9.6}$$

A trivial example of a slowly changing sequence is one which converges almost surely to a finite limit. The following example is a particularly important one.

EXAMPLE 9.7. Let η_n be as in Section 1—see especially (9.3) and (9.4). An easy consequence of the strong law of large numbers is that $n^{-1}\eta_n \to 0$ with probability one, so (9.5) holds. To see that (9.6) holds, consider first the special case where g is quadratic so (9.3) holds with g'' a constant. It suffices to verify (9.6) for the sequence $\eta_n = s_n^2/n$, where $s_n = \sum_1^n (x_k - \mu)$. By simple algebra, for $1 \leq k \leq n\delta$

$$|s_{n+k}^2/(n+k) - s_n^2/n| \leq \delta s_n^2/n + (s_{n+k} - s_n)^2/n + 2|s_n(s_{n+k} - s_n)|/n. \tag{9.8}$$

Let $A = \{s_n^2 \leq n\lambda^2\}$ and $B = \{\max_{1 \leq k \leq n\delta}(s_{n+k} - s_n)^2 \leq n\delta^{1/2}\}$. On $A \cap B$ the left hand side of (9.8) is smaller than $\delta\lambda^2 + \delta^{1/2} + 2\lambda\delta^{1/4}$. By Chebyshev's and Kolmogorov's inequalities $P(A^c \cup B^c) \leq P(A^c) + P(B^c) \leq \sigma^2(\lambda^{-2} + \delta^{1/2})$. By choosing $\lambda = \delta^{-1/5}$ and δ sufficiently small one verifies (9.6) in this special case. For general g the expansion (9.3) and the strong law of large numbers permits one to carry through the preceding argument with minor modifications to complete the proof of (9.6).

For $a > 0$, $m_0 = 1, 2, \ldots$ define

$$T = T_a = \inf\{n: n \geq m_0, Z_n \geq a\}, \tag{9.9}$$

$$\tau = \tau_a = \inf\{n: \tilde{S}_n \geq a\}, \tag{9.10}$$

and

$$\tau_+ = \inf\{n: \tilde{S}_n > 0\}.$$

Recall from Corollary 8.33 that if z_1 is nonarithmetic, then for all $x \geq 0$

$$\lim_{a \to \infty} P\{\tilde{S}_{\tau_a} - a \leq x\} = H(x),$$

where

$$H(x) = (E\tilde{S}_{\tau_+})^{-1} \int_{[0,x]} P\{\tilde{S}_{\tau_+} \geq u\}\, du. \tag{9.11}$$

The following result shows that under appropriate conditions $Z_{T_a} - a$ has the same limiting distribution quite generally.

Theorem 9.12. *Assume that the η_n are slowly changing and that z_1 is nonarithmetic. Then*

$$\lim_{a\to\infty} P\{Z_{T_a} - a \le x\} = H(x),$$

where H is defined in (9.11).

The proof of Theorem 9.12 is preceded by a preliminary result.

Lemma 9.13. *If* (9.5) *holds, then $P\{T_a < \infty\} = 1$ for all a, and as $a \to \infty$*

$$a^{-1}T_a \to \tilde{\mu}^{-1}$$

in probability.

Proof of Lemma 9.13. By the strong law of large numbers and (9.5), $n^{-1}Z_n \to \tilde{\mu}$ in probability. Hence along some subsequence n', $Z_{n'}/n' \to \tilde{\mu}$ with probability one, so $P\{T_a < \infty\} = 1$ for all a.

The strong law also implies that $n^{-1}\max_{1\le k\le n}\tilde{S}_k \to \tilde{\mu}$ with probability one, so by (9.5) $n^{-1}\max_{1\le k\le n} Z_k \to \tilde{\mu}$ in probability. Let $n_0 = a/\tilde{\mu}$. Fix $0 < \varepsilon < 1$ and let $n_1 = n_0(1+\varepsilon)$. Since $a = n_1\tilde{\mu}/(1+\varepsilon)$,

$$\begin{aligned} P\{T_a > n_0(1+\varepsilon)\} &= P\left\{\max_{1\le k\le n_1} Z_k < a\right\} \\ &= P\left\{\max_{1\le k\le n_1} Z_k < n_1\tilde{\mu}/(1+\varepsilon)\right\} \to 0 \end{aligned}$$

as $a \to \infty$. Similarly $P\{T_a < n_0(1-\varepsilon)\} \to 0$, which completes the proof. □

Proof of Theorem 9.12. Let $\varepsilon > 0$ be arbitrary and δ as in (9.6). Let $n_0 = a/\tilde{\mu}$, $n_1 = (1-\delta/4)n_0$, and $n_2 = (1+\delta/4)n_0$. The idea of the proof is very simple, although the execution involves some care. First we condition on $\mathscr{E}_{n_1}$ and then compare the (conditional) processes Z_n and $Z_{n_1} + \tilde{S}_n - \tilde{S}_{n_1}$ for $n \ge n_1$. According to (9.6), with probability at least $1-\varepsilon$ these processes will be within ε of each other for all $n_1 \le n \le n_2$; and according to Lemma 9.13, T_a will fall in this interval with probability close to one. Hence with probability close to one both processes will cross the level a at the same time and will have the same excess over the boundary to within ε. Since $Z_{n_1} + \tilde{S}_n - \tilde{S}_{n_1}$ for $n > n_1$ is (conditionally) a random walk, Corollary 8.33 may be applied to yield the desired result (see Figure 9.1).

Let $B_a = \{\max_{1\le k\le n_1} Z_k < a - a^{1/2}\} = \{T_{a-a^{1/2}} > n_1\}$. From Lemma 9.13 it follows that as $a \to \infty$

$$P(B_a^c) + P\{T_a > n_2\} \to 0. \tag{9.14}$$

On B_a define $\tau_a^{(\varepsilon)} = \inf\{n\colon n \ge n_1, \tilde{S}_n - \tilde{S}_{n_1} \ge a + \varepsilon - Z_{n_1}\}$. Let $x > 0$. Since $a - Z_{n_1} > a^{1/2}$ on B_a, for all a sufficiently large Corollary 8.33 yields

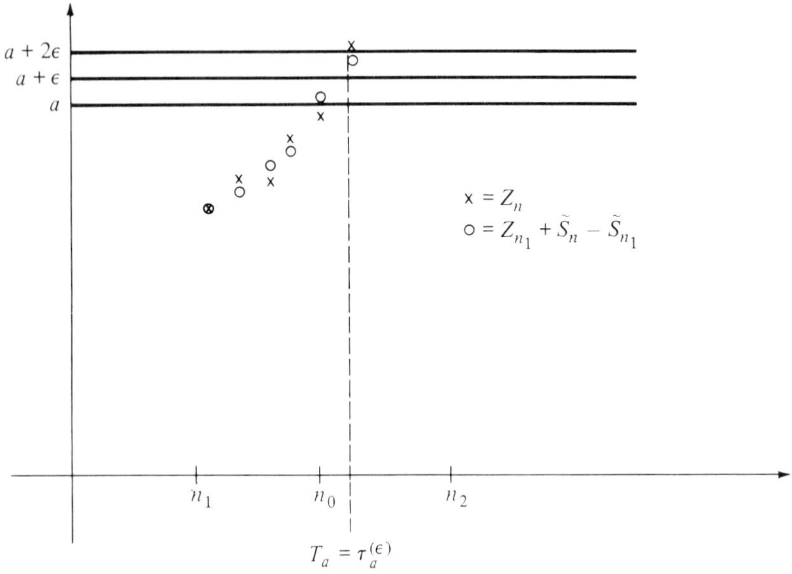

Figure 9.1.

(9.15) $$|P\{\tilde{S}_{\tau_a^{(\varepsilon)}} + \eta_{n_1} - (a + \varepsilon) \le x | \mathscr{E}_{n_1}\} - H(x)| < \varepsilon \text{ on } B_a,$$

where H is defined by (9.11). Let $x > 2\varepsilon$ and

$$C_a = B_a \cap \{T < n_2, \max_{n_1 \le k \le n_2} |\eta_k - \eta_{n_1}| < \varepsilon\}.$$

It is shown below that

(9.16) $$C_a \cap \{Z_{T_a} - a > x\} \subset \{\tau_a^{(\varepsilon)} = T_a, \tilde{S}_{\tau_a^{(\varepsilon)}} + \eta_{n_1} - (a + \varepsilon) > x - 2\varepsilon\}.$$

Then by (9.15) and (9.16),

$$\begin{aligned} P\{Z_{T_a} - a > x\} &\le P\{Z_{T_a} - a > x, C_a\} + P\{C_a^c\} \\ &\le P\{\tilde{S}_{\tau_a^{(\varepsilon)}} + \eta_{n_1} - (a + \varepsilon) > (x - 2\varepsilon), B_a\} + P(C_a^c) \\ &= E[P\{\tilde{S}_{\tau_a^{(\varepsilon)}} + \eta_{n_1} - (a + \varepsilon) > (x - 2\varepsilon) | \mathscr{E}_{n_1}\}; B_a] + P(C_a^c) \\ &\le E[1 - H(x - 2\varepsilon); B_a] + P(C_a^c) \\ &\le 1 - H(x - 2\varepsilon) + P(C_a^c). \end{aligned}$$

By (9.6) and (9.14), $\limsup_{a \to \infty} P(C_a^c) \le \varepsilon$, and hence since $\varepsilon > 0$ is arbitrary

$$\limsup_{a \to \infty} P\{Z_{T_a} - a > x\} \le 1 - H(x) \qquad (x > 0).$$

A similar argument proves

$$\liminf_{a \to \infty} P\{Z_{T_a} - a > x\} \ge 1 - H(x) \qquad (x > 0).$$

The main difference is that now one works with $\tau_a^{(-\varepsilon)} = \inf\{n: n \geq n_1, \tilde{S}_n - \tilde{S}_{n_1} \geq a - \varepsilon - Z_{n_1}\}$ and the relation (9.16) is replaced by

$$C_a \cap \{\tilde{S}_{\tau_a^{(-\varepsilon)}} + \eta_{n_1} - (a - \varepsilon) > x + 2\varepsilon\} \subset \{\tau_a^{(-\varepsilon)} = T_a, Z_{T_a} - a > x\}.$$

The details are omitted.

To verify (9.16) let $n_1 < k \leq n_2$. Observe that if $T_a = k$ and $\tilde{S}_k + \eta_k - a > x$, then on C_a, $\tilde{S}_k + \eta_{n_1} - (a + \varepsilon) > x - 2\varepsilon$ and hence $\tau_a^{(\varepsilon)} \leq k$. But if $\tau_a^{(\varepsilon)} = j < k$, then $\tilde{S}_j + \eta_{n_1} \geq a + \varepsilon$ so on C_a, $\tilde{S}_j + \eta_j \geq a$, which means that $T_a \leq j$. This contradicts $T_a = k$ and proves (9.16) (see Figure 9.1). □

A modification of the preceding argument along the lines of the proof of Theorem 8.34 yields the following result (cf. Problem 9.1).

Theorem 9.17. *Assume in addition to the conditions of Theorem* 9.12 *that* $\tilde{\sigma}^2 = \operatorname{var}(z_1) < \infty$ *and that* $n^{-1/2}\eta_n \to 0$ *in probability. Then for each* $-\infty < x < \infty$, $0 < y < \infty$

$$\lim_{a\to\infty} P\{T_a \leq a\tilde{\mu}^{-1} + x(a\tilde{\sigma}^2/\tilde{\mu}^3)^{1/2}, Z_{T_a} - a \leq y\} = \Phi(x)H(y).$$

Remarks. (i) There is a version of Theorem 9.12 when z_1 is arithmetic with span 1, say, but it requires the stronger hypothesis that η_n converge in law to a continuously distributed limit η. Then careful modification of the preceding argument shows that as $a \to \infty$ through integers, the limiting distribution of $Z_T - a$ is the same as the excess over a of the random walk $\tilde{S}_n = \tilde{S}_0 + z_1 + \cdots + z_n$, $n = 0, 1, \ldots$ where $\tilde{S}_0$ is distributed as η (independently of $z_1, z_2, \ldots$). However, the asymptotic independence of Theorem 9.17 is no longer valid except in very special cases. See Lalley (1982) for details.

(ii) The proof of Theorem 9.12 involved conditioning at the fixed time n_1 and random position Z_{n_1} which is close enough to a that η_n is essentially constant for $n_1 \leq n \leq T_a$ but far enough from a that the renewal theorem applied to $\tilde{S}_n - \tilde{S}_{n_1}$. An alternative possibility would be to condition on the random time $T_{a-\Delta(a)}$, where $\Delta(a) \to \infty$ but is small compared to a, e.g. $\Delta(a) = \log a$. This approach offers some advantages because the value of $Z_{T_{a-\Delta(a)}}$ is more nearly constant than Z_{n_1}. Something like it seems advisable if one wants to work directly with (9.2) to show that the limiting distribution of the excess of a random walk over a curved boundary is the same as the limiting distribution of the excess over an appropriate tangent (see Hogan, 1984).

Our next general result concerns $E(T_a)$. In the special case of (9.1), it is natural to use the expansion (9.3) and Wald's identity to obtain

$$E\left[Tg\left(T^{-1}\sum_1^T x_i\right)\right] = g(\mu)ET + \frac{1}{2}E\left[T^{-1}\left\{\sum_1^T (x_i - \mu)\right\}^2 g''(\zeta_T)\right] + o(1).$$

An obvious conjecture based on Theorems 2.40 and 9.12 is that

$$E\left[T^{-1}\left\{\sum_1^T (x_i - \mu)\right\}^2 g''(\zeta_T)\right] \to \sigma^2 g''(\mu)$$

and

$$E\left[Tg\left(T^{-1}\sum_1^T x_i\right)\right] = a + \int_{(0,\infty)} xH(dx) + o(1),$$

leading to

$$g(\mu)E(T_a) = a - \sigma^2 g''(\mu)/2 + E(\tilde{S}_{\tau_+}^2)/2E\tilde{S}_{\tau_+} + o(1) \tag{9.18}$$

as $a \to \infty$.

Unfortunately, in spite of the simplicity of the basic ideas and the individual calculations, the number of things to be checked in proving a result like (9.18) makes the argument rather tedious. Theorem 9.28 below is an illustrative, but not particularly general result. We begin with a simple preliminary approximation.

Lemma 9.19. *Assume* $\tilde{\sigma}^2 = \mathrm{var}(z_1) < \infty$, *that* (9.5) *holds, and for some* $0 < \varepsilon < \tilde{\mu}$

$$\sum_1^\infty P\{\eta_n \le -n\varepsilon\} < \infty. \tag{9.20}$$

Then $ET_a \sim a\tilde{\mu}^{-1}$ as $a \to \infty$.

Proof. Let $0 < \varepsilon < \tilde{\mu}$ be as in (9.20), and let $n_a = 2a/(\tilde{\mu} - \varepsilon)$. Then for $n \ge n_a$

$$\begin{aligned} P\{T > n\} &\le P\{\tilde{S}_n + \eta_n < a\} \le P\{\tilde{S}_n + \eta_n < n(\tilde{\mu} - \varepsilon)/2\} \\ &\le P\{\eta_n < -n\varepsilon\} + P\{\tilde{S}_n - n(\tilde{\mu} + \varepsilon)/2 < 0\} \\ &= q_n, \end{aligned}$$

say. By (9.20) and Problem 8.5, $\Sigma q_n < \infty$. Hence

$$\begin{aligned} E(T; T > 2n_a) &\le 2E(T - n_a; T > 2n_a) \\ &\le 2E(T - n_a; T > n_a) \le 2 \sum_{n \ge n_a} q_n \to 0 \end{aligned} \tag{9.21}$$

as $a \to \infty$. In particular T_a/a, $a \ge 1$, are uniformly integrable. Since by Lemma 9.13 $T_a/a \to \tilde{\mu}^{-1}$ in probability, the result follows. □

In order to formulate Theorem 9.28 below, consider the following conditions.

Assume there exist events $A_n \in \mathscr{E}_n$, $n = 1, 2, \ldots$ and $\mathscr{E}_n$ measurable random variables η_n^*, $n = 1, 2, \ldots$ such that

$$\sum_{n=1}^\infty P\left(\bigcup_{k=n}^\infty A_k^c\right) < \infty, \tag{9.22}$$

$$\eta_n = \eta_n^* \text{ on } A_n, \qquad n = 1, 2, \ldots, \tag{9.23}$$

$$\max_{0 \le k \le n} |\eta_{n+k}^*|,\ n = 1, 2, \ldots \text{ are uniformly integrable,} \tag{9.24}$$

(9.25) $$\sum_1^\infty P\{\eta_n^* \le -n\varepsilon\} < \infty \text{ for some } 0 < \varepsilon < \tilde{\mu},$$

(9.26) η_n^* converges in law to a random variable η,

and for some $0 < \delta < 1$

(9.27) $$aP\{T \le \delta a/\tilde{\mu}\} \to 0 \qquad (a \to \infty).$$

Theorem 9.28. *Suppose $\tilde{\sigma}^2 = \mathrm{var}(z_1) < \infty$ and z_1 is nonarithmetic. Assume also that (9.22)–(9.27) are satisfied and that η_n^*, $n = 1, 2, \ldots$ are slowly changing. Then as $a \to \infty$*

(9.29) $$E(T_a) = \tilde{\mu}^{-1}(a + \rho - E\eta) + o(1),$$

where $\rho = E(\tilde{S}_{\tau_+}^2)/2E\tilde{S}_{\tau_+}$ is the mean of the limiting distribution H of $\tilde{S}_{\tau_a} - a$.

In practice verifying (9.27) can cause some difficulty. Although the other conditions look complicated, the following result shows that they are satisfied quite generally.

Proposition 9.30. *Let $x_1, x_2, \ldots$ be independent and identically distributed with $\mu = Ex_1$, $\sigma^2 = \mathrm{var}(x_1)$, and $E|x_1|^\gamma < \infty$ for some $\gamma > 2$. Let g be twice continuously differentiable in a neighborhood U of μ with $g(\mu) > 0$, and suppose that $Z_n = ng(n^{-1}\sum_1^n x_k)$. Put $A_n = \{n^{-1}\sum_1^n x_k \in U\}$, $\tilde{S}_n = ng(\mu) + g'(\mu)\sum_1^n (x_k - \mu)$, and $\eta_n^* = \frac{1}{2}n^{-1}[\sum_1^n (x_k - \mu)]^2 g''(\zeta_n)I_{A_n}$, where ζ_n is defined as in (9.3). Then (9.22)–(9.26) are satisfied and η_n^*, $n = 1, 2, \ldots$ are slowly changing.*

Remarks. (i) Proposition 9.30 is easily generalized to vector valued x_n. Simple applications are the repeated significance tests of V.3 and V.4.

(ii) A useful generalization of Theorem 9.28 is to allow the decomposition $\eta_n = c(n) + \eta_n^*$ on A_n, where $c(n)$ is a sequence of constants which grow slowly compared to n. The stopping rule (4.3) can be put in this form with $c(n) = \log(n + 1)$. The conclusion of Theorem 9.28 must be modified by adding $-\tilde{\mu}^{-1}c(a\tilde{\mu}^{-1})$ to the right hand side of (9.29). Details of such a generalization with $c(n)$ limited to logarithmic growth are given by Hagwood and Woodroofe (1981). Lai and Siegmund (1979) have considered the case where $c(n)$ must grow more slowly than $n^{1/2}$. Their method imposes more severe moment conditions on z_1. In principle, Hogan's (1984) method should be still more general with regard to conditions on $c(\cdot)$.

Proof of Theorem 9.28. Let $n_0 = a\tilde{\mu}^{-1}$. It follows from (9.22) and (9.25) that (9.5) and (9.20) hold. Hence (9.21) holds. Together with (9.27) this implies that there exist $0 < \delta_1 < 1 < \delta_2$ such that

(9.31) $$aP\{T < n_1\} + E(T; T > n_2) \to 0,$$

where $n_i = \delta_i n_0$. Let

$$B_a = \{n_1 \le T \le n_2, \tau > n_1\} \cap \bigcap_{k \ge n_1} A_k,$$

where τ is defined in (9.10).

From (9.22), (9.31), and an easy calculation (cf. Problem 9.2) it follows that

$$aP(B_a^c) \to 0. \tag{9.32}$$

Hence

$$\begin{aligned} E(\tilde{S}_T; B_a) &= E(a + Z_T - a - \eta_T; B_a) \\ &= a + E(Z_T - a; B_a) - E(\eta_T; B_a) + o(1). \end{aligned} \tag{9.33}$$

An easy consequence of (9.23), (9.24), and (9.26) is that

$$E(\eta_T; B_a) \to E(\eta) \text{ as } a \to \infty. \tag{9.34}$$

Also by Wald's lemma

$$\begin{aligned} E(\tilde{S}_T; B_a) &= \tilde{\mu}ET - E(\tilde{S}_T; B_a^c) \\ &= \tilde{\mu}ET - \tilde{\mu}E(T; B_a^c) - E(\tilde{S}_T - \tilde{\mu}T; B_a^c). \end{aligned}$$

By (9.31) and (9.32) $E(T; B_a^c) \to 0$. By Wald's second moment identity, (9.32), and Lemma 9.19

$$\begin{aligned} |E(\tilde{S}_T - \tilde{\mu}T; B_a^c)| &\le \{P(B_a^c)E(\tilde{S}_T - \tilde{\mu}T)^2\}^{1/2} \\ &= \{P(B_a^c)\tilde{\sigma}^2 E(T)\}^{1/2} \to 0. \end{aligned}$$

Hence

$$E(\tilde{S}_T; B_a) = \tilde{\mu}E(T) + o(1). \tag{9.35}$$

Substitution of (9.34) and (9.35) into (9.33) yields

$$\tilde{\mu}E(T) = a + E(Z_T - a; B_a) - E\eta + o(1).$$

An appeal to Lemma 9.36 below completes the proof. □

Lemma 9.36. *Under the conditions of Theorem* 9.28, $E(Z_T - a; B_a) \to \rho = E(\tilde{S}_\tau^2)/2E\tilde{S}_{\tau_+}$.

Proof. By Theorem 9.12 it suffices to show that $(Z_T - a)I_{B_a}$, $a \ge 1$, are uniformly integrable. Let $x > 0$ and

$$C_a = C_a(x) = \left\{ \max_{n_1 \le k \le n_2} |\eta_k| I_{B_a} \le x \right\}.$$

Then

$$P\{B_a, Z_T - a > 3x\} \le P\{B_a \cap C_a, Z_T - a > 3x\} + P\{C_a^c\}, \tag{9.37}$$

and by an argument similar to the proof of Theorem 9.12

$$P\{B_a \cap C_a, Z_{T_a} - a > 3x\} \le P\{\tilde{S}_{\tau_{a+x}} - a > 2x\}$$
$$\le \sup_{b>0} P\{\tilde{S}_{\tau_b} - b > x\}.$$

The required uniform integrability now follows from (9.37), (9.24), and Problem 8.8. □

PROOF OF PROPOSITION 9.30. That η_n^*, $n = 1, 2, \ldots$ are slowly changing follows as in Example 9.7. Conditions (9.23) and (9.25) are trivially fulfilled, and (9.26) follows from the central limit theorem. To prove (9.24), let $s_n = \sum_1^n (x_k - \mu)$, and note that

$$P\left\{\max_{1 \le k \le n} s_{n+k}^2/(n+k) \ge x\right\} \le P\left\{\max_{1 \le k \le n} s_{n+k}^2 \ge xn\right\} \le x^{-\gamma/2} n^{-\gamma/2} E|s_{2n}|^\gamma.$$

Since

$$E|s_n|^\gamma = O(n^{\gamma/2}) \tag{9.38}$$

(e.g. Brillinger, 1962), (9.4) holds.

Condition (9.22) follows from a general result of Baum and Katz (1965). A direct argument which will be useful in other contexts is as follows. Let $\varepsilon > 0$ be such that $(\mu - \varepsilon, \mu + \varepsilon) \subset U$, so $P(\bigcup_{k=n}^\infty A_k^c) \le P\{|s_k| \ge k\varepsilon \text{ for some } k \ge n\}$. By the Hájek–Rényi–Chow inequality (e.g., Chow *et al.*, 1971, p. 25), this latter probability is majorized by

$$\varepsilon^{-\gamma}\left\{\sum_{k>n} k^{-\gamma}(E|s_k|^\gamma - E|s_{k-1}|^\gamma) + n^{-\gamma}E|s_n|^\gamma\right\}.$$

Summing by parts and using (9.38) yields (9.22). □

Remark 9.39. Theorem 9.17 leads to the conjecture

$$E(T - a\tilde{\mu}^{-1})^2 \sim a\tilde{\sigma}^2/\tilde{\mu}^3, \tag{9.40}$$

which can be verified by an argument similar to the proof of Theorem 9.28, provided some of the hypotheses are appropriately strengthened. The main idea is to write

$$(T - a\tilde{\mu}^{-1})I_{B_a} = -\tilde{\mu}^{-1}(\tilde{S}_T - T\tilde{\mu})I_{B_a} + \tilde{\mu}^{-1}(Z_T - a - \eta_T)I_{B_a}, \tag{9.41}$$

square, take expectations, and use Wald's second moment identity in conjunction with Lemma 9.19. Among other things it is necessary to verify for appropriate $n_a > a\tilde{\mu}^{-1}$ that

$$E(T^2; T > 2n_a) = o(a)$$

(cf. (9.21)). For this to be true, strengthening (9.20) to $\Sigma\, nP\{\eta_n \le -n\varepsilon\} < \infty$ and assuming that $E|z_1|^3 < \infty$ are more than sufficient. The details are omitted.

It appears that one should be able to improve (9.40) to

$$E(T - a\tilde{\mu}^{-1})^2 = a\tilde{\sigma}^2/\tilde{\mu}^3 + K + o(1)$$

for some constant K which can be identified. However, the appearance of the cross product term when (9.41) is squared makes this a substantially more difficult project. See Lai and Siegmund (1979) for the linear case.

For some applications (e.g. Theorem 9.54) it is useful to generalize Theorems 9.12 and 9.28 by allowing the slowly changing sequence to depend on a. A particularly simple example is Theorem 9.45 stated below. See Hogan (1984) for more profound results in this direction.

We continue to assume that $z_1, z_2, \ldots$ are independent and identically distributed with $E(z_1) = \tilde{\mu} > 0$, that $S_n = z_1 + \cdots + z_n$, and that $\mathscr{E}_n$ denotes the σ-algebra generated by $z_1, \ldots, z_n$. Suppose that for each $a > 0$, $\eta_n^{(a)}$ is adapted to $\mathscr{E}_n$ and define

$$T = T_a = \inf\{n: n \geq m_0, \tilde{S}_n + \eta_n^{(a)} \geq a\}. \tag{9.42}$$

Consider the following generalizations of (9.5) and (9.6). For each $\lambda > 0$

$$a^{-1} \max_{1 \leq n \leq a\lambda} |\eta_n^{(a)}| \to 0 \tag{9.43}$$

in probability. For each λ, $\varepsilon > 0$ there exists a $\delta = \delta(\lambda, \varepsilon)$ such that for all $n \leq a\lambda$

$$P\left\{\max_{1 \leq k \leq a\delta} |\eta_{n+k}^{(a)} - \eta_n^{(a)}| > \varepsilon\right\} < \varepsilon. \tag{9.44}$$

Theorem 9.45. *Assume that T is defined by* (9.42) *and that conditions* (9.43) *and* (9.44) *are satisfied. If z_1 is nonarithmetic, then as $a \to \infty$ for all $x \geq 0$*

$$P\{T < \infty, \tilde{S}_T + \eta_T^{(a)} - a \leq x\} \to H(x),$$

where H is defined by (9.11).

The proof requires only trivial modification of the proof of Theorem 9.12 and hence is omitted.

3. Applications to Repeated Significance Tests

In this section the nonlinear renewal theory of Theorems 9.12, 9.17, 9.28, and 9.45 is used to provide partial justification for the approximations suggested in Chapters IV and V.

Let $x_1, x_2, \ldots$ be independent and normally distributed with mean μ and unit variance. Put $S_n = x_1 + \cdots + x_n$ and let

$$(9.46)\qquad \begin{aligned} T &= \inf\{n: n \ge m_0, |S_n| \ge bn^{1/2}\} \\ &= \inf\{n: n \ge m_0, \tfrac{1}{2}n^{-1}S_n^2 \ge a\}, \end{aligned}$$

where $a = b^2/2$. This stopping rule is of the form (9.1) with $g(x) = \frac{1}{2}x^2$. In this special case (9.3) and (9.4) become

$$(9.47)\qquad \tilde{S}_n = \mu(S_n - n\mu/2) \text{ and } \eta_n = (S_n - n\mu)^2/2n.$$

It follows from Theorem 9.12 (cf. Example 9.7) that for $\mu \ne 0$

$$(9.48)\qquad \lim_{a\to\infty} P_\mu\{\tfrac{1}{2}T^{-1}S_T^2 - a \le x\} = \lim_{a\to\infty} P_\mu\{\mu(S_\tau - \tau\mu/2) - a \le x\},$$

where for $\mu > 0$

$$(9.49)\qquad \tau = \tau_a(\mu) = \inf\{n: S_n - n\mu/2 > a/\mu\}.$$

With the aid of Corollary 8.45 and (8.50) one can calculate various parameters of this limiting distribution. In particular,

$$(9.50)\qquad \lim_{a\to\infty} E_\mu[\exp\{-[\mu(S_\tau - \tau\mu/2) - a]\}] = v(\mu)$$

and

$$(9.51)\qquad \lim_{a\to\infty} E_\mu[\mu(S_\tau - \tau\mu/2) - a] = \rho(\mu),$$

where $v(\mu)$ is the quantity defined in (4.37) and

$$(9.52)\qquad \rho(\mu) = 1 + \mu^2/4 - \mu\sum_1^\infty n^{-1/2}[\phi(\tfrac{1}{2}\mu n^{1/2}) - \tfrac{1}{2}\mu n^{1/2}\Phi(-\tfrac{1}{2}\mu n^{1/2})].$$

Theorem 9.53. *Suppose T is defined by* (9.46). *For $\mu \ne 0$ as $b \to \infty$*

$$E_\mu(T) = (b^2 - 1)/\mu^2 + 2\mu^{-2}\rho(\mu) + o(1),$$

where $\rho(\mu)$ is given (for $\mu > 0$) by (9.52).

Proof. The proof is a direct application of Theorem 9.28 together with the identification of $\rho(\mu)$ given in (9.51). Because of Proposition 9.30 the only condition of Theorem 9.28 which must be checked is (9.27). By (9.47) $\tilde{\mu} = \mu^2/2$. Let $0 < \delta < 4^{-1}$ and $n_1 = \delta b^2/\mu^2$. Then for $\mu > 0$, say, $P_\mu\{T \le n_1\} \le P_\mu\{|S_n - n\mu| \ge bn^{1/2} - n\mu$ for some $n \le n_1\}$. It is easy to see that $bn^{1/2} - n\mu$ is increasing for $n \le n_1$, and an application of the Hájek–Rényi–Chow inequality as in the proof of Proposition 9.30 shows that $b^2 P_\mu\{T \le n_1\} \to 0$. This proves (9.27) and hence the theorem. (Note that except for identifying $\rho(\mu)$ the condition of normality was not used. It would have sufficed to assume that x_1, $x_2, \ldots$ are independent and identically distributed with mean $\mu \ne 0$, unit variance, and finite γth moment for some $\gamma > 2$.) □

Remark. Similarly (cf. Theorem 9.45), one may show that the correction for discrete time in Proposition 4.29 is $\frac{1}{2}(\mu - \frac{1}{2}\mu_1)^{-2}\Phi(-\Delta)\rho(2\mu - \mu_1)$, which in conjunction with Theorem 9.53 justifies the expansion (4.42) up to the terms converging to 0 as $b \to \infty$. Establishing rigorously whether the expansion in (4.42) contains the correct terms of order b^{-1} requires some additional effort. For the multidimensional case see Problem 9.8.

We now turn to an analysis of $P_0\{T < m | S_m = \xi\}$. Theorem 9.54 is a discrete time analogue of Theorem 4.14. For simplicity of exposition, it is assumed that m_0 is proportional to m as $b \to \infty$. Appropriate modifications for fixed m_0 are straightforward. Let

$$P_{\lambda,\xi}^{(m)}(A) = P_0(A | S_0 = \lambda, S_m = \xi) \qquad (A \in \mathscr{E}_m).$$

Theorem 9.54. *Assume $b \to \infty$, $m \to \infty$, and $m_0 \to \infty$ such that for some $0 < \mu_1 < \mu_0 < \infty$, $bm^{-1/2} = \mu_1$ and $bm_0^{-1/2} = \mu_0$. For each $0 < \xi_0 \le \mu_1$, $P_{0,m\xi_0}^{(m)}\{T < m\} \le (\mu_0/\mu_1)\exp\{-\tfrac{1}{2}m(\mu_1^2 - \xi_0^2)$. For each $\mu_1^2/\mu_0 < \xi_0 < \mu_1$*

$$P_{0,m\xi_0}^{(m)}\{T < m\} \sim \nu(\mu_1^2 \xi_0^{-1})\mu_1 \xi_0^{-1} \exp\{-\tfrac{1}{2}m(\mu_1^2 - \xi_0^2)\},$$

where ν is given by (4.37).

Corollary 9.55. *Suppose $bm^{-1/2} = \mu_1$ and $cm^{-1/2} = \gamma \in (\mu_1^2/\mu_0, \mu_1)$. Then for $\mu \ne 0$*

$$P_\mu\{T < m, |S_m| < cm^{1/2}\} \sim \frac{\phi[m^{1/2}(\mu_1 - |\mu|)]}{|\mu| m^{1/2}} \mu_1 \gamma^{-1} \nu(\mu_1^2 \gamma^{-1}) \exp[-m|\mu|(\mu_1 - \gamma)], \tag{9.56}$$

and

$$P_0\{T < m, |S_m| < cm^{1/2}\} \sim 2b\phi(b) \int_{\mu_1^2\gamma^{-1}}^{\mu_0} x^{-1}\nu(x)\,dx. \tag{9.57}$$

The relation (9.57) *also holds in the case $c = b$ ($\gamma = \mu_1$).*

Remarks. (i) Addition of $P_0\{|S_m| \ge bm^{1/2}\} = 2[1 - \Phi(b)]$ to the right hand side of (9.57) with $c = b$ gives the approximation suggested in (4.40). (ii) The approximation suggested in Problem 4.16 can be much more accurate than (9.56) when μ is close to zero. This case is important when computing a lower confidence bound for μ. A disadvantage is that one must perform a numerical integration, or alternatively, table a function of two variables.

Proof of Theorem 9.54. Taking over notation from the proof of Theorem 4.14 in the obvious way, we obtain by the same reasoning that led to (4.24).

$$
\begin{aligned}
P_{0,m\xi_0}^{(m)}\{T < m\} &= P_{0,m\xi_0}^{(m)}\{T^* \geq m_0\} \\
&= \exp[-m(\mu_1^2 - \xi_0^2)/2]\tilde{E}_{m\xi_0}^{(m)}\{(m/T^*)^{1/2} \\
&\quad \times \exp[-\tfrac{1}{2}(T^{*-1}S_{T^*}^2 - b^2)]; T^* \geq m_0\}.
\end{aligned} \tag{9.58}
$$

Since the $\tilde{P}_{m\xi_0}^{(m)}$ distributions of $S_n, n = m, m-1, \ldots$ running backwards from $S_m = m\xi_0$ are the same as the P_0 distributions of $m\xi_0 + S_n$, $n = 0, 1, \ldots$ running forwards, the expectation in (9.58) equals

$$
\begin{aligned}
&E_0\{[m/(m-\tau^*)]^{1/2}\exp\{-\tfrac{1}{2}[(m-\tau^*)^{-1}(m\xi_0 + S_{\tau^*})^2 - b^2]\}; \\
&\tau^* \leq m - m_0\},
\end{aligned} \tag{9.59}
$$

where $\tau^* = \inf\{n\colon n \geq 1,\ |m\xi_0 + S_n| \geq b(m-n)^{1/2}\}$. As in the proof of Theorem 4.14, it is easy to see that

$$
m^{-1}\tau^* \to 1 - (\xi_0/\mu_1)^2 \tag{9.60}
$$

in probability. To study the excess over the boundary in the exponential in (9.59) note that τ^* can be expressed

$$
\tau^* = \inf\{n\colon n \geq 1, \tfrac{1}{2}\mu_1^2\xi_0^{-1}n + S_n + \tfrac{1}{2}m^{-1}\xi_0^{-1}S_n^2 \geq \tfrac{1}{2}\xi_0^{-1}m(\mu_1^2 - \xi_0^2)\}.
$$

Hence by Theorem 9.45

$$
\lim_m P_0\{\tfrac{1}{2}\mu_1^2\xi_0^{-1}\tau^* + S_{\tau^*} + \tfrac{1}{2}m^{-1}\xi_0^{-1}S_{\tau^*}^2 - \tfrac{1}{2}\xi_0^{-1}m(\mu_1^2 - \xi_0^2) \leq x\} = H(x),
$$

where H is the limiting distribution of excess over a constant boundary for a normal random walk with mean $\frac{1}{2}\mu_1^2\xi_0^{-1}$ and variance 1 per step. From (9.60) and some simple algebra it follows that

$$
\lim_m P_0\{\tfrac{1}{2}(m-\tau^*)^{-1}(m\xi_0 + S_{\tau^*})^2 - \tfrac{1}{2}\mu_1^2 m \leq x\} = H(x\xi_0\mu_1^{-2}), \tag{9.61}
$$

which by (9.50) yields

$$
\int_{(0,\infty)} e^{-x}H(\xi_0\mu_1^{-2}\,dx) = \int_{(0,\infty)} \exp(-\mu_1^2\xi_0^{-1}x)H(dx) = v(\mu_1^2\xi_0^{-1}).
$$

Hence except for justification to take a limit inside an expectation, use of (9.61) and (9.60) in (9.59) and substitution of the result into (9.58) complete the proof. It is easy to see that for every $\varepsilon > 0$, $P_0\{\tau^* > m(1 - (\xi_0/\mu_1)^2)(1+\varepsilon)\}$ converges to 0 exponentially fast, so the interchange of limits causes no problems. □

Proof of Corollary 9.55. The proof of (9.56) is an immediate consequence of the theorem. To prove (9.57) in the case $\gamma < \mu_1$, let $\varepsilon > 0$. Since

$$
\begin{aligned}
P_0\{T < m, |S_m| < cm^{1/2}\} &= P_0\{T < m, |S_m| < (\mu_1^2/\mu_0)(1+\varepsilon)\mathrm{m}\} \\
&+ \int_{(\mu_1^2/\mu_0)(1+\varepsilon) < |\xi_0| < \gamma} P_{m\xi_0}^{(m)}\{T < m\}\phi(m^{1/2}\xi_0)m^{1/2}\,d\xi_0,
\end{aligned}
$$

in light of Theorem 9.54 it suffices to show

$$P_0\{T < m, |S_m| < (\mu_1^2/\mu_0)(1+\varepsilon)m\} \le \delta(\varepsilon)b\phi(b),$$

where $\delta(\varepsilon)$ does not depend on m and $\lim_{\varepsilon\to 0}\delta(\varepsilon) = 0$. Let $m' = m_0(1+2\varepsilon)^2$, so $b(m')^{1/2} = (\mu_1^2/\mu_0)(1+2\varepsilon)m$. Then

$$\begin{aligned}
P_0\{T \le m, |S_m| &< (\mu_1^2/\mu_0)(1+\varepsilon)m\} \\
&\le P_0\{T \le m'\} \\
&\quad + P_0\{m' < T \le m, |S_m| < (\mu_1^2/\mu_0)(1+\varepsilon)m\} \\
&\le 2(m' - m_0)(1 - \Phi(b)) \\
&\quad + \sum_{n=m'+1}^{m} P_0\{T = n\} P_0\{|S_m - S_n| > \varepsilon(\mu_1^2/\mu_0)m\},
\end{aligned}$$

from which existence of the required $\delta(\varepsilon)$ easily follows. The case $\gamma = \mu_1$ is handled by a similar argument. □

The asymptotic relation (9.56) does not hold when $c = b$ ($\gamma = \mu_1$), although (4.41) in effect suggests that it be used even in this case. There is an additional heuristic argument in support of (4.41), which although something of a digression is discussed next.

Hence consider the problem of computing

$$\begin{aligned}
P_\mu\{T < m, |S_m| < bm^{1/2}\} \\
= \int_{|\xi| < bm^{1/2}} P_\xi^{(m)}\{T < m\}\phi[(\xi - m\mu)/m^{1/2}]\,d\xi/m^{1/2}.
\end{aligned}$$

Suppose $\mu > 0$, and let $T_+ = \inf\{n: n \ge m_0, S_n \ge bn^{1/2}\}$. It is easy to see that

$$P_\mu\{T < m, |S_m| < bm^{1/2}\} \sim P_\mu\{T_+ < m, S_m < bm^{1/2}\},$$

and

$$\begin{aligned}
P_\mu\{T_+ < m, S_m < bm^{1/2}\} \\
= \int_{\xi < bm^{1/2}} P_\xi^{(m)}\{T_+ < m\}\phi[(\xi - m\mu)/m^{1/2}]\,d\xi/m^{1/2}.
\end{aligned} \tag{9.62}$$

Because of the positive drift it seems reasonable intuitively that if the random walk is to cross the boundary and then fall back below it, the crossing must take place at some n which is about equal to m, and ξ must be close to $bm^{1/2}$. (For a rigorous proof see Siegmund, 1978.) Hence we consider computation of $P_\xi^{(m)}\{T_+ < m\}$ for ξ close to $bm^{1/2}$—more precisely for $\xi = m\xi_0 = bm^{1/2} - x = m\mu_1 - x$ for some fixed $x > 0$. Then

$$\begin{aligned}
P_\xi^{(m)}\{T_+ < m\} = P_0\{S_n \ge bn^{1/2} \\
\text{for some } m_0 \le n < m \mid S_m = bm^{1/2} - x\}
\end{aligned}$$

$$= P_0\{S_m - S_n \le -x + b(m^{1/2} - n^{1/2}) \text{ for some } m_0 \le n < m | S_m = bm^{1/2} - x\}$$
$$= P_0\{S_i \le -x + \mu_1 m^{1/2}[m^{1/2} - (m-i)^{1/2}] \text{ for some } 1 \le i \le m - m_0 | m^{-1}S_m = \mu_1 - m^{-1}x\}.$$

A direct calculation shows that for each fixed n the joint distribution of $S_1, \ldots, S_n$ conditional on $m^{-1}S_m = \mu_1 - m^{-1}x$ converges as $m \to \infty$ to the joint distribution of $S_1, \ldots, S_n$ under P_{μ_1}. This in conjunction with the expansion

$$\mu_1 m^{1/2}[m^{1/2} - (m-i)^{1/2}] = \tfrac{1}{2}\mu_1 i + 0(i^2/m)$$

suggests that as $m \to \infty$, for $\xi = bm^{1/2} - x$

$$P_\xi^{(m)}\{T_+ < m\} \to P_{\mu_1}\{S_i \le -x + \tfrac{1}{2}i\mu_1 \text{ for some } i \ge 1\} = P_{\mu_1/2}\{\tilde{\tau}(-x) < \infty\}, \tag{9.63}$$

where $\tilde{\tau}(-x) = \inf\{n: S_n \le -x\}$. (An argument similar to that of Theorem 4.14 takes a slightly different route to the same conclusion; see Problem 9.10.)

Making the change of variable $\xi = \mu_1 m - x$ in (9.62) and using (9.63) yields

$$P_\mu\{T < m, |S_m| < bm^{1/2}\} \sim m^{-1/2}\phi[m^{1/2}(\mu_1 - \mu)] \int_{(0,\infty)} e^{(\mu_1 - \mu)x} P_{\mu_1/2}\{\tilde{\tau}(-x) < \infty\}\, dx. \tag{9.64}$$

The approximation in (9.64) is not a convenient one because the integral must be computed numerically and is a function of two variables, μ and μ_1.

However, by (8.49)

$$P_{\mu_1/2}\{\tilde{\tau}(-x) < \infty\} \sim \nu(\mu_1)e^{-\mu_1 x} \qquad (x \to \infty). \tag{9.65}$$

If we use for $P_{\mu_1/2}\{\tilde{\tau}(-x) < \infty\}$ the large x approximation given by (9.65), then (9.64) becomes

$$P_\mu\{T < m, |S_m| < bm^{1/2}\} \cong \nu(\mu_1)\phi[m^{1/2}(\mu_1 - \mu)]/m^{1/2}\mu, \tag{9.66}$$

which is no longer a bona fide asymptotic result, but is very simple to evaluate numerically with the aid of (4.38). The approximation (9.66) is that suggested in (4.41) and is tantamount to setting $c = b$ ($\gamma = \mu_1$) in (9.56).

Many of the preceding results can be generalized to a one parameter exponential family of distributions, and after substantially more work to a multiparameter exponential family. See Woodroofe (1982) and Lalley (1983). A useful method for generalizing (9.57) is sketched in IV.8 and Problem 4.13. This method is now developed in somewhat more detail to give an approximation for the significance level of the paired comparison test of V.2.

Let $y_1, y_2, \ldots$ be independent random variables assuming the values 1 and 0 with probabilities p and $q = 1 - p$ respectively. Let $S_n = y_1 + \cdots + y_n$ and $H(x) = x\log x + (1-x)\log(1-x) + \log 2$. To test $H_0: p = \frac{1}{2}$ against $H_1: p \ne \frac{1}{2}$, let

(9.67) $$T = \inf\{n\colon n \geq m_0, nH(n^{-1}S_n) \geq a\}.$$

Stop sampling at $\min(T, m)$ and reject H_0 if $T \leq m$ or $T > m$ and $mH(m^{-1}S_m) \geq d$ $(d \leq a)$. Let $Z_n = nH(n^{-1}S_n)$, $n = 1, 2, \ldots$. The power function of this test is

$$P_p\{T \leq m\} + P_p\{T > m, Z_m \geq d\} = P_p\{Z_m \geq d\} + P_p\{T < m, Z_m < d\}.$$
(9.68)

Theorem 9.69. *Assume that m_0, m, a, and $d \to \infty$ in such a way that for some $\frac{1}{2} < p_1 < p_2 < p_0 < 1$, $H(p_0) = a/m_0$, $H(p_1) = a/m$, and*

(9.70) $$H\{p_2 H(p_1)/H(p_2) + \tfrac{1}{2}[1 - H(p_1)/H(p_2)]\} = d/m.$$

Then for T defined by (9.67)

$$P_{1/2}\{T < m, Z_m < d\} \sim (2a/\pi)^{1/2} e^{-a} \int_{p_2}^{p_0} [pqH(p)]^{-1/2} v^*(p)\, dp,$$

where

(9.71) $$v^*(p) = \lim_{a \to \infty} E_p[\exp\{-(Z_T - a)\}].$$

Remark 9.72. The stopping rule (9.67) if of the form (9.1), so by (9.3), (9.4), and Theorem 9.12, the P_p $(p \neq \frac{1}{2})$ limiting distribution of excess over the stopping boundary for the process $Z_n = nH(n^{-1}S_n)$ is related to that of the random walk

$$\tilde{S}_n = \tilde{S}_n(p) = nH(p) + (S_n - np)H'(p) \qquad (n = 1, 2, \ldots).$$

Although the random walk S_n, $n = 1, 2, \ldots$ is obviously arithmetic, a simple analysis shows that $\tilde{S}_n(p)$ can be arithmetic for at most a denumerable set of values of p. Hence except for this denumerable set of p values, by Theorem 9.12 the limit in (9.71) exists and can be evaluated according to the prescription worked out in Corollary 8.45 and Theorem 8.51. Note, however, that the random walk $\tilde{S}_n(p)$ does not fulfill the strongly nonarithmetic condition of Theorem 8.51. See Remark 8.53 for the appropriate generalization.

Proof of Theorem 9.69. Let $P = \int_0^1 P_p\, dp$, and let E denote expectation with respect to P. An easy calculation shows that the likelihood ratio of $y_1, \ldots, y_n$ under P relative to $P_{1/2}$ is

(9.73) $$L_n = 2^n \int_0^1 p^{S_n} q^{n - S_n}\, dp = \left[(n+1)\binom{n}{S_n} 2^{-n}\right]^{-1}.$$

Hence by a slight variation on the usual form of Wald's likelihood ratio identity

$$P_{1/2}\{m_0 < T < m, Z_m < d\} = E\{L_T^{-1} P_{1/2}\{Z_m < d \mid \mathscr{E}_T\}; m_0 < T < m\}.$$

(It is easy to show that $P_0\{T = m_0\} = O(a^{-1/2} e^{-a})$ and hence can be neglected.)

Let $\hat{p}_n = n^{-1}S_n$ and $\hat{q}_n = 1 - \hat{p}_n$. From (9.73) and Stirling's formula it follows after some computation that

$$P_{1/2}\{m_0 < T < m, Z_m < d\}$$
$$\sim (2a/\pi)^{1/2}e^{-a}\int_{1/2}^{1} E_p\{(T/a\hat{p}_T\hat{q}_T)^{1/2}\exp[-(Z_T - a)]P_{1/2}\{Z_m < d|\mathscr{E}_T\};$$
$$m_0 < T < m\}\,dP. \tag{9.74}$$

By the usual arguments as $a \to \infty$

$$P_p\{a^{-1}T \to [H(p)]^{-1}\} = 1 \text{ for } p < p_0, \tag{9.75}$$

$$P_p\{m_0 < T < m\} \to 1 \text{ for } p_1 < p < p_0 \qquad \to 0 \text{ for } p \notin [p_1, p_0], \tag{9.76}$$

$$P_p\{\hat{p}_T \to p\} = 1; \tag{9.77}$$

and as noted above

$$E_p\{\exp[-(Z_T - a)]\} \to v^*(p) \tag{9.78}$$

for almost all values of p. The proof is completed by substituting (9.75)–(9.78) in conjunction with Lemma 9.79 below into (9.74). □

Lemma 9.79. *Under the conditions of Theorem* 9.69, *for each $p_1 < p < p_0$ there exists an event $B_a \in \mathscr{E}_T$ such that $P_p(B_a) \to 1$ and on B_a*

$$P_{1/2}\{Z_m < d|\mathscr{E}_T\} \to \begin{cases} 1 & \text{if } p > p_2 \\ 0 & \text{if } p < p_2, \end{cases}$$

where p_2 is defined by (9.70).

Proof. First note that since H is convex and $H(\frac{1}{2}) = 0$, the function taking p into $\frac{1}{2} + (p - \frac{1}{2})H(p_1)/H(p)$ is decreasing for $p > \frac{1}{2}$. Hence p_2 is uniquely defined by (9.70) and $H[pH(p_1)/H(p) + \frac{1}{2}\{1 - H(p_1)/H(p)\}]$ is $>d/m$ or $<d/m$ according as $p < p_2$ or $p > p_2$. On $\{T < m\}$, by Taylor's theorem

$$\begin{aligned} m^{-1}Z_m &= H(m^{-1}S_m) \\ &= H[(m^{-1}T)(T^{-1}S_T) + \tfrac{1}{2}(1 - m^{-1}T) \\ &\quad + m^{-1}\{S_m - S_T - \tfrac{1}{2}(m - T)\}] \\ &= H[(m^{-1}T)(T^{-1}S_T) + \tfrac{1}{2}(1 - m^{-1}T)] \\ &\quad + m^{-1}[S_m - S_T - \tfrac{1}{2}(m - T)]H'(\zeta_m) \end{aligned}$$

for a suitable intermediate value ζ_m. It follows from (9.75) and (9.77) that

$$(m^{-1}T)(T^{-1}S_T) + \tfrac{1}{2}(1 - m^{-1}T) \to pH(p_1)/H(p) + \tfrac{1}{2}[1 - H(p_1)/H(p)]$$

Table 9.1. Approximations for $\alpha = P_{1/2}\{Z_m \geq d\} + P_{1/2}\{T < m, Z_m < d\}$

					α		
$b = (2a)^{1/2}$	$c = (2d)^{1/2}$	m_0	m	k	Theorem 9.69	Normal Approx.	Monte Carlo
3.05	2.05	10	50	1	.043	.050	.043
3.05	3.05	10	50	1	.019	.022	.019
2.84	2.03	10	50	5	.044	.050	.043
2.84	2.84	10	50	5	.025	.019	.022
2.75	2.03	10	50	10	.044	.050	.044
2.75	2.75	10	50	10	.024	.020	.022
2.69	2.05	16	80	16	.045	.050	.043
2.69	2.69	16	80	16	.028	.024	.023

with P_p probability one. By Chebyshev's inequality

$$P_{1/2}\{m^{-1}|S_m - S_T - \tfrac{1}{2}(m - T)| > \delta | \mathscr{E}_T\} \leq (4m\delta^2)^{-1}.$$

The lemma follows easily by putting $B_a = \{|a^{-1}T - 1/H(p)| < \varepsilon, |T^{-1}S_T - p| < \varepsilon\}$ for sufficiently small $\varepsilon > 0$. □

It is straightforward to extend Theorem 9.69 to cover group sequential tests of group size $k > 1$.

Table 9.1 compares approximations to

$$\alpha = P_{1/2}\{Z_m \geq d\} + P_{1/2}\{T < m, Z_m < d\} \tag{9.80}$$

derived from Theorem 9.69 with normal approximations as suggested in V.2 (and corrected for excess over the boundary whenever $k > 1$). Monte Carlo estimates are also included. For the Monte Carlo experiment importance sampling as suggested in Remark 4.45 was used. The number of repetitions was 5000 whenever $d < a$ and 2500 whenever $d = a$. These choices give standard errors roughly equal to .001.

The approximation based on Theorem 9.69 is generally quite good and noticeably better than the normal approximation. It is not as good as the approximations of Chapter IV, however. A possible explanation is that Theorem 9.69 suggests a smooth function of a to approximate a discontinuous one, but it is not easy to see how to make a continuity correction. (The overall approximation to (9.80) is discontinuous because of discontinuities in the evaluation of $P_{1/2}\{Z_m \geq d\}$—either by exact calculation or a continuity corrected normal approximation.)

For $p \neq \frac{1}{2}$ a normal approximation with mean and variance given by (5.6) is usually quite good. For example, for $b = 2.75$, $c = 2.03$, $m_0 = 10$, $m = 50$, $k = 10$, and $p = .75$, the approximation yields $P_p\{T \leq m\} \cong .85$, $P_p\{\text{Reject } H_0\} \cong .96$, and $E_p(T \wedge m) \cong 33$ compared to Monte Carlo estimates (2500 repetitions) of .86, .95, and 32 respectively.

4. Application to Fixed Width Confidence Intervals for a Normal Mean

The following argument provides more precise justification for Theorem 7.10. We use the notation of VII.2. Let $g(x) = x^{-1}$. The stopping rule T of (7.7) equals $\tilde{T} + 1$, where in terms of $s_n = \sigma^{-2} \sum_1^n y_i^2$

(9.81) $$\tilde{T} = \inf\{n: n \geq m_0 - 1, a_{n+1}[(n+1)/n]^2 ng(n^{-1}s_n) \geq n_0\}.$$

By (7.9)

$$a_{n+1}[(n+1)/n]^2 = 1 + n^{-1}(2 - a_0) + n^{-1}\varepsilon_n$$

for some $\varepsilon_n \to 0$ $(n \to \infty)$. It is shown below that Theorem 9.28 and (9.40) are applicable with

$$\tilde{S}_n = ng(1) + (s_n - n)g'(1) = 2n - s_n,$$

and η_n asymptotically equal to

$$(2 - a_0)g(n^{-1}s_n) + (s_n - n)^2 g''(1)/2n = (2 - a_0)n/s_n + n^{-1}(s_n - n)^2.$$

A simple calculation shows that (9.29) yields (7.11) (keep in mind that $T = \tilde{T} + 1$); and (9.29), (9.40), and the proof of Theorem 7.10 yield (7.12). Moreover, ρ is identified as $E(2\tau_+ - s_{\tau_+})^2/2E(2\tau_+ - s_{\tau_+})$, where $\tau_+ = \inf\{n: 2n - s_n > 0\}$.

An argument along the lines of Proposition 9.30 shows that the conditions of Theorem 9.28 and of (9.40) are satisfied except possibly for (9.27), where as always a special argument is required.

To verify (9.27) note that since s_n has a χ^2 distribution with n degrees of freedom, for all $n \geq m_0 - 1$

$$\begin{aligned} P\{\tilde{T} = n\} &\leq P\{s_n \leq a_{n+1}(n+1)^2/n_0\} \\ &= 2^{-n/2}[\Gamma(n/2)]^{-1} \int_0^{a_{n+1}(n+1)^2/n_0} x^{n/2-1} e^{-x/2}\, dx \qquad (9.82) \\ &\leq C_n n_0^{-n/2}, \end{aligned}$$

where $C_n = [a_{n+1}(n+1)^2]^{n/2}/2^{n/2}\Gamma(n/2+1)$. Hence for any fixed $n \geq m_0 - 1$

$$\begin{aligned} P\{\tilde{T} \leq n\} \leq \sum_{k=m_0-1}^{n} P\{\tilde{T} = k\} &= O(n_0^{-(m_0-1)/2}) \\ &= o(n_0^{-1}) \qquad (n_0 \to \infty), \end{aligned}$$

provided $(m_0 - 1)/2 > 1$, i.e. $m_0 \geq 4$. From Stirling's formula it is easy to see that $C_k \leq (ke)^{k/2}$ for all k sufficiently large, so by (9.82)

$$P\{n < \tilde{T} \leq n_0^{1/2}\} = O(n_0^{1/2}(en_0^{-1/2})^{n/2}) = o(n_0^{-1})$$

provided n is large enough. Finally the Hájek–Rényi–Chow inequality as in the proof of Proposition 9.30 shows that for $\delta < \frac{1}{2}$

$$P\{n_0^{1/2} < \tilde{T} \le \delta n_0\} \le P\{s_k \le 2k^2/n_0 \text{ for some } n_0^{1/2} < k \le \delta n_0\}$$
$$\le P\{|k - s_k| \ge k(1 - 2\delta) \text{ for some } n_0^{1/2} < k \le \delta n_0\}$$
$$\le \text{const.} \sum_{k \ge n_0^{1/2}} k^{-4} = o(n_0^{-1}).$$

Hence (9.27) holds, so Theorem 9.28 and (9.40) are indeed applicable as indicated above.

5. Woodroofe's Method

In a number of beautiful papers Woodroofe has developed an alternative to the method presented in this book for computing approximations to the error probabilities of various sequential tests. His method appears to have definite advantages if one wants to obtain second order corrections for approximations like that given in (9.57) (cf. Takahashi and Woodroofe, 1981, and Woodroofe and Takahashi, 1982), and it plays an important role in Lalley's (1983) study of repeated significance tests for multiparameter exponential families. A disadvantage with regard to the didactic goals of this book is that in continuous time Woodroofe's method is not easy to use and loses its intuitive appeal, so one has difficulty applying it to derive the simple exact results of Chapter III and the related but more complicated results of Chapter X. This section contains an informal introduction to Woodroofe's method in the context of repeated significance tests for a normal mean.

Let $x_1, x_2, \dots$ be independent and normally distributed with mean μ and variance 1. Let $S_n = x_1 + \cdots + x_n$ and define the stopping rule $T = \inf\{n: n \ge m_0, |S_n| \ge bn^{1/2}\}$. In order to calculate an approximation to

$$P_\mu\{T \le n\} = \sum_{k=m_0}^{n} P_\mu\{T = k\}, \tag{9.83}$$

Woodroofe first calculates an approximation to the individual terms $P_\mu\{T = k\}$, from which an approximation to the cumulative probability (9.83) follows by summation (and approximation of the sum by an integral).

For the most part, the terms $P_\mu\{T = k\}$ are of no interest in themselves, but to provide some focus for the subsequent discussion, here is an example in which they are. If T is used to define a repeated significance test of H_0: $\mu = 0$ against H_1: $\mu \ne 0$, and if stopping occurs at $T = n$ ($>m_0$), the attained significance level of the test is defined in IV.5 to be $P_0\{T \le n\}$. This definition in effect neglects excess over the stopping boundary and orders the boundary points $\pm bn^{1/2}$ in terms of $|S_T|/T \cong b/T^{1/2}$, with large values considered to be more evidence against H_0. In the case of a group sequential test with a small number of moderately large groups, this excess may not be completely negligible, and a perhaps more appealing definition of attained significance level is the following. If $T = n$ and $|S_n| = bn^{1/2} + x$, the attained level is defined to be

(cf. Fairbanks and Madsen, 1983)

$$P_0\{T \le n-1\} + P_0\{T = n, |S_n| \ge bn^{1/2} + x\}. \tag{9.84}$$

(If the test is truncated after m observations the definition remains the same as in IV.5.) The first probability in (9.84) can be approximated by Corollary 9.55. The second can be studied as follows (cf. Woodroofe, 1976, 1982).

Let $T_+ = \inf\{n: n \ge m_0, S_n \ge bn^{1/2}\}$. Suppose $n - m_0 \to \infty$, $b \to \infty$, and $bn^{-1/2} \to \mu^*$ for some fixed $\mu^* > 0$. For fixed $x \ge 0$,

$$P_0\{T = n, |S_n| \ge bn^{1/2} + x\} \sim 2P_0\{T_+ = n, S_n \ge bn^{1/2} + x\}$$

so it suffices to consider the latter probability. By conditioning on S_n, one obtains the identity

$$\begin{aligned} &P_0\{T_+ = n, S_n \ge bn^{1/2} + x\} \\ &= \int_{bn^{1/2}+x}^{\infty} P_\xi^{(n)}\{T_+ > n-1\}\phi(\xi/n^{1/2})\, d\xi/n^{1/2}, \end{aligned} \tag{9.85}$$

where as usual $P_\xi^{(n)}(A) = P_0(A \mid S_n = \xi)$ $(A \in \mathscr{E}_n)$.

Reasoning now proceeds along a similar path to that leading to (9.64) and (9.66). Because of the relation $bn^{1/2} \sim n\mu^*$ and the rate of decrease of the tail of the normal distribution, the only values of ξ which contribute asymptotically to the integral in (9.85) are $bn^{1/2} + x + O(1)$. Hence consider the computation of $P_\xi^{(n)}\{T_+ > n-1\}$ for $\xi = bn^{1/2} + y$ for fixed $y > 0$. Then, since $bn^{-1/2} \to \mu^*$,

$$\begin{aligned} P_\xi^{(n)}\{T_+ > n-1\} &= P_0\{S_k < bk^{1/2} \\ &\qquad \text{for all } m_0 \le k \le n-1 \mid S_n = bn^{1/2} + y\} \\ &= P_0\{S_n - S_k > y + b(n^{1/2} - k^{1/2}) \\ &\qquad \text{for all } m_0 \le k < n \mid S_n = bn^{1/2} + y\} \\ &= P_0\{S_i > y + b[n^{1/2} - (n-i)^{1/2}] \\ &\qquad \text{for all } 1 \le i \le n - m_0 \mid n^{-1}S_n = \mu^* + o(1)\}. \end{aligned} \tag{9.86}$$

Given that $S_n = \xi$ the expected value of S_k for $k < n$ is $k\xi/n$. This means that under the conditioning $S_n = bn^{1/2} + y$ the expected path of S_k for $m_0 \le k < n$ is far from the stopping boundary $bk^{1/2}$ except for k about equal to n. Hence it seems plausible that the range of values of i in (9.86) can be restricted to, say, $1 \le i \le K$ for some large constant K, or $1 \le i \le \log n$, without changing the probability by much. Also, a direct calculation shows that for each fixed i the joint distribution of $(S_1, \ldots, S_i)$ given that $n^{-1}S_n \to \mu^*$ converges to the unconditional joint P_{μ^*}-distribution of $(S_1, \ldots, S_i)$. In conjunction with the Taylor series expansion

$$b[n^{1/2} - (n-i)^{1/2}] = \tfrac{1}{2}bn^{-1/2}i + O(bn^{-3/2}i^2) \to \tfrac{1}{2}\mu^* i,$$

the preceding argument suggests that

(9.87)
$$P_{\xi}^{(n)}\{T_+ > n-1\} \to P_{\mu^*}\{S_i > y + \tfrac{1}{2}\mu^* i \text{ for all } i \geq 1\}$$
$$= P_{\mu^*/2}\{S_i > y \text{ for all } i \geq 1\}.$$

This last probability can be rewritten (Problem 8.13) to yield

Theorem 9.88. *Suppose $n - m_0 \to \infty$ and $b \to \infty$ so that $bn^{-1/2} \to \mu^*$ for some $0 < \mu^* < \infty$. Then for arbitrary $x \geq 0$*

(9.89)
$$P_0\{T = n, |S_n| \geq bn^{1/2} + x\}$$
$$\sim 2b^{-1}\phi(b)\mu^*[E_{\mu^*/2}(\tau_+)]^{-1}\int_x^\infty e^{-\mu^* y}P_{\mu^*/2}\{S_{\tau_+} \geq y\}\,dy,$$

where $\tau_+ = \inf\{k\colon S_k > 0\}$.

Remarks. (i) Putting $x = 0$ in (9.89) and using Corollary 8.45 yield

(9.90)
$$P_0\{T = n\} \sim bn^{-1}\nu(b/n^{1/2})\phi(b)$$

provided $n - m_0 \to \infty$ and $bn^{-1/2}$ converges to a finite, positive limit. Summing (9.90) from m_0 to m shows that if $bm^{-1/2} = \mu_1 < \mu_0 = bm_0^{-1/2}$

$$P_0\{T \leq m\} \sim b\phi(b)\sum_{n=m_0+m_0^{1/2}}^{m} b^2 n^{-1}\nu(bn^{-1/2})b^{-2}$$
$$\sim b\phi(b)\int_{\mu_0^{-2}}^{\mu_1^{-2}} y^{-1}\nu(y^{-1/2})\,dy,$$

in agreement with (9.57).

(ii) It is obviously possible to adapt the argument preceding Theorem 9.88 to obtain an asymptotic approximation for $P_\mu\{T = n\}$ for $\mu > 0$. However, some reflection shows that summing the resulting expression cannot possibily yield reasonable approximations for $P_\mu\{T \leq m\}$, at least not for values of μ close to $\mu_1 = b/m^{1/2}$. With some additional computation one can obtain an asymptotic approximation for $P_\mu\{T = n, |S_m| < bm^{1/2}\}$, which can then be summed on n to yield a result equivalent to (9.64).

Hence Theorem 9.88 or, more precisely, its essential ingredient, (9.87), can be used as a basis for deriving Theorem 9.54 and a variety of related results. Ironically, however, it is not clear whether (9.89) solves the problem posed at the beginning of this section because one must still evaluate the integral on the right hand side of (9.89); and this appears to be a formidable problem involving the actual distribution of S_{τ_+}—not just its first moment.

After some manipulation the right hand side of (9.89) can be shown to equal

(9.91)
$$2n^{-1/2}\phi(b)\int_0^\infty e^{\mu^* y}P_{-\mu^*/2}\{\tau(y) = \infty\}[1 - \Phi(x + \tfrac{1}{2}\mu^* + y)]\,dy,$$

where as usual $\tau(x) = \inf\{n\colon S_n \geq x\}$. If we use for $P_{-\mu^*/2}\{\tau(y) < \infty\}$ the large y approximation given by (9.65) then (9.91) can be approximately evaluated as

$$(9.92)\quad \begin{aligned} &2n^{-1/2}\phi(b)\{\mu^{*-1}[\Phi(\tfrac{1}{2}\mu^* - x)\exp(-\mu^* x) - \Phi(-\tfrac{1}{2}\mu^* - x)] \\ &\quad - \nu(\mu^*)[\phi(\tfrac{1}{2}\mu^* + x) - (\tfrac{1}{2}\mu^* + x)\Phi(-\tfrac{1}{2}\mu^* - x)]\}, \end{aligned}$$

which is easily computed.

For a direct derivation of (9.91) which does not go via Theorem 9.88, see Problem 9.12.

Problems

9.1. Prove Theorem 9.17.

9.2. Show that if $\tilde{\sigma}^2 = \text{var}(z_1) < \infty$, then τ defined by (9.9) satisfies (9.27), i.e. for all $0 < \delta < 1$, $aP\{\tau_a < \delta a/\tilde{\mu}\} \to 0$ $(a \to \infty)$.

9.3. Consider the test of power one defined by (4.2). Find asymptotic expressions for $P_0\{T < \infty\}$ and $E_\mu(T)$ as $B \to \infty$ (cf. Problems 4.5 and 4.6). What can be said in favor of a truncated version of this test as a competitor to a repeated significance test?

9.4. Fill in details of the proof of Theorem 5.29. Use Theorem 8.51 to derive a computable expression for ν^*. Study the test of V.4 for grouped data.

9.5. Under the conditions of Theorem 9.54, but with $m_0 = o(m)$, prove that

$$P_0\{T \le m\} \sim 2b\phi(b)\int_{\mu_1}^{\infty} x^{-1}\nu(x)\,dx.$$

9.6. Suggest approximations for the significance level, power, and expected sample size for a discrete time version of the sequential test obtained by truncating the stopping rule (4.66).

9.7. Prove (9.56) for $\mu_1^2\mu_0^{-1} < \gamma < \mu_1$ and (9.57) for $\gamma = \mu_1$.

9.8. For T defined by (5.11), show that as $b \to \infty$, for $\mu \ne 0$,

$$E_\mu(T) = (b^2 - r_1)/\theta^2 + 2\theta^{-2}\rho(\theta) + o(1),$$

where $\theta = [\sum_{i=1}^{r}(\mu_i - \bar{\mu}.)^2]^{1/2}$, $r_1 = r - 1$, and $\rho(\theta)$ is given by (9.52).

9.9. Under the assumptions of Theorem 9.17 prove that $Z_T - a$ and $(\tilde{S}_T - T\tilde{\mu})/T^{1/2}$ are aymptotically independent.

9.10. Let $x_1, x_2, \ldots$ be independent and normally distributed with mean μ and variance 1. Let $S_n = x_1 + \cdots + x_n$, and $T_+ = \inf\{n: n \ge m_0, S_n \ge bn^{1/2}\}$. Suppose $b \to \infty, m \to \infty$ with $bm^{-1/2} = \mu_1 > 0$, and let $\xi = bm^{1/2} - x$ for some fixed $x \ge 0$. Show by a likelihood ratio calculation along the lines of the proof of Theorem 4.14 (see especially the remark following the proof) that $P_\xi^{(m)}\{T_+ < m\} \to P_{\mu_1/2}\{\tilde{\tau}(-x) < \infty\}$, where $\tilde{\tau}(-x) = \inf\{n: S_n \le -x\}$.

9.11.† Formulate and prove a version of Theorem 9.69 for a one-parameter exponential family.

9.12. Give an argument in support of (9.91) starting from

$$P_0\{T_+ = n, S_n \geq bn^{1/2} + x\}$$
$$= (n-1)^{-1/2} \int_0^\infty \phi[b - y/(n-1)^{1/2}]$$
$$\times P_\xi^{(n-1)}\{T_+ > n-1\}[1 - \Phi(bn^{1/2} + x - b(n-1)^{1/2} + y)]\,dy,$$

where $\xi = b(n-1)^{1/2} - y$.

9.13. Give an approximation for the power of a repeated likelihood ratio test in a one-parameter exponential family having a smooth density (cf. Woodroofe, 1982).

9.14. Find an approximation to the significance level of a repeated significance test for a normal mean when observations are made at irregularly spaced intervals. More precisely, suppose $T = \inf\{n_i: n_i \geq m_0, |S_{n_i}| \geq bn_i^{1/2}\}$, where for some $k \geq 1$, $n_i = \pi(ik/m)m$, $i = 0, 1, \ldots, [m/k]$ and π: $[0, 1] \to [0, 1]$ is differentiable, strictly increasing, and satisfies $\pi(0) = 0$, $\pi(1) = 1$. Find an approximation for $P_0\{m_0 < T \leq m\}$ and check that it reduces to the known result for group repeated significance tests with fixed group size equal to k.

9.15. Consider a repreated significance test for a normal mean μ. Define a confidence interval for μ taking excess over the boundary into account, and give appropriate generalizations of (9.91) and (9.92) for the required probability calculations.

9.16. Use the results of Section 4, Problem 2.1, and Problem 8.14 to show that the correction $\tilde{\rho}$ for excess under the boundary in Problem 7.5 is about equal to $\frac{3}{2}$.

9.17.† Assuming that the $o(1)$ in Theorem 9.53 is in fact $o(b^{-1})$ (presumably it is $O(b^{-2})$), justify heuristically the approximation (4.42). Note that the term involving $\nu(\mu_1)$ in (4.42) results from an approximation similar to the one leading from (9.64) to (9.66). See also Problem 4.20.

CHAPTER X

Corrected Brownian Approximations

1. $P_{\mu_0}\{\tau(b) < \infty\}$ Revisited

Although we have informally thought of Brownian motion as a continuous approximation to random walks in discrete time, the approximations of Chapters IV and V, which are developed more completely in Chapter IX, do not actually utilize this idea. For example, Theorem 9.54 and Corollary 9.55, which are used as partial justification for (4.40), (4.41), (4.49), and (4.50), involve probabilities of large deviations, which can be quite different for Brownian motion and for random walk in discrete time.

This chapter is concerned with problems which are scaled so that Brownian motion appears as a first order approximation, and we obtain a correction term to improve the accuracy of this approximation. The results are in the spirit of Edgeworth expansions to supplement the central limit theorem. Although the method seems generally less adequate for dealing with nonlinear stopping boundaries than an approach based on large deviations, for some linear problems it is easier to apply and/or more appropriate. See especially Theorems 10.13, 10.16, and 10.41.

For illustrative purposes we begin with the one-sided first passage time discussed in VIII.1–4.

Through Sections 1–3 we assume that $z_1, z_2, \ldots$ are independent random variables with distribution of the form

$$dF_\mu(x) = \exp\{\theta x - \psi(\theta)\}\, dF(x), \tag{10.1}$$

where $\mu = \psi'(\theta)$ is a $1-1$ function, so one can indifferently consider μ to be a function of θ or θ a function of μ, say $\theta(\mu)$ (cf. (8.56)). It will be convenient to assume that F is standardized by $\int x\, dF(x) = 0$ and $\int x^2\, dF(x) = 1$. In particular, $F_0 = F$.

Let $S_n = z_1 + \cdots + z_n$, and for $b > 0$ define

$$\tau = \tau(b) = \inf\{n\colon S_n \geq b\}.$$

Let μ_0 and μ_1 be conjugate in the sense that $\mu_0 < 0 < \mu_1$ and $\psi(\theta_0) = \psi(\theta_1)$, where $\theta_i = \theta(\mu_i)$. By (10.1) and Proposition 2.24

$$P_{\mu_0}\{\tau(b) < \infty\} = e^{-\Delta b} E_{\mu_1}\{\exp[-\Delta(S_\tau - b)]\}, \tag{10.2}$$

where $\Delta = \theta_1 - \theta_0$. (In a different notation this is just (8.3) or (8.9).)

Suppose now that $b \to \infty$ and $\mu_0 \to 0$ in such a way that $\mu_0 b$ remains bounded away from 0 and $-\infty$. The normalization $\int x\, dF(x) = 0, \int x^2\, dF(x) = 1$ implies that $\theta(0) = \psi'(0) = 0(= \psi(0))$, so

$$\psi(\theta) = \tfrac{1}{2}\theta^2 + \kappa\theta^3/6 + 0(\theta^4), \tag{10.3}$$

where $\kappa = E_0 z_1^3 = \int x^3\, dF(x)$. From (10.3) it follows easily that $\mu_1 \to 0$ and Δb remains bounded away from 0 and ∞. For simplicity assume that Δb is some fixed number $2\delta \in (0, \infty)$.

Expansion of (10.2) yields

$$P_{\mu_0}\{\tau < \infty\} = e^{-\Delta b}[1 - \Delta E_{\mu_1}(S_\tau - b) + \tfrac{1}{2}\Delta^2 E_{\mu_1}(S_\tau - b)^2 + \cdots]. \tag{10.4}$$

This together with Corollary 8.33 suggests heuristically

Theorem 10.5. *Suppose $b \to \infty$ and $\Delta \to 0$ so that $\Delta b = 2\delta \in (0, \infty)$. If F is nonarithmetic*

$$P_{\mu_0}\{\tau < \infty\} = e^{-2\delta}[1 \quad 2\delta b^{-1}\rho_+ + o(b^{-1})],$$

where $\rho_+ = E_0(S_{\tau_+}^2)/2E_0 S_{\tau_+}$. A similar result holds in the arithmetic case if $b \to \infty$ through multiples of the span of F.

Remark 10.6. Since $1 - x = e^{-x} + o(x)$ as $x \to 0$, and e^{-x} is always positive whereas $1 - x$ is not necessarily so, it seems reasonable for numerical purposes to re-express the conclusion of Theorem 10.5 as

$$P_{\mu_0}\{\tau < \infty\} = \exp[-\Delta(b + \rho_+)] + o(\Delta). \tag{10.7}$$

This has a particularly suggestive interpretation, since the right hand side of (10.7) is the probability that a Brownian motion with drift $-\tfrac{1}{2}\Delta$ ever reaches the level $b + \rho_+$, i.e. to correct for discrete time the boundary at b is replaced by a boundary at $b + \rho_+$ and then the appropriate result for Brownian motion is used.

Additional justification for (10.7) is given by Siegmund (1979), who shows that if F is strongly nonarithmetic, the error term in (10.7) is actually $o(\Delta^2)$. Unlike Theorem 10.5, this result is not intuitively apparent from (10.4), which *au contraire* suggests that the coefficient of Δ^2 should involve $\lim E_{\mu_1}(S_\tau - b)^2 = E_0 S_{\tau_+}^3/3E_0 S_{\tau_+}$ (cf. Problem 10.3 and Appendix 4).

Remark 10.8. A reasonably general method for calculating ρ_+ is discussed in Section 4. For the standard normal case ρ_+ is the constant $\rho \cong .583$ which was introduced in II.3, III.5, and IV.3. In comparison with the approximation (8.58), note that in (10.7) there is only this one constant to evaluate numerically, whereas ν_+ in (8.58) is a function of μ_0. Not surprisingly, a local expansion of ν_+ for small μ_0 yields $\nu_+ = 1 - \rho_+\Delta + \frac{1}{2}\rho_+^2\Delta^2 + o(\Delta^2)$ (see Proposition 10.37).

The following lemmas will prove useful. Let $W(t)$, $0 \le t \le \infty$, denote Brownian motion with drift μ and put

$$\tau_W = \tau_W(b) = \inf\{t\colon W(t) \ge b\}.$$

Let $G(t;\mu,b) = P_\mu\{\tau_W(b) \le t\}$ (cf. (3.15)). Suppose to be definite that F is nonarithmetic and let

$$H(x) = (E_0 S_{\tau_+})^{-1}\int_{(0,x)} P_0\{S_{\tau_+} \ge y\}\,dy. \tag{10.9}$$

Lemma 10.10 *Suppose $b \to \infty$ and $m \to \infty$ so that for some fixed $0 < \zeta < \infty$, $b = \zeta m^{1/2}$. Then for all $0 \le t, x \le \infty$*

$$\lim P_0\{\tau(b) \le mt, S_{\tau(b)} - b \le x\} = G(t;0,\zeta)H(x).$$

Proof. The limiting marginal distribution of $\tau(b)/m$ is discussed in Appendix 1. That of $S_{\tau(b)} - b$ follows from the renewal theorem as in Corollary 8.33 once it is known that $P_0\{\tau_+ < \infty\} = 1$ and $E_0 S_{\tau_+} < \infty$ (cf. Problem 8.9). The asymptotic independence of $\tau(b)/m$ and $S_{\tau(b)} - b$ follows by an argument similar to the proof of Theorem 8.34 (cf. Problem 10.5).

Lemma 10.11. *Suppose $b \to \infty$ and $\mu \to 0$ in such a way that for some $-\infty < \xi < \infty$, $\mu b \to \xi$. Then for all $0 \le t, x \le \infty$, and $r > 0$*

$$\lim P_\mu\{\tau(b) \le b^2 t, S_{\tau(b)} - b \le x\} = G(t;\xi,1)H(x)$$

and

$$\lim E_\mu\{(S_\tau - b)^r; \tau < \infty\} = E_0 S_{\tau_+}^{r+1}/(r+1)E_0(S_{\tau_+}).$$

Proof. Let $m = b^2 t$. By Proposition 2.24

$$\begin{aligned} &P_\mu\{\tau(b) \le m, S_{\tau(b)} - b \le x\} \\ &\quad = E_0[\exp\{\theta S_\tau - \tau\psi(\theta)\}; \tau \le m, S_\tau - b \le x] \\ &\quad = \exp(\theta b)E_0[\exp\{\theta(S_\tau - b) - \tau\psi(\theta)\}; \tau \le m, S_\tau - b \le x]. \end{aligned} \tag{10.12}$$

From (10.3) it follows that $\theta b \to \xi$ and $\psi(\theta) \sim \frac{1}{2}\theta^2 \sim \frac{1}{2}\xi^2/b^2$. Hence at least for all finite x, Lemma 10.10 shows that the right hand side of (10.12) converges to

$$\begin{aligned} &\exp(\xi)E_0[\exp\{-\tfrac{1}{2}\xi^2\tau_W(1)\}; \tau_W(1) \le t]H(x) \\ &\quad = E_0[\exp\{\xi W(\tau_W(1)) - \tfrac{1}{2}\xi^2\tau_W(1); \tau_W(1) \le t]H(x) \\ &\quad = P_\xi\{\tau_W(1) \le t\}H(x) = G(t;\xi,1)H(x). \end{aligned}$$

That this calculation is also valid when $x = \infty$ follows from the renewal theorem once it is known that $E_0\{\exp(\lambda S_{\tau_+})\} < \infty$ for some $\lambda > 0$ (Problem 8.9).

The indicated convergence of $E_\mu\{(S_\tau - b)^r\}$ follows by a similar calculation and is omitted.

PROOF OF THEOREM 10.5. The indicated expansion follows immediately from (10.2), the inequality $1 - x \leq e^{-x} \leq 1 - x + x^2/2$ $(x \geq 0)$, and Lemma 10.11.

2. Sequential Probability Ratio Tests and Cusum Tests

Since the constant v_+ in (8.58) is numerically very close to $\exp(-\Delta\rho_+)$, where $\rho_+ = E_0(S_{\tau_+}^2)/2E_0(S_{\tau_+})$ and $\Delta = \theta_1 - \theta_0$ (cf. 10.37), the approximations (8.58) and (10.7) are almost the same for computational purposes. Of course (10.7) is easier to use because it requires the numerical evaluation of only one constant whereas v_+ is actually a function of μ_0.

For the stopping rule (8.57) of a sequential probability ratio test in the exponential family (8.56), (8.64) gives an heuristic approximation for $P_{\mu_0}\{S_N \geq b\}$. By scaling this problem as in Theorem 10.5 one obtains an approximation similar to (8.64) which has the same precise justification as (10.7).

Theorem 10.13. *Let $z_1, z_2, \ldots$ be independent random variables with distribution function of the form* (10.1) *for some nonarithmetic F standardized to have mean* 0 *and variance* 1. *Let N be defined by*

$$N = \inf\{n: S_n \notin (a, b)\}.$$

Assume $\mu_0 < 0 < \mu_1$ are such that $\psi(\theta_1) = \psi(\theta_0)$ $(\theta_i = \theta(\mu_i))$. Let $\Delta = \theta_1 - \theta_0$. Suppose $b \to \infty$, $a \to -\infty$, and $\Delta \to 0$ in such a way that $\Delta b = 2\xi$ and $b/|a| = \zeta$ for some fixed $0 < \xi < \infty$, $0 < \zeta < \infty$. Then for $i = 0$ or 1

$$P_{\mu_i}\{S_N \geq b\} = \frac{1 - \exp[(-1)^i\Delta(a + \rho_-)]}{\exp[(-1)^i\Delta(b + \rho_+)] - \exp[(-1)^i\Delta(a + \rho_-)]} + o(\Delta), \tag{10.14}$$

where $\rho_\pm = E_0(S_{\tau_\pm}^2)/2E_0(S_{\tau_\pm})$. Also

$$E_{\mu_i}(N) = \mu_i^{-1}[(a + \rho_-) + (b - a + \rho_+ - \rho_-)p^*] + o(\Delta^{-1}) \tag{10.15}$$

where p^* denotes the right hand side of (10.14).

Note that (10.14) and (10.15) are just the Wald approximations of Chapter II but with b and a replaced by $b + \rho_+$ and $a + \rho_-$. The proof of Theorem 10.13 combines in an obvious way the arguments of VIII.5 and X.1, and consequently is omitted. For the corresponding results when $\mu = 0$, see Problems 8.3 and 8.4.

For a more novel example, consider the cusum tests of II.6. In this case one would like to compute $E_\mu(N)/P_\mu\{S_N \geq b\}$ for N defined as in Theorem 10.13 with $a = 0$; see (2.52)

Theorem 10.16. *Assume the conditions of Theorem* 10.13, *except that* $a = 0$ *in the definition of* N. *As* $b \to \infty$ *and* $\Delta \to 0$ *so that* $\Delta b = 2\xi \in (0, \infty)$,

$$E_{\mu_0}(N)/P_{\mu_0}\{S_N \geq b\} = (\Delta|\mu_0|)^{-1}\{\exp[\Delta(b + \rho_+ - \rho_-)] - \Delta[b + \rho_+ - \rho_-] - 1\} + o(\Delta^{-1}), \tag{10.17}$$

and

$$E_{\mu_1}(N)/P_{\mu_1}\{S_n \geq b\} = (\Delta\mu_1)^{-1}\{\exp[-\Delta(b + \rho_+ - \rho_-)] + \Delta[b + \rho_+ - \rho_-] - 1\} + o(\Delta^{-1}), \tag{10.18}$$

where ρ_+ *and* ρ_- *are as defined in Theorem* 10.13.

Remark. Simple algebra shows that (10.17) and (10.18) are of the form of (2.56) with b replaced by $b + \rho_+ - \rho_-$.

Proof of Theorem 10.16. Since the proof of (10.17) is comparatively straightforward, only the proof of (10.18) is given here.

The reasoning leading to (8.61) and (8.69) together with an application of Theorem 10.5 yields

$$\begin{aligned} P_{\mu_0}\{S_N \geq b\} &= e^{-\Delta(b+\rho_+)}P_{\mu_1}\{S_N \geq b\} + o(\Delta) \\ &= e^{-\Delta(b+\rho_+)}P_{\mu_1}\{\tau_- = \infty\}/[1 - e^{-\Delta(b+\rho_+-\rho_-)}] + o(\Delta) \end{aligned} \tag{10.19}$$

and

$$\begin{aligned} P_{\mu_1}\{S_N \geq b\} &= P_{\mu_1}\{\tau_- = \infty\} + e^{\Delta\rho_-}P_{\mu_0}\{S_N \geq b\} + o(\Delta) \\ &= P_{\mu_1}\{\tau_- = \infty\}/[1 - e^{-\Delta(b+\rho_+-\rho_-)}] + o(\Delta). \end{aligned} \tag{10.20}$$

By Wald's identity

$$\begin{aligned} \frac{\mu_1 E_{\mu_1}(N)}{P_{\mu_1}\{S_N \geq b\}} &= \frac{E_{\mu_1}(S_N)}{P_{\mu_1}\{S_N \geq b\}} \\ &= b + E_{\mu_1}(S_N - b|S_N \geq b) + \frac{E_{\mu_1}(S_N; S_N \leq 0)}{P_{\mu_1}\{S_N \geq b\}}. \end{aligned} \tag{10.21}$$

An easy argument along established lines yields

$$E_{\mu_1}(S_N - b|S_N \geq b) \to \rho_+, \tag{10.22}$$

so it remains to examine the third term on the right hand side of (10.21). It is straightforward to see that

$$
\begin{aligned}
E_{\mu_1}(S_N; S_N \le 0) &= E_{\mu_0}[S_N \exp(\Delta S_N); S_N \le 0] \\
&= E_{\mu_0}[S_{\tau_-} \exp \Delta S_{\tau_-})] \\
&\quad - E_{\mu_0}\{E_{\mu_0}[S_{\tau_-} \exp(\Delta S_{\tau_-})|S_N]; S_N \ge b\} \\
&= E_{\mu_0}(S_{\tau_-} + \Delta S_{\tau_-}^2) - P_{\mu_0}\{S_N \ge b\}\rho_- + o(\Delta);
\end{aligned} \tag{10.23}
$$

and from (10.19) it follows immediately that

$$
P_{\mu_0}\{S_N \ge b\}\rho_-/P_{\mu_1}\{S_N \ge b\} = \rho_- e^{-\Delta(b+\rho_+)} + o(1). \tag{10.24}
$$

From Lemma 10.27 below one obtains

$$
\begin{aligned}
E_{\mu_0}(S_{\tau_-} + \Delta S_{\tau_-}^2) &= E_0(S_{\tau_-}) + \tfrac{1}{2}\theta_0 E_0(S_{\tau_-}^2) + \Delta E_0(S_{\tau_-}^2) + o(\Delta) \\
&= E_0(S_{\tau_-})(1 + 3\Delta\rho_-/2) + o(\Delta)
\end{aligned} \tag{10.25}
$$

and similarly

$$
\begin{aligned}
P_{\mu_1}\{\tau_- = \infty\} &= E_{\mu_1}\{1 - e^{\Delta S_{\tau_-}}\} \\
&= -\Delta E_0(S_{\tau_-})(1 + \tfrac{1}{2}\Delta\rho_-) + o(\Delta^2).
\end{aligned} \tag{10.26}
$$

□

Putting (10.20)–(10.26) together yields (10.18).

The following lemma justifies (10.25) and (10.26).

Lemma 10.27. *Let* $r > 0$, $\mu_1 > 0$. *Then*

$$
\lim_{\mu_1 \downarrow 0} \mu_1 E_{\mu_1}\{\tau_+ S_{\tau_+}^r\} = E_0 S_{\tau_+}^{r+1}/(r+1) \tag{10.28}
$$

and consequently as $\mu_1 \downarrow 0$

$$
E_{\mu_1} S_{\tau_+}^r = E_0 S_{\tau_+} + \theta_1 r E_0 S_{\tau_+}^{r+1}/(r+1) + o(\theta_1), \tag{10.29}
$$

where $\theta_1 = \theta(\mu_1)$.

Proof. Suppose initially that (10.28) holds. From

$$
E_{\mu_1}(S_{\tau_+}^r) = E_0\{S_{\tau_+}^r \exp[\theta_1 S_{\tau_+} - \tau_+ \psi(\theta_1)]\} \tag{10.30}
$$

and the dominated convergence theorem it follows that

$$
\lim_{\mu_1 \downarrow 0} E_{\mu_1}(S_{\tau_+}^r) = E_0(S_{\tau_+}^r). \tag{10.31}
$$

Also (10.30) implies that $f(\theta) = E_\mu(S_{\tau_+}^r)$ is continuously differentiable for small positive μ and that

$$
f'(\theta) = E_\mu\{S_{\tau_+}^r[S_{\tau_+} - \tau_+ \psi'(\theta)]\}. \tag{10.32}
$$

Since $f(\theta_1) = f(\theta) + (\theta_1 - \theta)f'(\theta) + \int_\theta^{\theta_1} [f'(y) - f'(\theta)]\,dy$, (10.29) follows by letting $\theta \to 0$ and appealing to (10.28), (10.31), and (10.32).

It remains to prove (10.28). Obviously

$$\mu_1 E_{\mu_1}(\tau_+ S_{\tau_+}^r) = \mu_1 \sum_{n=0}^{\infty} E_{\mu_1}(S_{\tau_+}^r; \tau_+ > n)$$

(10.33)

$$= \mu_1 \sum_0^{\infty} \int_{[0,\infty)} E_{\mu_1}[S_{\tau(x)} - x]^r P_{\mu_1}\{\tau_+ > n, -S_n \in dx\}.$$

Let $M = \min_{0 \le n < \infty} S_n$ and let $\tau_-^{(j)}$ denote the jth weak descending ladder epoch (cf. (8.39)). By Proposition 8.38 and Corollary 8.39

$$P_{\mu_1}\{\tau_+ > n, S_n \le -x\} = E_{\mu_1}(\tau_+) \sum_{j=0}^{\infty} P_{\mu_1}\{\tau_-^{(j)} = n, S_{\tau_+^{(j)}} \le -x\} P_{\mu_1}\{\tau_- = \infty\}$$

$$= E_{\mu_1}(\tau_+) \sum_{j=0}^{\infty} P_{\mu_1}\{\tau_-^{(j)} = n, \tau_-^{(j+1)} = \infty, S_{\tau_+^{(j)}} \le -x\}$$

$$= E_{\mu_1}(\tau_+) P_{\mu_1}\{M \le -x\}.$$

Hence by Wald's identity and a change of variable (10.33) may be rewritten

$$\mu_1 E_{\mu_1}(\tau_+ S_{\tau_+}^r) = E_{\mu_1}(S_{\tau_+}) \int_{[0,\infty)} E_{\mu_1}\{S_{\tau(x/\theta_1)} - (x/\theta_1)\}^r P_{\mu_1}\{-M \in dx/\theta_1\}.$$

(10.34)

It follows from Theorem 10.5 that

$$P_{\mu_1}\{M \le -x/\theta_1\} \to e^{-2x} \qquad (10.35)$$

and from Lemma 10.11 that

$$E_{\mu_1}\{S_{\tau(x/\theta_1)} - (x/\theta_1)\}^r \to E_0 S_{\tau_+}^{r+1}/(r+1) E_0 S_{\tau_+}. \qquad (10.36)$$

The relation (10.28) follows formally from substitution of (10.31), (10.35), and (10.36) in (10.34). To justify this formal substitution it suffices to note that the convergence indicated in (10.35) and (10.36) is uniform on $[\varepsilon, 1/\varepsilon]$ for every $\varepsilon > 0$, that the left hand side of (10.36) is bounded by a constant multiple of e^x for all x, and that for small μ_1, $P_{\mu_1}\{M \le -x/\theta_1\} \le \exp(-3x/2)$ for all x. The details of this justification are omitted. □

Of independent interest is the following consequence of Lemma 10.27, which justifies the local approximation of $\nu(\mu)$ in (4.38) and, more generally, of ν_+ in Remark 10.8. For the similar expansion of $\rho(\mu)$ given in (4.39) see Problem 10.2.

Proposition 10.37. *Under the distributional assumptions* (10.1), *the constant* ν_+ *of* (8.59) (*cf. also* (8.48) *and* (8.49)) *equals*

$$P_{\mu_0}\{\tau_+ = \infty\}/\Delta E_{\mu_1}(S_{\tau_+}) \qquad (10.38)$$

and has the local expansion

$$1 - \rho_+ \Delta + \tfrac{1}{2}\rho_+^2 \Delta^2 + o(\Delta^2) \qquad \Delta \to 0, \qquad (10.39)$$

where $\rho_+ = \frac{1}{2} E_0 S_{\tau_+}^2 / E_0 S_{\tau_+}$ *and* $\Delta = \theta_1 - \theta_0$.

Proof. That v_+ in (8.59) equals the expression in (10.38) is an immediate consequence of Corollary 8.39 and Wald's identity. To obtain (10.39) one uses Lemma 10.27 to calculate as follows.

$$\begin{aligned} P_{\mu 0}\{\tau_+ = \infty\}/\Delta E_{\mu_1} S_{\tau_+} &= (1 - E_{\mu_1} e^{-\Delta S_{\tau_+}})/\Delta E_{\mu_1} S_{\tau_+} \\ &= E_{\mu_1}(S_{\tau_+} - \tfrac{1}{2}\Delta S_{\tau_+}^2 + \Delta^2 S_{\tau_+}^3/6)/E_{\mu_1}(S_{\tau_+}) + o(\Delta^2) \\ &= 1 - \frac{\Delta(E_0 S_{\tau_+}^2 + \frac{2}{3}\theta_1 E_0 S_{\tau_+}^3 + o(\Delta))}{2(E_0 S_{\tau_+} + \frac{1}{2}\theta_1 E_0 S_{\tau_+}^2 + o(\Delta))} + \frac{\Delta^2 E_0 S_{\tau_+}^3}{6 E_0 S_{\tau_+}} + o(\Delta^2) \\ &= 1 - \rho_+ \Delta + \tfrac{1}{2}\rho_+^2 \Delta^2 + o(\Delta^2). \qquad \square \end{aligned}$$

3. Truncated Tests

We continue to use the notation and assumptions of Sections 1 and 2. Let $c \le b$ and consider

$$P_\mu\{\tau < m, S_m < c\}, \tag{10.40}$$

where as usual $\tau = \inf\{n: S_n \ge b\}$. The corresponding probability for Brownian motion is an important ingredient in studying the tests defined in Chapter III—even (perhaps surprisingly) the truncated sequential probability ratio test (cf. (3.36)). In this section we attempt to improve upon the simple Brownian approximation to (10.40) given in (3.14). The principal result, Theorem 10.41, justifies and generalizes (3.28).

The approximations of Sections 1 and 2 lend themselves to the following interpretation. In (10.7), Theorem 10.13, and Theorem 10.16, one corrects the Brownian motion approximation by displacing the stopping boundary a constant amount to account for excess over the boundary and by using $\pm\frac{1}{2}\Delta = \pm\frac{1}{2}(\theta_1 - \theta_0)$ in place of the expected value μ to correct for non-normality. The probability (10.40) requires a much more elaborate correction for non-normality, in which the third moment $\kappa = E_0(z_1^3)$ enters explicitly.

Theorem 10.41. *Let $z_1, z_2, \ldots$ be independent random variables with distribution function of the form* (10.1) *with F standardized to have mean* 0 *and variance* 1. *Assume also that for some $n_0 \ge 1$ and all μ in some neighborhood of* 0

$$\int_{-\infty}^{\infty} |E_\mu e^{i\lambda z_1}|^{n_0}\, d\lambda < \infty. \tag{10.42}$$

Assume that $b = m^{1/2}\zeta$ and $\xi = m^{1/2}\xi_0$ for some $\zeta > 0$ and $-\infty < \xi_0 < \zeta$. Then as $m \to \infty$

$$P_\xi^{(m)}\{\tau < m\} = \exp\{-2(b + \rho_+)(b + \rho_+ - \xi - \kappa/3)/(m + \kappa\xi/3)\} + o(m^{-1/2}), \tag{10.43}$$

where $\kappa = E_0 z_1^3$ *and* $\rho_+ = E_0 S_{\tau_+}^2 / 2E_0 S_{\tau_+}$. *If in addition,* $c = m^{1/2}\gamma$ *for some* $\gamma \le \zeta$,

$$\text{(10.44)} \qquad P_0\{\tau < m, S_m < c\} = \Phi\left(\frac{c + \kappa/3 - 2(b + \rho_+)}{(m + \kappa c/3)^{1/2}}\right) + o(m^{-1/2})$$

as $m \to \infty$. *Now suppose* $\mu_0 < 0 < \mu_1$ *are such that* $\psi(\theta_0) = \psi(\theta_1)$, *where* $\theta_j = \theta(\mu_j)$. *Let* $\Delta = (\theta_1 - \theta_0)$. *Assume as above that* $b = m^{1/2}\zeta$ *and* $c = m^{1/2}\gamma$ *for some* $\zeta > 0, \gamma \le \zeta$, *and also that* $m^{1/2}\Delta = \delta$ *is a fixed positive constant. Then as* $m \to \infty$, *for* $j = 0$ *or* 1

$$\text{(10.45)} \qquad \begin{aligned} P_{\mu_j}\{\tau < m, S_m < c\} &= \exp[-(-1)^j \Delta(b + \rho_+)] \\ &\times \Phi\left(\frac{c + \kappa/3 - 2(b + \rho_+)}{(m + \kappa c/3)^{1/2}} + \tfrac{1}{2}(-1)^j \Delta(m + \kappa c/3)^{1/2}\right) + o(m^{-1/2}). \end{aligned}$$

Remark. Note that (10.43)–(10.45) are of the form of the corresponding Brownian motion formulae with b replaced by $b + \rho_+$, $\xi(c)$ replaced by $\xi + \kappa/3$ $(c + \kappa/3)$, m replaced by $m + \kappa\xi/3$ $(m + \kappa c/3)$, and the drift of Brownian motion replaced by $\pm\frac{1}{2}\Delta$ (cf. (3.13) and (3.14)). Customarily Edgeworth type expansions are written as a sum of several terms (e.g. Feller, 1972, p. 538), which in each of (10.43)–(10.45) have been compressed into a single probability. In addition to illustrating the relation with Brownian motion (10.43)–(10.45) have the advantages of computational simplicity and in special cases agree with approximations obtained elsewhere. (For example, for $m = \infty$ (10.45) is consistent with Theorem 10.5; and for normal variables (10.43) is the same as (8.79).)

For a numerical example suppose that $x_1, x_2, \ldots$ are independent with probability density function $\lambda e^{-\lambda x}$ for $x \ge 0$. A sequential probability ratio test of H_0: $\lambda = \lambda^{(0)}$ against H_1: $\lambda = \lambda^{(1)}$ $(\lambda^{(0)} < \lambda^{(1)})$ is defined by a stopping rule of the form

$$\text{(10.46)} \qquad N = \inf\{n: S_n \notin (a, b)\},$$

where $S_n = \sum_1^n z_k = \sum_1^n (1 - \lambda^* x_k)$ with $\lambda^* = (\lambda^{(1)} - \lambda^{(0)})/\log(\lambda^{(1)}/\lambda^{(0)})$ (cf. Problem 2.4). Suppose that the test is truncated after m observations, and at truncation H_0 is rejected if $S_m \ge c$ for suitable $a \le c \le b$. (There are several, roughly equivalent, versions of this test. Since $S_n - S_{n-1} \le 1$, it is possible without changing the error probabilities, to curtail the test with acceptance of H_0 if for some $n \le m$, $\sum_1^n x_k \ge (m - c)/\lambda^*$ (see Figure 10.1). A second variation is to assume that observation takes place continuously in time, so instead of $S_n = n - \lambda^* \sum_1^n x_k$ one observes $X(t) - \lambda^* t$, where $X(t)$ is a Poisson process with intensity λ. For this version to be equivalent to the previous one, the a in (10.46) should be replaced by $a - 1$.)

The test described above is the sequential test of MIL-STD-781C, whose exact operating characteristics have been computed numerically by Aroian (1963), Woodall and Kurkjian (1962), and Epstein *et al.* (1963). Suppose that

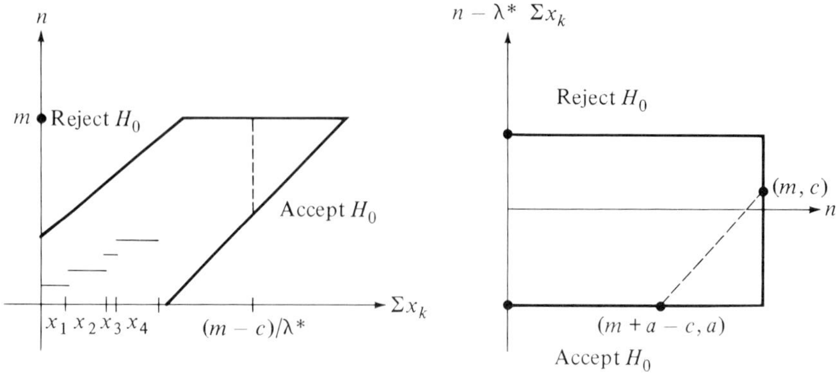

Figure 10.1.

$\lambda^{(0)} = 1$, $\lambda^{(1)} = 1.5$, $b = 5.419$, $a = -4.419$, $m = 41$, and $c = .311$. For the continuous time version of the test without truncation (hence with $a = -5.519$) the no excess approximations of Chapter II yield Type I and Type II error probabilities of .1 (cf. Problem 2.4(b)). Theorem 10.13 and Problem 8.10 yield the approximations $P_{\mu_0}\{S_N \geq 5.419\} \cong .087$ and $P_{\mu_1}\{S_N \leq -4.419\} \cong .101$, which according to Epstein *et al.* are correct to the accuracy given. For the truncated test Epstein *et al.* obtain $P_{\mu_0}\{S_{N \wedge m} \geq c\} = .115$ and $P_{\mu_1}\{S_{N \wedge m} < c\} = .125$. An application of (3.36), (10.45), and Problem 8.10 yields the approximations .115 and .126 respectively. Epstein *et al.* also give $P_\mu\{N \leq n, S_N \geq b\}$ and $P_\mu\{N \leq n, S_N \leq a\}$ for $n < m$ and $\mu = \mu_0$, 0, and μ_1. These values are of interest in computing attained significance levels and (for general μ) in computing confidence intervals. In particular, they obtain $P_{\mu_0}\{N \leq 30, S_N \geq b\} = .060$ and $P_{\mu_1}\{N \leq 30, S_N \geq b\} = .624$, for which (3.36), (10.45), and Problem 8.10 yield the approximations .058 and .630.

The proof of Theorem 10.41 sketched below proceeds by a likelihood ratio calculation along the lines of numerous other arguments appearing in this monograph. It is particularly interesting to note the following points of contrast with the superficially similar proof of Theorem 8.72. (i) The basic likelihood ratio identity in the proof of Theorem 8.72 involves the conditional probability $P_\xi^{(m)}$ and the unconditional probability P_{μ_1} (cf. (8.83)). Here there is an intervening step which requires that one first introduce the likelihood ratio of $P_\xi^{(m)}$ relative to $P_{\xi'}^{(m)}$, for a suitable ξ' as in Chapter III—especially (3.13). After some manipulation one introduces the likelihood ratio of $P_{\xi'}^{(m)}$ relative to an unconditional probability, which for convenience is taken to be P_0. It does not seem reasonable to try to combine the two steps. (ii) In the proof of Theorem 8.72, under the new probability τ/m converges to a point, but here its distribution remains nondegenerate. This has important implications for extensions to nonlinear stopping boundaries. In formulations where the distribution of τ/m concentrates at a point, one can expand the stopping boundary in a Taylor series about that point. Otherwise one must deal "honestly" with the nonlinear boundary, which is a much more difficult problem.

Proof of Theorem 10.41. The following argument yields (10.43). In principle (10.44) and (10.45) are direct consequences, although a complete justification is tricky (see the remarks below).

Let f_n denote the P_0 probability density function of S_n (which by (10.42) exists for all $n \geq n_0$). Repeated use is made of the expansion

$$f_n(xn^{1/2})n^{1/2} = \phi(x)\{1 + (\kappa/6n^{1/2})(x^3 - 3x)\} + (1 + |x|^3)^{-1}o(n^{-1/2}), \tag{10.47}$$

where the $o(\cdot)$ is uniform in x (cf. Petrov, 1972, p. 207).

Let $P_{\lambda,\xi}^{(m)}(A) = P_0(A|S_0 = \lambda, S_m = \xi)$ and $P_{\lambda,\mu}(A) = P_\mu(A|S_0 = \lambda)$. Let $\lambda = m^{1/2}\lambda_0$ and $\xi = m^{1/2}\xi_0$ for some λ_0, $\xi_0 < \zeta$, and set $\xi' = m^{1/2}(2\zeta - \xi_0)$. It is not difficult to use (10.47) to show for $m_1 = m[1 - (\log m)^{-2}]$ and for some $\varepsilon_m \to 0$

$$P_{\lambda,\xi}^{(m)}\{\tau < m\} - P_{\lambda,\xi}^{(m)}\{\tau < m_1, S_\tau - b < \varepsilon_m m^{1/2}\} = o(m^{-1/2}) \tag{10.48}$$

and

$$P_{\lambda,\xi'}^{(m)}\{\tau < m\} - P_{\lambda,\xi'}^{(m)}\{\tau < m_1, S_\tau - b < \varepsilon_m m^{1/2}\} = o(m^{1/2}). \tag{10.49}$$

Set $A_m = \{\tau < m_1, S_\tau - b < \varepsilon_m m^{1/2}\}$, and let $l^{(m)}(n, S_n)$ denote the likelihood ratio of $z_1, \ldots, z_n$ under $P_{\lambda,\xi}^{(m)}$ relative to $P_{\lambda,\xi'}^{(m)}$.

Then for all $n \leq m - n_0$

$$l^{(m)}(n, S_n) = \frac{f_{m-n}(\xi - S_n)f_m(\xi' - \lambda)}{f_m(\xi - \lambda)f_{m-n}(\xi' - S_n)}, \tag{10.50}$$

and by (10.48)

$$P_{\lambda,\xi}^{(m)}\{\tau < m\} = E_{\lambda,\xi'}^{(m)}\{l^{(m)}(\tau, S_\tau); A_m\} + o(m^{-1/2}). \tag{10.51}$$

Substitution of (10.50) into (10.51) and expansion via (10.47) easily yield the anticipated first order result

$$P_{\lambda,\xi}^{(m)}\{\tau < m\} \to \exp\{-2(\zeta - \lambda_0)(\zeta - \xi_0)\},$$

which suggests rewriting (10.51) in the form (cf. (10.49))

$$\begin{aligned} &P_{\lambda,\xi}^{(m)}\{\tau < m\} - \exp\{-2(\zeta - \lambda_0)(\zeta - \xi_0)\} + o(m^{-1/2}) \\ &\quad = E_{\lambda,\xi'}^{(m)}\{l^{(m)}(\tau, S_\tau) - \exp[-2(\zeta - \lambda_0)(\zeta - \xi_0)]; A_m\}. \end{aligned} \tag{10.52}$$

The likelihood ratio of $z_1, \ldots, z_n$ under $P_{\lambda,\xi'}^{(m)}$ relative to $P_{\lambda,0}$ is

$$f_{m-n}(\xi' - S_n)/f_m(\xi' - \lambda),$$

and hence by (10.50) the right hand side of (10.52) equals

$$E_{\lambda,0}\left\{\frac{f_{m-\tau}(\xi - S_\tau)}{f_m(\xi - \lambda)} - \exp[-2(\zeta - \lambda_0)(\zeta - \xi_0)]\frac{f_{m-\tau}(\xi' - S_\tau)}{f_m(\xi' - \lambda)}; A_m\right\}. \tag{10.53}$$

In principle the derivation of (10.43) is completed by using (10.47) to expand the integrand in (10.53) and by applying Lemma 10.11 to evaluate the resulting expression. To see what happens in the simplest case, which also illustrates the pitfalls of the most general, suppose $f_n(x) = n^{-1/2}\phi(xn^{-1/2})$. Some algebra shows that the integrand in (10.53) equals

$$[(1-\tau/m)^{1/2}\phi(\xi_0-\lambda_0)]^{-1}\left\{\phi\left(\frac{\zeta-\xi_0+R/m^{1/2}}{(1-\tau/m)^{1/2}}\right)-\phi\left(\frac{\zeta-\xi_0-R/m^{1/2}}{(1-\tau/m)^{1/2}}\right)\right\},$$

where $R = S_\tau - m^{1/2}\zeta$. This can be expanded to give

$$\begin{aligned}(10.54)\qquad &-2(1-\tau/m)^{-1/2}\exp\{\tfrac{1}{2}(\xi_0-\lambda_0)^2-\tfrac{1}{2}(\zeta-\xi_0)^2/(1-\tau/m)\}\\ &\times[(\zeta-\xi_0)R/m^{1/2}(1-\tau/m)]+O([1+R^2]/m)\end{aligned}$$

uniformly on A_m. For $\xi_0 \neq \zeta$ (10.54) is a bounded continuous function of τ/m on A_m. Hence it can be substituted into (10.53) and evaluated via Lemma 10.11. Putting $\lambda = 0$ and performing the required calculus yields (10.43) for the special case of normally distributed z_n. Details of the general case are omitted.

In order to integrate (10.43) to obtain (10.44) and (10.45), one would like to know the extent to which (10.43) holds uniformly in $\xi_0 \le \zeta$. Some care with the preceding argument shows that (10.43) is uniform in $-\log m \le \xi_0 \le \zeta - \varepsilon$ for each $\varepsilon > 0$, and this, together with (10.47), suffices to prove (10.44) and (10.45) for $\gamma = cm^{-1/2} < \zeta$.

However, the argument completely collapses when ξ_0 can be arbitrarily close to ζ. (For example, (10.54) is not a uniformly bounded function of τ/m for $\tau < m_1$, nor can one expect (10.48) and (10.49) to hold uniformly.) To handle this case define

$$\tau^* = \sup\{n: n \le m, S_n \ge b\}$$

and observe that for $\xi < b$

$$P_{0,\xi}^{(m)}\{\tau < m\} = P_{0,\xi}^{(m)}\{\tau^* > 0\} = P_{\xi,0}^{(m)}\{\tau < m\}.$$

Hence it remains to investigate

$$P_{\lambda,\xi}^{(m)}\{\tau < m\}$$

for $\xi = 0$ and λ in the range $m^{1/2}(\zeta - \varepsilon) \le \lambda < m^{1/2}\zeta$. By easy modifications of the preceding argument one can show that

$$P_{\lambda,0}^{(m)}\{\tau < m\} = \exp\{-2\zeta(\zeta-\lambda_0)\} + O(m^{-1/2})$$

uniformly in $\zeta - \varepsilon \le m^{-1/2}\lambda < \zeta$ for each $\varepsilon > 0$, which suffices to complete the proof of (10.44) and (10.45) when $\gamma = \zeta$. □

4. Computation of $E_0(S_{\tau_+}^2)/2E_0(S_{\tau_+})$

Let $z_1, z_2, \ldots$ be independent and identically distributed with mean 0, variance 1, $\kappa = Ez_1^3$, and $g(\lambda) = E\{\exp(i\lambda z_1)\}$. Let $S_n = z_1 + \cdots + z_n$ and $\tau_+ = \inf\{n: S_n > 0\}$. By Problem 8.9 $P\{\tau_+ < \infty\} = 1$ and $E(S_{\tau_+}^2) < \infty$. In order to apply the results of Sections 1–3 one must be able to evaluate $E(S_{\tau_+}^2)/2E(S_{\tau_+})$. The following theorem gives an expression suitable for numerical computation.

For the special case of normal variables it yields the constant $\rho \cong .583$ which was used throughout Chapters II–V.

Theorem 10.55. *Assume that the distribution of z_1 is continuous. Then*

$$E(S_{\tau_+}^2)/2E(S_{\tau_+}) = \kappa/6 - \pi^{-1}\int_0^\infty \lambda^{-2}\,\mathrm{Re}\log\{2[1-g(\lambda)]/\lambda^2\}\,d\lambda.$$

PROOF. Suppose initially that

$$\int_{-\infty}^{\infty} |g(\lambda)|\,d\lambda < \infty. \tag{10.56}$$

This assumption will be removed later.

The proof proceeds by first showing that

$$\begin{aligned} E(S_{\tau_+}) &= \lim_{\alpha\to 0} \alpha^{-1}[1 - E\exp(-\alpha S_{\tau_+})] \\ &= 2^{-1/2}\exp\left\{\pi^{-1}\int_0^\infty \lambda^{-1}\,\mathrm{Im}\log[1-g(\lambda)]\,d\lambda\right\}. \end{aligned} \tag{10.57}$$

and then evaluating

$$\lim_{\alpha\to 0}\alpha^{-1}\log\{[1 - E\exp(-\alpha S_{\tau_+})]/\alpha E S_{\tau_+}\} = -E(S_{\tau_+}^2)/2E(S_{\tau_+}). \tag{10.58}$$

Note (essentially by the change of variable $\lambda = \alpha e^u$) that

$$\log\alpha = (2\pi)^{-1}\int_{-\infty}^{\infty}[\alpha/(\alpha^2+\lambda^2)]\log\lambda^2\,d\lambda. \tag{10.59}$$

By Theorem 8.45, (10.56), and a calculation similar to that in the proof of Theorem 8.51,

$$\log\{1 - E[\exp(-\alpha S_{\tau_+})]\} = (2\pi)^{-1}\int_{-\infty}^{\infty}(\alpha + i\lambda)^{-1}\log[1-g(\lambda)]\,d\lambda,$$

so by (10.59)

$$\begin{aligned} \log\{[1 - E\exp(-\alpha S_{\tau_+})]/\alpha\} &= (2\pi)^{-1}\int_{-\infty}^{\infty}[\lambda/(\alpha^2+\lambda^2)]\,\mathrm{Im}\log[1-g(\lambda)]\,d\lambda \\ &\quad + (2\pi)^{-1}\int_{-\infty}^{\infty}[\alpha/(\alpha^2+\lambda^2)]\{\mathrm{Re}\log[1-g(\lambda)] \\ &\quad - \log(\tfrac{1}{2}\lambda^2)\}\,d\lambda - \tfrac{1}{2}\log 2. \end{aligned} \tag{10.60}$$

Since $\mathrm{Im}\log[1-g(\lambda)]$ behaves like λ near 0 and is integrable at ∞, the first term on the right hand side of (10.60) converges to $\pi^{-1}\int_0^\infty \lambda^{-1}\,\mathrm{Im}\log[1 - g(\lambda)]\,d\lambda$ as $\alpha \to 0$. Since $\mathrm{Re}\log[1-g(\lambda)] - \log(\frac{1}{2}\lambda^2)$ is continuous, equals 0 at 0, is well behaved away from 0, and since $\pi^{-1}\alpha/(\alpha^2+\lambda^2)$ acts like a Dirac

δ-function as $\alpha \to 0$, the second integral on the right hand side of (10.60) converges to 0. This gives (10.57).

Subtraction of the logarithm of (10.57) from (10.60) yields

$$\alpha^{-1} \log\{[1 - E\exp(-\alpha S_{\tau_+})]/\alpha E(S_{\tau_+})\}$$

(10.61)
$$= -(2\pi)^{-1} \int_{-\infty}^{\infty} [\alpha/(\alpha^2 + \lambda^2)]\lambda^{-1} \operatorname{Im}\log[1 - g(\lambda)]\, d\lambda$$

$$+ (2\pi)^{-1} \int_{-\infty}^{\infty} (\alpha^2 + \lambda^2)^{-1} \{\operatorname{Re}\log[1 - g(\lambda)] - \log(\tfrac{1}{2}\lambda^2)\}\, d\lambda.$$

Since $\operatorname{Re}\log[1 - g(\lambda)] - \log(\frac{1}{2}\lambda^2) \sim \lambda^2$ as $\lambda \to 0$, the second integral on the right hand side of (10.61) converges as $\alpha \to 0$ to

$$(2\pi)^{-1} \int_{-\infty}^{\infty} \lambda^{-2} \{\operatorname{Re}\log[1 - g(\lambda)] - \log(\tfrac{1}{2}\lambda^2)\}\, d\lambda.$$

Since $\lambda^{-1} \operatorname{Im}\log[1 - g(\lambda)]$ is continuous and equals $\kappa/3$ at 0, is well behaved away from 0, and since $\pi^{-1}\alpha/(\alpha^2 + \lambda^2)$ again behaves like a δ-function, the first integral on the right hand side of (10.61) converges to $-\frac{1}{2}\kappa/3 = -\kappa/6$. This completes the evaluation of (10.58) and yields the asserted result under the assumption (10.56).

To remove the assumption (10.56), introduce the smoothed random walk $z_n^{(\sigma)} = z_n + \sigma y_n$ as in the proof of Remark 8.53. Let $S_n^{(\sigma)}$, $\tau_+^{(\sigma)}$, $g^{(\sigma)}$, etc. be defined in the obvious way. Then $g^{(\sigma)}(\lambda) = g(\lambda)\exp(-\frac{1}{2}\lambda^2\sigma^2)$ satisfies (10.56), so (10.61) holds for the random walk $S_n^{(\sigma)}$, $n = 1, 2, \ldots$, $\tau_+^{(\sigma)}$, $g^{(\sigma)}$, etc.

Now let $\sigma \to 0$. By the continuity of the distribution of z_n, $\tau_+^{(\sigma)} \to \tau_+$ and $S_{\tau_+^{(\sigma)}}^{(\sigma)} \to S_{\tau_+}$ with probability one. Hence

$$1 - E\{\exp(-\alpha S_{\tau_+^{(\sigma)}}^{(\sigma)})\} \to 1 - E\{\exp(-\alpha S_{\tau_+})\}.$$

A somewhat more elaborate argument (cf. Problem 10.6) shows that

(10.62)
$$E(S_{\tau_+^{(\sigma)}}^{(\sigma)}) \to E(S_{\tau_+}).$$

It is easy to see that

$$\int_{-\infty}^{\infty} [\alpha/(\alpha^2 + \lambda^2)]\lambda^{-1} \operatorname{Im}\log[1 - g^{(\sigma)}(\lambda)]\, d\lambda$$

$$\to \int_{-\infty}^{\infty} [\alpha/(\alpha^2 + \lambda^2)]\lambda^{-1} \operatorname{Im}\log[1 - g(\lambda)]\, d\lambda,$$

and the argument in the proof of Remark 8.53 shows that

$$\int_{-\infty}^{\infty} (\alpha^2 + \lambda^2)^{-1} \{\operatorname{Re}\log[1 - g^{(\sigma)}(\lambda)] - \log(\tfrac{1}{2}\lambda^2)\}\, d\lambda$$

$$\to \int_{-\infty}^{\infty} (\alpha^2 + \lambda^2)^{-1} \{\operatorname{Re}\log[1 - g(\lambda)] - \log(\tfrac{1}{2}\lambda^2)\}\, d\lambda.$$

Table 10.1. $P_p\{\tau \leq m\}$ ($b = 5.25; \eta = .3219; \rho_+ = .32$)

p	m	Brownian Approximation	Exact (Armitage, 1957)	Corrected Approximation
.5	44	.0327	.0205	.0206
.8	44	.953	.947	.956
.5	20	.0251	.0153	.0161
.5	14	.0189	.0110	.0126

This brings us back to (10.61) (without the σ), and from there the proof proceeds as before. □

Remark. The hypothesis that z_1 have a continuous distribution is used only to guarantee that $S_{\tau_+^{(\sigma)}}^{(\sigma)} \to S_{\tau_+}$ in the preceding unsmoothing argument. It seems plausible that the condition might be weakened or eliminated entirely (see Problem 10.7).

EXAMPLE 10.63. Let $x_1, x_2, \ldots$ be independent random variables with $P_p\{x_i = 1\} = p = 1 - P_p\{x_i = -1\}$. Let $S_n' = x_1 + \cdots + x_n$ and define for $b, \eta > 0$

$$\tau = \inf\{n: S_n' \geq b + \eta n\}. \tag{10.64}$$

For the Bernoulli matched pair model of V.2, (10.64) is the stopping rule of a one-sided sequential probability ratio test of $p = \frac{1}{2}$ against $p = p_1$ for a particular $p_1 = p_1(\eta) > \frac{1}{2}$ (cf. Problem 5.6). In order to approximate $P_p\{\tau \leq m\}$ it is irresistibly tempting to apply Theorem 10.41 to the random walk with increments $z_i = (x_i - \eta)/(1 - \eta^2)^{1/2}$ (which have mean 0 and variance 1 for $p = (1 + \eta)/2 = p^*$, say), even though the condition (10.42) is not satisfied. As always, one can write

$$P_p\{\tau \leq m\} = P_p\{S_m' \geq b + \eta m\} + P_p\{\tau < m, S_m' < b + \eta m\}.$$

The first term can be evaluated exactly; and to approximate the second using (10.45), one need only calculate $\rho_+ = E_{p^*}(S_{\tau_+}^2)/2E_{p^*}(S_{\tau_+})$, which is an easy numerical integration based on Theorem 10.55 as extended in Problem 10.7. Table 10.1 compares a Brownian approximation with mean $2p - 1$ and variance $4p(1 - p)$, the corrected Brownian approximation described above, and some numerically computed results reported by Armitage (1957).

PROBLEMS

10.1. Use Theorem 10.13 and Problem 8.10 to justify the approximations suggested in Problem 2.4.

10.2. Prove that as $\theta \downarrow 0$,

$$E_\mu(S_{\tau_+}^2)/2E_\mu(S_{\tau_+}) = \rho_+ - \tfrac{1}{2}\rho_+^2\theta + \theta E_0(S_{\tau_+}^3)/E_0(S_{\tau_+}) + o(\theta),$$

where as usual $\rho_+ = E_0(S_{\tau_+}^2)/2E_0(S_{\tau_+})$. For the case of normal (variance 1) random variables this becomes

$$\rho_+ + \mu/4 + o(\mu),$$

which implies (4.41).

10.3. Suppose it is known that as $b \to \infty$, *uniformly* in $\mu \in (0, \varepsilon)$

$$E_\mu(S_\tau - b) = E_\mu(S_{\tau_+}^2)/2E_\mu(S_{\tau_+}) + o(b^{-1}). \tag{10.65}$$

Show that under the conditions of Theorem 10.5 the remainder in (10.7) is $o(\Delta^2)$. (Siegmund, 1979, observes that (10.65) holds if z_1 is strongly nonarithmetic in the sense of Theorem 8.51.)

10.4. Use Theorem 10.16 and Problem 8.10 to justify the improved approximations suggested in Problems 2.17 and 2.18.

10.5.* By an argument similar to the proof of Theorem 8.34 (or otherwise), show that under P_0, $b^{-2}\tau_b$ and $S_\tau - b$ are asymptotically independent.

10.6. Prove (10.62).

10.7. Consider Theorem 10.55, but without assuming that z_1 has a continuous distribution. Let $\bar{\tau} = \inf\{n: S_n \geq 0\}$. Show that for $j = 1, 2$

$$E(S_{\tau_+}^j) = E(S_{\tau_+}^j; \tau_+ = \bar{\tau})/P\{\tau_+ = \bar{\tau}\}$$

and hence that

$$E(S_{\tau_+}^2)/E(S_{\tau_+}) = \lim_{\sigma \to 0} E[(S_{\tau_+^{(\sigma)}}^{(\sigma)})^2]/E(S_{\tau_+^{(\sigma)}}^{(\sigma)}).$$

Complete the proof of this generalization of Theorem 10.55.

10.8.† Develop corrected Brownian approximations for $E_\mu(\tau \wedge m)$ and for the stopping time of a truncated sequential probability ratio test. Compare these theoretically (insofar as possible) and numerically with an alternative approximation along the lines of Theorems 4.27 and 9.53.

CHAPTER XI

Miscellaneous Boundary Crossing Problems

The subject of this chapter is several results which can be obtained by refinements of the methods developed in Chapters IV and IX. Some of them have already been used in Chapters IV and V.

1. Proof of Theorem 4.21

Let $W(t)$, $0 < t < \infty$, denote Brownian motion and $T = \inf\{t: t \geq m_0, |W(t)| \geq bt^{1/2}\}$. The following generalizes Theorem 4.21.

Theorem 11.1. *Suppose b, c, m_0, and m all tend to ∞ in such a way that for some $0 < \gamma \leq \mu_1 < \mu_0$, $cm^{-1/2} = \gamma$, $bm^{-1/2} = \mu_1$, and $bm_0^{-1/2} = \mu_0$. Then*

$$P_0\{T < m, |W(m)| < cm^{1/2}\} = (b - b^{-1})\phi(b)\log(mc^2/m_0 b^2) + b^{-1}\phi(b)\{3 - (bc^{-1})^2\} + o(b^{-1}\phi(b)).$$

PROOF. The following calculation is an extension of the proof of Theorem 4.14 and uses the notation developed in that argument. By (4.25)

$$P_0\{T < m, |W(m)| < cm^{1/2}\} \tag{11.2}$$
$$= 2\mu_1^{-1} b\phi(b) \int_{[0,\gamma]} \tilde{E}_{m\xi_0}^{(m)}\{(m/T^*)^{1/2}; T^* \geq m_0\}\, d\xi_0.$$

In contrast to the proof of Theorem 4.14 which uses only the fact that $m^{-1} T^* \to (\xi_0 \mu_1^{-1})^2$ in $\tilde{P}_{m\xi_0}^{(m)}$-probability, the present argument analyzes the central limit behavior of $m^{-1}T^*$. The ideas are simple but their implementation rather lengthy. Particular attention must be paid to values of ξ_0 which are in the range $\mu_1^2/\mu_0 + O(m^{-1/2})$, for then $\tilde{P}_{m\xi_0}^{(m)}\{T^* \geq m_0\}$ does not converge to 0 nor to 1. (See Figure 4.2 and Lemma 11.9.)

Let $\xi_0 \in (0, \mu_1]$ and define

$$\tau^* = \tau_{\xi_0}^* = \inf\{t\colon |m\xi_0 + W(t)| \geq b(m-t)^{1/2}\},$$

so the P_0-distribution of $m - \tau^*$ is the same as the $\tilde{P}_{m\xi_0}^{(m)}$-distribution of T^*, and hence

$$\text{(11.3)}\quad \tilde{E}_{m\xi_0}^{(m)}\{(m/T^*); T^* \geq m_0\} = E_0\{[m/(m-\tau_{\xi_0}^*)]^{1/2}; \tau_{\xi_0}^* \leq m - m_0\}.$$

Lemma 11.4. *Let $\mu_1^2\mu_0^{-1} \leq \xi_0 \leq \mu_1$. As $m \to \infty$*

$$\text{(11.5)}\quad E_0\{[m/(m-\tau^*)]^{1/2}\} = \mu_1\xi_0^{-1} + \{1 - (\xi_0/\mu_1)^2\}\mu_1/m\xi_0^3 + o(m^{-1}).$$

Let $\delta > 0$. The relation (11.5) is valid uniformly on $[\mu_1^2\mu_0^{-1}, \mu_1 - \delta]$; the first term of (11.5) with a remainder of order m^{-1} is valid uniformly on $[\mu_1^2\mu_0^{-1}, \mu_1 - \delta m^{-1}]$.

Proof. The proof is given for fixed $\xi_0 \in [\mu_1^2\mu_0^{-1}, \mu_1)$. The uniformity follows by similar calculations with considerably more attention to details.

Let $t^* = 1 - (\xi_0\mu_1^{-1})^2$. An easy law of large numbers argument shows that $m^{-1}\tau^* \to t^*$ in probability as $m \to \infty$. Hence by a Taylor series expansion

$$\text{(11.6)}\quad \begin{aligned} E_0\{[m/(m-\tau^*)]^{1/2}\} &= \mu_1\xi_0^{-1} + \tfrac{1}{2}(\mu_1/\xi_0)^3 E_0(m^{-1}\tau^* - t^*) \\ &\quad + \tfrac{3}{8}(\mu_1/\xi_0)^5 E_0(m^{-1}\tau^* - t^*)^2 + o(m^{-1}). \end{aligned}$$

Since $m^2\xi_0^2 + 2m\xi_0 W(\tau^*) + [W(\tau^*)]^2 = b^2(m - \tau^*)$, by Wald's identities

$$\text{(11.7)}\quad E_0(\tau^*) = mt^*/(1 + 1/\mu_1^2 m) = mt^* - \mu_1^{-2}t^* + O(m^{-1}).$$

Also

$$\text{(11.8)}\quad E_0(\tau^* - mt^*)^2 \sim 4mt^*\xi_0^2/\mu_1^4$$

(cf. Remark 9.39). Now (11.5) follows by substituting (11.7) and (11.8) into (11.6). □

Lemma 11.9. *Let $\xi_0 = \mu_1^2\mu_0^{-1} + \Delta m^{-1/2}$ and $t^{**} = 1 - (\mu_1\mu_0^{-1})^2$. Uniformly in $|\Delta| \leq \log m$*

$$\begin{aligned} &E_0\{[m/(m-\tau^*)]^{1/2}; \tau^* \leq m - m_0\} \\ &= \mu_0\mu_1^{-1}\Phi(\Delta t^{**-1/2}) + m^{-1/2}\{\mu_1^{-1}\phi(\Delta t^{**-1/2})/t^{**1/2} \\ &\quad - \mu_0^2\mu_1^{-3}\Delta\Phi(\Delta t^{**-1/2}) - \mu_0^2\mu_1^{-3}t^{**1/2}\phi(\Delta t^{**-1/2})\} \\ &\quad + O(m^{-3/4}) + o(m^{-1/2}\phi(\Delta)). \end{aligned}$$

PROOF. By a Taylor series expansion (cf. (11.6))

$$
\begin{aligned}
&E_0\{[m/(m-\tau^*)]^{1/2}; \tau^* \le m - m_0\} \\
&\quad = \mu_1 \xi_0^{-1} P_0\{\tau^* \le m - m_0\} + \tfrac{1}{2}(\mu_0/\mu_1)^3 E_0\{(m^{-1}\tau^* - t^*); \tau^* \le m - m_0\} \\
&\qquad + O(\Delta m^{-1}).
\end{aligned}
\tag{11.10}
$$

Note that $m - m_0 = mt^{**}$ and $bm_0^{1/2} - m\xi_0 = -\Delta m^{1/2}$, so by an argument similar to the proof of Corollary 4.19

$$
\begin{aligned}
P_0\{\tau^* \le m - m_0\} &= P_0\{W(m - m_0) \ge bm_0^{1/2} - m\xi_0\} \\
&\quad + P_0\{\tau^* < m - m_0, W(m - m_0) < bm_0^{1/2} - m\xi_0\} \\
&= 1 - \Phi(-\Delta t^{**-1/2}) \\
&\quad + \int_0^{(\log m)^2} \phi\{[-\Delta - xm^{-1/2}]t^{**-1/2}\} e^{-\mu_0 x}\, dx/(mt^{**})^{1/2} \\
&\quad + O(m^{-3/4}) + o(m^{-1/2}\phi(\Delta)) \\
&= \Phi(\Delta t^{**-1/2}) + m^{-1/2}\phi(\Delta t^{**-1/2})/\mu_0 t^{**1/2} \\
&\quad + O(m^{-3/4}) + o(m^{-1/2}\phi(\Delta)).
\end{aligned}
\tag{11.11}
$$

A similar calculation shows that

$$P_0\{\tau^* - mt^* \le xm^{1/2}\} = \Phi(\mu_0 x/2t^{**1/2}) + O(m^{-1/2});$$

and from (11.8) it is easily seen that

$$m^{-1/2}E_0(\tau^* - mt^*; \tau^* - mt^* \le -m^{3/4}) = O(m^{-1/4}).$$

Using these results and integrating by parts shows that

$$
\begin{aligned}
&m^{-1}E_0\{\tau^* - mt^*: \tau^* \le m - m_0\} \\
&\quad = m^{-1}E_0\{\tau^* - mt^*; \tau^* - mt^* \le 2\mu_0^{-1}\Delta m^{1/2} + O(1)\} \\
&\quad = 2m^{-1/2}\mu_0^{-1}t^{**1/2} \int_{(-m^{1/4}, \Delta t^{**-1/2})} z\phi(z)\, dz + O(m^{-3/4}) \\
&\quad = -2m^{-1/2}\mu_0^{-1}t^{**1/2}\phi(\Delta t^{**-1/2}) + O(m^{-3/4}).
\end{aligned}
$$

Substituting this and (11.11) into (11.10) yields the lemma. □

It is now straightforward to evaluate (11.2). It is easy to see that the integral over $[0, \mu_1^2\mu_0^{-1} - m^{-1/2}\log m]$ is $o(m^{-1})$; and in the range $\mu_1^2\mu_0^{-1} + m^{-1/2}\log m \le \xi_0 \le \mu_1$ the expectation over $\tau^* \le m - m_0$ can be replaced by the expectation over the entire sample space with an error which is $o(m^{-1})$. Hence by (11.3) the integral in (11.2) equals

$$\int_{\mu_1^2\mu_0^{-1}-m^{-1/2}\log m}^{\mu_1^2\mu_0^{-1}} E_0\{[m/(m-\tau^*)]^{1/2}; \tau^* \le m - m_0\}\, d\xi_0$$
$$- \int_{\mu_1^2\mu_0^{-1}}^{\mu_1^2\mu_0^{-1}+m^{-1/2}\log m} E_0\{[m/(m-\tau^*)]^{1/2}; \tau^* > m - m_0\}\, d\xi_0$$
$$+ \int_{\mu_1^2\mu_0^{-1}}^{\gamma} E_0\{[m/(m-\tau^*)]^{1/2}\}\, d\xi_0 + o(m^{-1}).$$

Lemmas 11.4 and 11.9 can be used to evaluate these integrals and complete the proof of Theorem 11.1. □

2. Expected Sample Size in the Case of More than Two Treatments

The following calculation justifies (5.15). The argument is not completely rigorous but seems capable of being made so. The notation of V.2 is used. In particular $\gamma_1 = bm^{-1/2}$ and $\theta = \gamma_1 + \Delta m^{-1/2}$. As in Theorem 4.27 we write

$$E_\mu(T \wedge m) = E_\mu(T) - E_\mu(T - m; T > m)$$

and approximate the two terms separately. Writing (5.4) in the form

$$T = \inf\{n: n \ge m_0, n^{-1}\|S_n\|^2 \ge b^2\}, \tag{11.12}$$

we can apply Theorem 9.28 (cf. Problem 9.8) to show that as $b \to \infty$

$$E_\mu(T) = m\gamma_1^2/\theta^2 + [2\rho(\theta) - r + 1]/\theta^2 + o(1),$$

where ρ is given exactly in (9.52) and approximately in (4.39). However, a more elaborate argument is required in order to generalize Proposition 4.29.

To approximate $E_\mu(T - m; T > m)$ it is possible to generalize the proof of Proposition 4.29 or the calculation suggested in Remark 4.34 (see Problem 4.8). The first approach seems more geometric and less computational, and hence is followed here.

Assume now that T is defined by (11.12) with r_1 dimensional Brownian motion $W(t)$ instead of S_n. Without loss of generality assume that $\mu = (\theta, 0, \ldots, 0)'$.

Proposition 11.13. *Let* $\gamma_1 = bm^{-1/2}$ *and* $\theta = \gamma_1 + \Delta m^{-1/2}$. *As* $b \to \infty$

$$E_\mu(T - m; T > m) = m^{1/2}(\theta - \tfrac{1}{2}\gamma_1)^{-1}[\phi(\Delta) - \Delta\Phi(-\Delta)]$$
$$+ \gamma_1^{-2}[(r_1 - 1)\Phi(-\Delta) + (1 + \Delta^2)\Phi(-\Delta) - \Delta\phi(\Delta)] + o(1).$$

PROOF. For simplicity assume that $r_1 = 2$. (See Remark 11.18 for general r_1.) Let ω_m denote the angle made by $W(m)$ with the x_1-axis and write

$$E_\mu(T - m; T > m) = E_\mu\{E_\mu[T - m | W(m)]; T > m\}$$

$$(11.14) \qquad = \iint E_\mu(T - m | \|W(m)\| = \rho, \omega_m = w, T > m) \times P_\mu\{T > m, \|W(m)\| \in d\rho, \omega_m \in dw\},$$

where the integral is over $-\pi < w < \pi, 0 < \rho < \gamma_1 m$.

As in the proof of Proposition 4.29, the important range of values of ρ is $\gamma_1 m - O(m^{1/2})$, so it is convenient to write $\rho = \gamma_1 m - zm^{1/2}$. Also

$$P_\mu\{T > m, \|W(m)\| \in d\rho, \omega_m \in dw\} = P_\mu\{\|W(m)\| \in d\rho, \omega_m \in dw\}(1 - P_0\{T < m | \|W(m)\| = \rho, \omega_m = w\}),$$

and as in one dimension the conditional probability is $o(m^{-1})$ for the important values of ρ. Hence integration in (11.14) can be taken with respect to

$$(11.15) \qquad P_\mu\{\|W(m)\| \in d\rho, \omega_m \in dw\} = \exp[\theta\rho \cos w - \tfrac{1}{2}\theta^2 m - \tfrac{1}{2}\rho^2 m^{-1}]\rho \, d\rho \, dw/2\pi m.$$

Since ρ is of order m and $\cos w = 1 - (w^2/2) + O(w^4)$, the effective range of w is $O(m^{-1/2})$.

Now consider the conditional expectation in (11.14). For $\rho = \gamma_1 m - zm^{1/2}$ and w of order $m^{-1/2}$, the value of u in Figure 11.1 is of order $m^{1/2}$. In the $O(m^{1/2})$ units of time it takes the Brownian path to reach the stopping boundary by drifting at rate θ in the x_1 direction from the point $(\rho \cos w, \rho \sin w)$, it will diffuse a distance $O(m^{1/4})$ in the x_2 direction. An easy calculation shows that the boundary is asymptotically effectively flat, so diffusion in the x_2 direction can be neglected. Hence the conditional expectation in (11.14) can be computed according to the one-dimensional result (4.33) to yield

$$(11.16) \qquad E_\mu(T - m | \|W(m)\| = \rho, \omega_m = w, T > m) = u/(\theta - \tfrac{1}{2}\gamma_1) - u^2/m\gamma_1^2 + o(u^2/m).$$

By a Taylor series expansion (cf. Figure 11.1)

$$(11.17) \qquad \begin{aligned} u &= [(\gamma_1 m)^2 - (\rho \sin w)^2]^{1/2} - \rho \cos w \\ &= \gamma_1 m - (\rho \sin w)^2/2\gamma_1 m + O(mw^4) - (\gamma_1 m - zm^{1/2}) \cos w \\ &= zm^{1/2}(1 - \tfrac{1}{2}w^2) - \tfrac{1}{2}\gamma_1 mw^2 + \tfrac{1}{2}\gamma_1 mw^2 + O(mw^4) + O(w^2). \end{aligned}$$

Use of (11.17) in (11.16) and substitution of the result together with (11.15) into (11.14) reduce the problem to one of calculating

$$\int_0^\infty \int_{-\pi}^{\pi} \left\{ \frac{zm^{1/2}(1 - \frac{1}{2}w^2)}{(\theta - \frac{1}{2}\gamma_1)} - \gamma_1^{-2} z^2 \right\} \rho \exp\{\theta\rho \cos w - \tfrac{1}{2}\theta^2 m - \tfrac{1}{2}\rho^2 m^{-1}\} \frac{dw \, dz}{2\pi m^{1/2}}.$$

Asymptotic evaluation of this integral is tedious but straightforward and yields Proposition 11.13 in the special case $r_1 = 2$. □

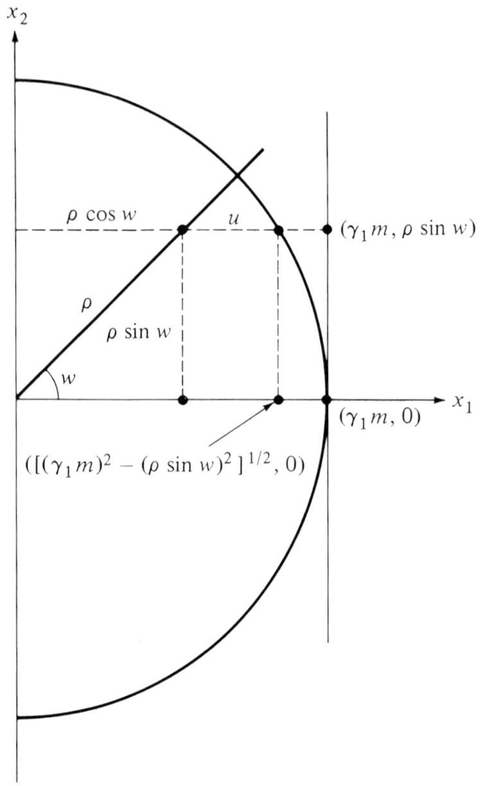

Figure 11.1.

Remark 11.18. The corresponding calculation for general r_1-dimensional Brownian motion is quite similar. The only difference is that the measure in (11.15) is, after integrating out $r_1 - 2$ variables,

$$C_{r_1-1}(\sin w)^{r_1-2}\rho^{r_1-1}\exp[-\tfrac{1}{2}\rho^2 m^{-1} + \theta\rho\cos w - \tfrac{1}{2}\theta^2 m]\, dw\, d\rho/(2\pi m)^{r_1/2}$$

for $0 < w < \pi$, $0 < \rho < m\gamma_1$ (see the hint for Problem 5.1).

3. The Discrete Brownian Bridge

Let $x_1, x_2, \ldots$ be independent and normally distributed with mean 0 and variance 1. Set $S_n = x_1 + \cdots + x_n$, and for $b > 0$, $m = 2, 3, \ldots$ define the stopping rule

(11.19) $$T = \inf\{n: n \geq m_0, |S_n| \geq b[n(1 - n/m)]^{1/2}\}.$$

This section is concerned with obtaining an asymptotic expression for

(11.20) $$P_0^{(m)}(T \leq m_1)$$

as $m \to \infty$. As always

$$P_\xi^{(m)}(A) = P_0(A \mid S_m = \xi).$$

Remark 11.21. A Brownian motion process $W(t)$, $0 < t < 1$, conditioned by $W(1) = \xi$ is called a Brownian bridge with drift ξ. It is easily verified by direct computation of means and covariances that the $P_\xi^{(1)}$ joint distributions of $W(t)$ are the same as the unconditional joint distributions of

(11.22) $$W(t) - tW(1) + \xi t \qquad (0 < t < 1).$$

Alternatively if $W(t)$ itself has drift ξ these distributions are the same as those of

(11.23) $$(1 - t)W[t/(1 - t)] \qquad (0 < t < 1).$$

The continuous analogue of (11.19) is

(11.24) $$\tilde{T} = \inf\{t: t \geq t_0, |W(t)| \geq b[t(1 - t)]^{1/2}\},$$

and from (11.23) it follows for $0 < t_0 < t_1 < 1$ that

$$\begin{aligned} P_0^{(1)}\{\tilde{T} \leq t_1\} &= P_0\{(1 - t)|W[t/(1 - t)]| \\ &\qquad \geq b[t(1 - t)]^{1/2} \text{ for some } t_0 \leq t \leq t_1\} \\ &= P_0\{|W(s)| \geq bs^{1/2} \text{ for some } t_0/(1 - t_0) \leq s \leq t_1/(1 - t_s)\}. \end{aligned}$$

Hence according to Theorem 4.21 (cf. Remark (i) following Theorem 4.21) as $b \to \infty$

$$\begin{aligned} P_0^{(1)}(\tilde{T} \leq t_1) = (b - b^{-1})\phi(b)\log[t_1(1 - t_0)/t_0(1 - t_1)] \\ + 4b^{-1}\phi(b) + o(b^{-1}\phi(b)). \end{aligned}$$

However, one does not obtain the analogous discrete time result by a simple transformation.

Below are several examples in which something like (11.20) arises as the approximate significance level of a statistical test. The particularly interesting Example 11.28 is not a problem of sequential analysis.

Example 11.25. Suppose that balls are drawn sequentially without replacement from an urn containing r red and w white balls, and that we would like to test whether the proportion $p = r/(r + w)$ of red balls equals some specified constant p_0. Let $m = r + w$ and let v_n denote the number of red balls in the first n draws. Then as $m \to \infty$

$$(v_{[mt]} - [mt]p)/m^{1/2} \qquad (0 < t < 1)$$

converges to a Brownian bridge process with drift 0. Alternatively, if $p - p_0 = \xi/m^{1/2}$, then

(11.26) $$(v_{[mt]} - [mt]p_0)/m^{1/2}$$

is approximately a Brownian bridge with drift ξ; and testing $H_0: p = p_0$ is asymptotically equivalent to testing $H_0: \xi = 0$ for a Brownian bridge (see Problem 3.13). For this limiting problem the stopping rule (11.24) but with (11.26) in place of $W(t)$ can be used to define a repeated significance test, the significance level of which is given approximately by $P_0^{(1)}(\tilde{T} \le t_1)$. If balls are drawn in groups of size m/k, so the observed values of the process v_n are $v_{[m/k]}$, $v_{[2m/k]}, \ldots$, the significance level for a repeated significance test when m is large is given approximately by (11.20).

EXAMPLE 11.27. Consider the same model as Example 11.25 with the interpretation that the red balls represent defective items in a lot subject to sampling inspection. It may be reasonable to assume that r is fixed and m is large, so that in the limit $v_{[mt]}$ converges to a Poisson process $N(t)$, $0 < t < 1$, conditional on $N(1) = r$. The situation is somewhat similar to Example 11.25, but there may be more appropriate tests of $H_0: r = r_0$ than a repeated significance test, and the approximate evaluation of error probabilities is quite different in detail.

EXAMPLE 11.28. The following is one simple example from the quite considerable literature on "change point" problems. Assume that $x_1, x_2, \ldots, x_m$ are independent normally distributed random variables and that x_i has the mean value μ_i and variance 1. Suppose we are interested in testing the hypothesis $H_0: \mu_1 = \cdots = \mu_m$ against the alternative H_1: There exists k, $1 < k < m - 1$ such that $\mu_1 = \cdots = \mu_k \neq \mu_{k+1} = \mu_{k+2} = \cdots = \mu_m$. If k were known, the problem would be a two sample test of the equality of the means of the first sample $x_1, x_2, \ldots, x_k$ and the second sample $x_{k+1}, x_{k+2}, \ldots, x_m$. For this problem the log likelihood ratio statistic would be

$$\Lambda_{k,m} = k(m-k)(\bar{x}_k - \bar{x}_{k,m})^2/m,$$

where $\bar{x}_k = k^{-1}\sum_1^k x_i$ and $\bar{x}_{k,m} = (m-k)^{-1}\sum_{k+1}^m x_i$. Since k is in fact unknown, the log likelihood ratio statistic is

$$\max_{1 \le n \le m-1} \Lambda_{n,m}, \tag{11.29}$$

and the significance level of the likelihood ratio test is the probability under H_0 that the random variable (11.29) exceeds some constant C.

Let $S_n = \sum_1^n x_i$. Simple algebra shows that

$$\Lambda_{n,m} = (S_n - nS_m/m)^2/n(1 - n/m).$$

It is easy to see that under H_0 the random variables

$$(n/m)(1 - n/m)\Lambda_{n,m} = m^{-1}(S_n - nS_m/m)^2, \qquad n = 1, 2, \ldots, m-1$$

have the same joint distribution as the process (11.22) with $\xi = 0$ observed at the time points $t_n = n/m$, $n = 1, 2, \ldots, m-1$. Hence the probability under H_0 that the random variable (11.29) exceeds C is just (11.20) with $m_0 = 1$, $m_1 = m - 1$, and $b = C^{1/2}$.

The principal result of this section is

Theorem 11.30. *Let T be defined by* (11.19), *and assume that $b \to \infty$, $m_0 \to \infty$, $m_1 \to \infty$, and $m \to \infty$ in such a way that for some $0 < t_0 < t_1 < 1$ and $\mu_1 > 0$*

$$m_i/m \to t_i (i = 0, 1) \text{ and } b/m^{1/2} = \mu_1.$$

Then as $m \to \infty$

$$(11.31)\quad P_0^{(m)}\{m_0 < T \le m_1\} \sim 2b\phi(b) \int_{\mu_1(t_1^{-1}-1)^{1/2}}^{\mu_1(t_0^{-1}-1)^{1/2}} x^{-1} v(x + \mu_1^2 x^{-1})\, dx,$$

where v is given by (4.37).

Remark 11.32. Theorem 11.30 suggests the approximation

$$P_0^{(m)}(T \le m_1) \cong 2b\phi(b) \int_{b(m_1^{-1}-m^{-1})^{1/2}}^{b(m_0^{-1}-m^{-1})^{1/2}} x^{-1} v(x + b^2/mx)\, dx + 2[1 - \Phi(b)],$$

(11.33)

which is shown in Table 11.1 below to be reasonably accurate even when $m_0 = 1$ and $m_1 = m - 1$, although it does not seem to be nearly as accurate as (4.40).

Remark 11.34. For studying the power of the tests suggested in Examples 11.25 and 11.28 it is useful to have an approximation for $P_{m\xi_0}^{(m_1)}\{T < m_1\}$. It can be shown that for

$$\mu_1(1 - t_1)[t_0/(1 - t_0)]^{1/2} < |\xi_0| < \mu_1[t_1(1 - t_1)]^{1/2}$$

this probability is

$$\sim [t_1(1 - t_1)]^{1/2} \mu_1 \xi_0^{-1}$$

$$\exp\{-\tfrac{1}{2}m[\mu_1^2 - \xi_0^2/t_1(1 - t_1)]\} v[\mu_1^2(1 - t_1)/\xi_0 + \xi_0/(1 - t_1)].$$

Since we shall not attempt a complete discussion of the statistical problems mentioned above, the proof of this result is omitted (see Siegmund, 1985).

Table 11.1 gives numerical examples to illustrate the accuracy of (11.33). In some cases exact computations from Worsley (1983) are available for comparison. In others the comparisons are with the results of a Monte Carlo experiment, which are given $\pm$ an estimated standard error to distinguish them from the exact results.

Proof of Theorem 11.30. The following argument combines the ideas in IV.8 and the proof of Theorem 8.72. Let $Q^{(m)} = \int_{-\infty}^{\infty} P_\xi^{(m)}\, d\xi/(2\pi)^{1/2}$. An easy calculation shows that the likelihood ratio of $x_1, \ldots, x_n$ $(n < m)$ under $Q^{(m)}$ relative to $P_0^{(m)}$ is

$$[m(m - n)/n]^{1/2} \exp[\tfrac{1}{2}S_n^2/n(1 - n/m)],$$

from which it follows by familiar arguments that

Table 11.1. $P_0^{(m)}\{T \le m_1\}$

				Probability	
b	m_0	m_1	m	Approximation	Exact or Monte Carlo
2.50	1	5	6	.052	.05
2.99	1	4	5	.010	.01
2.55	1	19	20	.106	.10
3.38	1	24	25	.010	.01
2.97	1	49	50	.054	.05
2.30	1	5	10	.074	.073 ± .001
2.30	2	8	10	.095	.093 ± .001
2.30	2	6	10	.069	.070 ± .001
2.30	6	15	30	.075	.079 ± .001
2.30	6	24	30	.128	.128 ± .002

Exact values from Worsley (1983).

$$\begin{aligned} &P_0^{(m)}\{m_0 < T \le m_1\} \\ &= m^{-1}\int_{-\infty}^{\infty} E_\xi^{(m)}\{[T/(1-T/m)]^{1/2}\exp[-\tfrac{1}{2}S_T^2/T(1-T/m)]; \\ &\qquad m_0 < T \le m_1\}\, d\xi/(2\pi)^{1/2} \\ &= \int_{-\infty}^{\infty} E_{m\xi}^{(m)}\{[T/(1-T/m)]^{1/2}\exp[-\tfrac{1}{2}S_T^2/T(1-T/m)]; \\ &\qquad m_0 < T \le m_1\}\, d\xi/(2\pi)^{1/2}. \end{aligned} \tag{11.35}$$

Let P_μ denote the probability measure under which $x_1, x_2, \ldots$ are independent and normally distributed with mean μ and variance 1. It is easy to see that the likelihood ratio of $x_1, \ldots, x_n$ under $P_{m\xi}^{(m)}$ relative to P_ξ is $\exp[-\frac{1}{2}(S_n - n\xi)^2/(m-n)]/(1-n/m)^{1/2}$, and hence (11.35) may be rewritten

$$\begin{aligned} &P_0^{(m)}\{m_0 < T \le m_1\}/2m^{1/2}\phi(b) \\ &= \int_0^{\infty} E_\xi\left[\frac{(T/m)^{1/2}}{(1-T/m)}\exp\left\{-\frac{1}{2}\left[\frac{S_T^2}{T(1-T/m)} - \mu_1^2 m\right] - \frac{1}{2}\frac{(S_T-\xi T)^2}{m-T}\right\};\right. \\ &\qquad \left. m_0 < T \le m_1\right] d\xi. \end{aligned} \tag{11.36}$$

It is easy to see that under P_ξ

$$T/m \to \mu_1^2/(\mu_1^2 + \xi^2) \tag{11.37}$$

with probability one. In particular

$$(T/m)^{1/2}/(1-T/m) \to \mu_1(\mu_1^2+\xi^2)^{1/2}/\xi^2; \tag{11.38}$$

and the range of integration in (11.35) asymptotically equals

(11.39) $$\mu_1(t_1 - 1)^{1/2} < |\xi| < \mu_1(t_0^{-1} - 1)^{1/2}.$$

By a variation on Theorem 9.17 (cf. Corollary 8.37) the excess over the boundary, $S_T^2/T(1 - T/m) - \mu_1^2 m$, and $(S_T - \xi T)^2/(m - T)$ are asymptotically independent, so it suffices to evaluate separately the expectations of the different factors in the integrand of (11.36).

By (11.37) and Theorem 2.40

$$E_\xi\{\exp[-\tfrac{1}{2}(S_T - \xi T)^2/(m - T)]\} = E_\xi\left[\exp\left\{-\frac{1}{2}\frac{(S_T - \xi T)^2}{T}\frac{T/m}{1 - T/m}\right]\right\}$$

(11.40) $$\to \int_{-\infty}^{\infty} \exp(-\tfrac{1}{2}z^2\mu_1^2/\xi^2)\phi(z)\,dz = \xi/(\mu_1^2 + \xi^2)^{1/2}.$$

If we rewrite the definition (11.19) in the form

$$T = \inf\{n: n \ge m_0, n^{-1}S_n^2 + \mu_1^2 n \ge \mu_1^2 m\},$$

it is easy to see that Theorem 9.12 applies, and that the random walk which determines the limiting distribution of $\frac{1}{2}(T^{-1}S_T^2 + \mu_1^2 T - \mu_1^2 m)$ has normally distributed increments with mean $\frac{1}{2}(\xi^2 + \mu_1^2)$ and variance ξ^2. Some calculation now yields

(11.41) $$E_\xi\left\{\exp\left[\frac{-\frac{1}{2}S_T^2}{T(1 - T/m)} - \tfrac{1}{2}\mu_1^2 m\right]\right\} \to \nu(\mu_1^2/\xi + \xi).$$

Substitution of (11.38)–(11.41) into (11.36) completes the proof of the theorem. □

Problems†

11.1. Let $W(t)$ denote d-dimensional Brownian motion and $\tau = \inf\{t: \|W(t)\| \ge m\zeta\}$. Show that

$$P\{\tau \le m | W(m) = 0\} = \exp(-2m\zeta^2)\{(2\pi^{1/2}/\Gamma(d/2))(2m\zeta^2)^{(d-1)/2} \times [1 - (d - 1)/8m\zeta^2 + o(m^{-1})]\}$$

as $m \to \infty$. More generally, find an asymptotic approximation for

$$P\{\tau \le m | W(m) = m\xi_0\} \qquad (\|\xi_0\| < \zeta).$$

Compare these approximations numerically with the exact computations of Kiefer (1959). Give (first order) asymptotic approximations for the corresponding results in discrete time.

Hint: Consider the likelihood ratio of $W(s)$, $s \le t$, under $P_\xi^{(m)}$ relative to

$$\tilde{P}^{(m)} = \int_{S_{d-1}} P_\eta^{(m)}\, d\sigma(w)/\sigma(S_{d-1}),$$

where $\|\eta\| = 2m\zeta - \|\xi\|$, w is the angle between η and some fixed direction in d-dimensional space, and σ denotes surface measure.

11.2. Prove Theorem 11.30 as an application of Problem 9.14.

11.3. Derive (11.31) as an application of Woodroofe's method described in IX.5 and/or by the method of proof of Theorem 9.54. (For a rigorous development of the appropriate nonlinear renewal theorem for the latter method, see Hogan, 1984.)

11.4. The following change point problem (cf. Example 11.28) is the starting point for the investigations of Levin and Kline (1984). Suppose that $x_1, \dots, x_m$ are independent observations from an exponential family of the form (8.56). Let $\tilde{\mu}_i$ denote the expected value of x_i $(i = 1, 2, \dots, m)$. Let $S_n = x_1 + \cdots + x_n$ and $\mu_0 < \mu_1$.

(a) Show that the log likelihood ratio statistic for testing H_0: $\tilde{\mu}_1 = \cdots = \tilde{\mu}_m = \mu_0$ against the "square wave" alternative H_1: $\exists 1 \le n_1 < n_2 < m$ such that $\tilde{\mu}_1 = \cdots = \tilde{\mu}_{n_1} = \mu_0, \tilde{\mu}_{n_1+1} = \cdots = \tilde{\mu}_{n_2} = \mu_1, \tilde{\mu}_{n_2+1} = \cdots = \tilde{\mu}_m = \mu_0$ is

$$\max_{0 \le k \le n \le m} \{(\theta_1 - \theta_0)(S_n - S_k) - (n - k)[\psi(\theta_1) - \psi(\theta_0)]\}, \tag{11.42}$$

where $\theta_i = \theta(\mu_i)$ $(i = 0, 1)$.

(b) Assume for notational simplicity that $\mu_0 < 0 < \mu_1$ and $\psi(\theta_1) = \psi(\theta_0)$. (This can always be achieved by redefining x_i to be $x_i - [\psi(\theta_1) - \psi(\theta_0)]/(\theta_1 - \theta_0)$ and changing the value of μ accordingly.) Let $\tau = \inf\{n: S_n - \min_{0 \le k \le n} S_k \ge b\}$. The significance level of the test of H_0 based on (11.42) is (essentially) $P_{\mu_0}\{\tau \le m\}$. Let $\sigma = \max\{k: S_k = \min_{0 \le j \le \tau} S_j\}$ and $T = \inf\{n: S_n \notin (0, b)\}$. Show that

$$P_{\mu_0}\{\sigma = k, \tau \le m\} = P_{\mu_0}\left\{\tau \ge k, S_k = \min_{i \le k} S_i\right\} P_{\mu_0}\{T \le m - k, S_T \ge b\}.$$

(c) Suppose that $b \to \infty$, and $m \to \infty$ in such a way that for some $0 < \varepsilon < 1$, $\varepsilon < bm^{-1} < \mu_1(1 - \varepsilon)$. Show that

$$P_{\mu_0}\{\tau \le m\} \sim (\theta_1 - \theta_0)(m\mu_1 - b)v_+^2 \exp[-(\theta_1 - \theta_0)b],$$

where v_+ is given by (8.59).

Hint: Use Proposition 8.38 and (a slight generalization of) (8.61).

APPENDIX 1

Brownian Motion

The purpose of this appendix is to give the reader not already familiar with the Brownian motion process some feeling for Brownian motion as an asymptotic approximation to random walks and hence to log likelihood ratios. Also, the basic likelihood ratio (3.11) is computed.

Let $x_1, x_2, \ldots$ be independent, identically distributed random variables with mean 0 and variance σ^2, which without loss of generality may be taken equal to 1. Let $S_n = x_1 + \cdots + x_n$ $(n = 0, 1, \ldots)$. By the central limit theorem $m^{-1/2}S_m$ converges in law to a normally distributed random variable with mean 0 and variance 1. More generally, for each fixed t, $m^{-1/2}S_{[mt]}$ converges in law to a normal $(0, t)$ random variable. (Here $[x]$ denotes the largest integer $\leq x$.)

We now ask if in some sense the random function taking t into $m^{-1/2}S_{[mt]}$ converges in law as $m \to \infty$ simultaneously for all t. Although a complete answer is beyond the scope of this book (cf. Billingsley, 1968), assuming that the limit exists, we can see what kind of stochastic process to limit must be.

Denote the limiting random function by $W(t)$, $0 \leq t < \infty$. The preceding remarks show that $W(t)$ is normally distributed with mean 0 and variance t. Since the convergence predicated in the preceding paragraph must entail convergence at any fixed finite number of time points, for each $k = 1, 2, \ldots$ and $0 \leq s_1 < t_1 \leq s_2 < t_2 \leq \cdots \leq s_k < t_k$ the random variables

$$m^{-1/2}(S_{[mt_j]} - S_{[ms_j]}), \qquad j = 1, 2, \ldots, k$$

must converge to $W(t_j) - W(s_j), j = 1, \ldots, k$, which therefore must be independent and normally distributed with mean 0 and variance $t_j - s_j$. Similarly, for $s < t$ the covariance of $W(s)$ and $W(t)$ must be given by

$$E\{W(s)W(t)\} = \lim_{m\to\infty} m^{-1}E(S_{[ms]}S_{[mt]})$$

$$= \lim_{m\to\infty} m^{-1}E\{S_{[ms]}^2 + S_{[ms]}(S_{[mt]} - S_{[ms]})\}$$

$$= \lim_{m\to\infty} m^{-1}[ms] = s = \min(s, t).$$

Thus the process $W(t)$, $0 \le t < \infty$, must have all the properties of Brownian motion as defined in III.1, except possibly for property (iv): that $W(t)$ is a continuous function of t. This property is much more technical and will be assumed without further comment.

Now let $\tilde{\tau}(b) = \inf\{n: S_n \ge b\}$ and $\tau(b) = \inf\{t: W(t) \ge b\}$ for $b > 0$. Suppose $b = m^{1/2}\zeta$ and $n = [mt]$. From the assumed convergence of $m^{-1/2}S_{[mt]}$ to $W(t)$, one expects to find that

$$P\{\tilde{\tau}(b) \le n\} = P\left\{\max_{1\le k\le n} S_k \ge b\right\}$$

(A.1) $$= P\left\{\max_{0\le s\le t} m^{-1/2}S_{[ms]} \ge \zeta\right\} \to P\left\{\max_{0\le s\le t} W(s) \ge \zeta\right\}$$

$$= P\{\tau(\zeta) \le t\}.$$

Moreover, it follows easily from the definition that if $W(t)$, $0 \le t < \infty$, is Brownian motion (with 0 drift), then for any m the process $W(mt)/m^{1/2}$, $0 \le t < \infty$, is also Brownian motion. Hence

$$P\{\tau(\zeta) \le t\}P\left\{\lim_{0\le s\le t} m^{-1/2}W(ms) > \zeta\right\}$$

$$= P\left\{\lim_{0\le s\le mt} W(s) \ge m^{1/2}\zeta\right\} = P\{\tau(m^{1/2}\zeta) \le mt\}.$$

In conjunction with (A.1) this suggests the approximation

(A.2) $$P\{\tilde{\tau}(b) \le n\} \cong P\{\tau(b) \le n\}$$

provided n and b are large and b is proportional to $n^{1/2}$.

Of course there are many similar approximations, of which (A.2) is only one example. Another is that for $\zeta_1 < 0 < \zeta_2$, the probability that S_k, $k = 0, 1, \ldots$ crosses $m^{1/2}\zeta_1$ before $m^{1/2}\zeta_2$ and before time mt is approximately the probability that $W(s)$, $0 \le s < \infty$, crosses ζ_1 before ζ_2 and before time t.

The situation is more complicated if $\mu = E(x_1) \ne 0$. Since $ES_m = m\mu$, in order that $m^{-1/2}S_{[mt]}$ converge in law it is necessary that μ be proportional to $m^{-1/2}$ as $m \to \infty$.

More precisely, assume $x_{m1}, x_{m2}, \ldots$ are independent and identically distributed with mean $\mu = \mu_m$ and unit variance. Let $S_{m,n} = \sum_{i=1}^n x_{mi}$. If x_{m1}^2 is uniformly (in m) integrable, and if for some fixed real number ξ, $\mu_m \sim \xi m^{-1/2}$ as $m \to \infty$, then $S_{m,[mt]}$, $0 \le t < \infty$, converges in law to Brownian motion with drift ξ. Considerations like those above again lead to approximations like (A.2), but now a complete justification requires in addition to b becoming infinitely large at rate $m^{1/2}$ that μ tend to 0 at rate $m^{-1/2}$, so that μb tends to a finite limit.

In light of these requirements it should not be surprising that the Brownian approximation is often not an especially good one (cf. III.5).

For an example, suppose that $x_1, x_2, \ldots$ are independent with probability density function of the form

$$f_\theta(x) = \exp[\theta x - \psi(\theta)]f(x)$$

relative to some fixed probability density function $f(x)$, and assume (without loss of generality) that $\psi(0) = 0$. Recall from II.3 that $\psi'(\theta) = E_\theta(x_1)$ and $\psi''(\theta) = \text{var}_\theta(x_1)$. The stopping rule of a sequential probability ratio test of H_0: $\theta = 0$ against H_1: $\theta = \theta_1 > 0$ can be expressed

$$N = \text{first } n \geq 1 \text{ such that } S_n - n\psi(\theta_1)/\theta_1 \notin (\zeta_1/\theta_1, \zeta_2/\theta_1),$$

where $S_n = x_1 + \cdots + x_n$ and to avoid considering special cases we assume that $\zeta_1 < 0 < \zeta_2$. Now suppose that $\theta_1 \downarrow 0$. Note that

$$E_0(x_1 - \psi(\theta_1)/\theta_1) = \psi'(0) - [\psi(\theta_1) - \psi(0)]/\theta_1 = -\tfrac{1}{2}\theta_1\psi''(0) + O(\theta_1^2)$$

and

$$\begin{aligned} E_{\theta_1}(x_1 - \psi(\theta_1)/\theta_1) &= \psi'(\theta_1) - [\psi(\theta_1) - \psi(0)]/\theta_1 \\ &= \psi'(\theta_1) - \psi'(0) - \tfrac{1}{2}\theta_1\psi''(0) + O(\theta_1^2) \\ &= \tfrac{1}{2}\theta_1\psi''(0) + O(\theta_1^2). \end{aligned}$$

Hence under both H_0 and H_1 (or more generally if $\theta = r\theta_1$ for an arbitrary real number r) the mean increment of the random walk $S_n - n\psi(\theta_1)/\theta_1$ converges to 0 at the same rate that the stopping boundaries $|\zeta_1|/\theta_1$ and ζ_2/θ_1 move out to ∞. In light of the discussion following (A.2) one expects to find (and it can be proved) that for $j = 0$ or 1

$$P_{\theta_j}\{\theta_1[S_N - N\psi(\theta_1)] \geq \zeta_2\} \to P\{W(t) \text{ hits } \zeta_2 \text{ before hitting } \zeta_1\},$$

where $W(t)$, $0 \leq t < \infty$, is Brownian motion with drift $-(-1)^j\tfrac{1}{2}\psi''(0)$ and variance $\psi''(0)$.

It is readily verified that under these conditions the Wald approximations calculated in Problem 2.13 converge to the appropriate Brownian motion results given in Theorem 3.6. (Of course, in the case of a normal mean the Wald approximations are exactly the corresponding Brownian motion results.)

We now turn to a verification of (3.11). The proof of Proposition 3.1 is similar but somewhat simpler. Let $0 < t_1 < \cdots < t_k = t < m$, and let $-\infty < w_i < \infty$ $(i = 1, \ldots, k)$. Put $w_k = w$. Then

$$\begin{aligned} &P_\xi^{(m)}\{W(t_1) \in dw_1, \ldots, W(t_k) \in dw_k\} \\ &= P_0\{W(t_1) \in dw_1, \ldots, W(t_k) \in dw_k, W(m) \in d\xi\}/P_0\{W(m) \in d\xi\} \\ &= \phi(w_1/t_1^{1/2})(dw_1/t_1^{1/2})\phi\{(w_2 - w_1)/(t_2 - t_1)^{1/2}\}(dw_2/(t_2 - t_1)^{1/2})\ldots \\ &\quad \phi\{(\xi - w_k)/(m - t_k)\}(d\xi/(m - t_k)^{1/2})/\phi(\xi/m^{1/2})(d\xi/m^{1/2}). \end{aligned}$$

In the ratio of this expression for $\xi = \xi_0$ to the same expression for $\xi = \xi_1$, all factors except those involving ξ_i ($i = 0, 1$) cancel, which leaves

$$
\begin{aligned}
&P_{\xi_0}^{(m)}\{W(t_1)\in dw_1,\ldots,W(t_k)\in dw_k\}/P_{\xi_1}^{(m)}\{W(t_1)\in dw_1,\ldots,W(t_k)\in dw_k\}\\
&\quad = \phi\{(\xi_0 - w)/(m-t)^{1/2}\}\phi(\xi_1/m^{1/2})/\phi\{(\xi_1 - w)/(m-t)^{1/2}\}\phi(\xi_0/m^{1/2})\\
&\quad = \exp\left\{\left[(\xi_0 - \xi_1)w - \frac{t}{2m}(\xi_0^2 - \xi_1^2)\right]\Big/(m-t)\right\}.
\end{aligned}
$$

This can be reexpressed as

$$
\text{(A.3)}\qquad
\begin{aligned}
P_{\xi_0}^{(m)}(A) = E_{\xi_1}^{(m)}\Bigg[\exp\Bigg\{\Bigg[&(\xi_0 - \xi_1)W(t)\\
&- \frac{t}{2m}(\xi_0^2 - \xi_1^2)\Bigg]\Big/(m-t)\Bigg\}; A\Bigg]
\end{aligned}
$$

for every event A defined in terms of $W(t_1), \ldots, W(t_k)$ for some $k = 1, 2, \ldots$ and arbitrary $0 < t_1 < \ldots < t_k = t < m$, which is equivalent to the assertion (3.11). More precisely, (3.11) is equivalent to the validity of (A.3) for every $A \in \mathscr{E}_t$, but this follows from (A.3) for every A determined by $W(t_1), \ldots, W(t_k)$, $k = 1, 2, \ldots$ and the fact that a probability measure on $\mathscr{E}_t$ is determined by its value on cylinder sets (cf. Loève, 1977, p. 94).

APPENDIX 2

Queueing and Insurance Risk Theory

The one-sided first passage problem as discussed in detail in VIII.1–4 and again in X.1 does not have an important role in statistical applications. The justification for its inclusion in this book is that its mathematical simplicity provides a convenient pedagogical device. In the theory of the single server queue and in the ruin problem of insurance risk theory, however, it arises naturally as a problem of primary importance. The purpose of this appendix is to point out the pertinent relations and also to illustrate the connection between the single server queue and the cusum process of II.6.

Assume that customers arrive at a server at times $u_1 < u_1 + u_2 < u_1 + u_2 + u_3 < \dots$ and are served in order of their arrival. Let v_n $(n = 1, 2, \dots)$ denote the service time and W_n the waiting time of the nth customer. It is not difficult to derive a relation between W_n and W_{n-1}. The waiting time of the nth customer equals that of the $(n-1)$th increased by the service time of the $(n-1)$th customer and decreased by the amount of time between the arrival of the $(n-1)$th and the nth customer, during which the $(n-1)$th customer is waiting but the nth is not. Hence $W_n = W_{n-1} + v_{n-1} - u_n$, provided of course that this latter expression is non-negative. Otherwise the nth customer finds an unoccupied server when he arrives and $W_n = 0$. It is convenient to assume that a customer indexed by 0 arrives at time 0 to an empty queue and has a service time of length v_0. Hence

$$W_0 = 0$$

and

$$W_n = (W_{n-1} + v_{n-1} - u_n)^+ \qquad (n = 1, 2, \dots)$$

describe recursively the waiting time process. Let $z_n = v_{n-1} - u_n$, so these equations can be rewritten

(A.4) $$W_0 = 0,\ W_n = (W_{n-1} + z_n)^+ \qquad (n = 1, 2, \ldots).$$

A simple stochastic model is to assume (i) that the arrival times form a renewal process, so $u_1, u_2, \ldots$ are independent and identically distributed, and (ii) that the service times $v_0, v_1, v_2, \ldots$ are likewise independent and identically distributed. Then the z's are independent and identically distributed, so the process W_n defined by (A.4) is a random walk which is reset to 0 whenever it enters $(-\infty, 0)$.

Let $S_n = z_1 + \cdots + z_n$. By recursion from n backwards, one obtains from (A.4)

(A.5)
$$\begin{aligned} W_n &= \max(0, W_{n-1} + z_n) = \max(0, (W_{n-2} + z_{n-1})^+ + z_n) \\ &= \max(0, W_{n-2} + z_{n-1} + z_n, z_n) \\ &= \max(0, (W_{n-3} + z_{n-2})^+ + z_{n-1} + z_n, z_n) \\ &= \max(0, W_{n-3} + z_{n-2} + z_{n-1} + z_n, z_{n-1} + z_n, z_n) \\ &= \cdots = \max(0, z_1 + \cdots + z_n, z_2 + \cdots + z_n, \ldots, z_n) \\ &= \max_{0 \le j \le n} (S_n - S_j) = S_n - \min_{0 \le j \le n} S_j. \end{aligned}$$

This shows that W_n is the same process as that introduced in II.6, although here the natural object of study is not the stopping rule (2.47), but the distribution of W_n and its limit as $n \to \infty$.

Since $(z_1, \ldots, z_n)$ have the same joint distribution as $(z_n, \ldots, z_1)$, it follows from (A.5) that W_n has the same distribution as

$$\max(0, z_n + \cdots + z_1, z_{n-1} + \cdots + z_1, \ldots, z_1) = \max_{0 \le k \le n} S_k.$$

Hence

(A.6) $$P\{W_n \ge b\} = P\{\tau(b) \le n\},$$

where $\tau(b) = \inf\{n\colon S_n \ge b\}$, and $\lim_{n\to\infty} P\{W_n \ge b\} = P\{\tau(b) < \infty\}$. Consequently, one can interpret the probability $P\{\tau(b) < \infty\}$ as the steady state probability that the waiting time exceeds b in a single server queue with recurrent input and independent, identically distributed service times. Equation (A.6) shows that not only $P\{\tau(b) < \infty\}$ but also the complete distribution of $\tau(b)$ is of interest; and Theorem 10.41 provides a useful approximation.

Remarks. (i) Although the marginal distributions of W_n and $\max_{0 \le k \le n} S_k$ are the same, their joint distributions manifestly are not; and hence one cannot study the first passage times of II.6 by the "time reversal" technique of the preceding paragraph. (ii) The arguments of II.6 make use of the renewal property of $S_n - \min_{0 \le k \le n} S_k$, to wit that the process starts from scratch upon reaching the state 0. This fact is particularly transparent from the representation (A.4).

Now consider a portfolio of an insurance company which has an initial reserve u_0, which increases deterministically and linearly at rate $\rho > 0$ due to receipt of premiums and decreases randomly due to the payment of claims. Let $N(t)$ denote the number of claims during the time interval $[0, t]$, and let v_i denote the amount of the ith claim. The reserve of the portfolio at time t is

$$X(t) = u_0 + \rho t - \sum_{i=1}^{N(t)} v_i,$$

provided that $X(s) \geq 0$ for all $s < t$. At time $\tilde{\tau} = \inf\{t: X(t) < 0\}$ the claims against the portfolio exceed its reserves and it is ruined. Quantities of particular interest are the probability of ultimate ruin. $P\{\tilde{\tau} < \infty\}$ and the probability of ruin before time t, $P\{\tilde{\tau} \leq t\}$.

Suppose that $v_1, v_2, \ldots$ are independent, identically distributed random variables and that $N(t)$ is a renewal process. Let $\sigma_1 < \sigma_2 < \ldots$ denote the times of increase of $N(t)$ (i.e. times at which claims occur), so $u_1 = \sigma_1, u_2 = \sigma_2 - \sigma_1$, $u_3 = \sigma_3 - \sigma_2, \ldots$ are independent, identically distributed, positive random variables. Obviously if ruin occurs, it occurs at the time of some claim. Note that $X(\sigma_n) = u_0 + \sum_{i=1}^{n} (\rho u_i - v_i) = u_0 - S_n$, say. Then $P\{\tilde{\tau} < \infty\} = P\{\tau < \infty\}$, where $\tau = \inf\{n: S_n > u_0\}$, so under some further restrictions on the distribution of $\rho u_1 - v_1$ the probability of ultimate ruin can be approximated by the results of VIII.1–4 or X.1. The methods of Theorem 10.41 can presumably be adapted to provide approximations for $P\{\tilde{\tau} \leq t\}$. (See Asmussen, 1984, for a closely related approach.)

APPENDIX 3

Martingales and Stochastic Integrals

This appendix gives a heuristic discussion of the martingale theory underlying the results of V.5. Although the processes of V.5 evolve in continuous time, the discussion here is limited to discrete time, which involves the same concepts without the technicalities. For a complete development of the appropriate mathematical foundations in continuous time, see Jacobsen (1982).

Let $\mathscr{E}_1 \subset \mathscr{E}_2 \subset \dots$ be an increasing sequence of σ-fields. Typically we are observing some stochastic processes at times $n = 1, 2, 3, \dots$, and $\mathscr{E}_n$ denotes the class of all events which can be defined in terms of the observations up to and including the nth one. It is convenient to let $\mathscr{E}_0$ denote the trivial σ-field which consists of the empty set and the entire sample space. A sequence of random variables $u_1, u_2, \dots$ is said to be adapted to $\mathscr{E}_n$, $n = 1, 2, \dots$ (or just adapted when no confusion can arise) if u_n is $\mathscr{E}_n$-measurable, i.e. is defined in terms of events in $\mathscr{E}_n$, for every $n = 1, 2, \dots$. Thus if the sequence $u_1, u_2, \dots$ is adapted, the value of u_n becomes known at the same time as the events in $\mathscr{E}_n$. A sequence $v_1, v_2, \dots$ is called predictable if it is adapted to $\mathscr{E}_{n-1}$, $n = 1, 2, \dots$. Thus the value v_n of a predictable sequence becomes known with the events in $\mathscr{E}_{n-1}$. A sequence of random variables M_n, $n = 1, 2, \dots$ which is adapted to $\mathscr{E}_n$, $n = 1, 2, \dots$ is called an $\mathscr{E}_n$-martingale (or more simply a martingale) if $E(|M_n|) < \infty$ and for each $n \geq 2$ the conditional expectations of M_n given the past satisfy

$$E(M_n | \mathscr{E}_{n-1}) = M_{n-1} \tag{A.7}$$

with probability one.

Remark A.8. Note that if $M_n, n = 1, 2, \dots$ is an $\mathscr{E}_n$-martingale, $\mathscr{E}_n' \subset \mathscr{E}_n$, and M_n is adapted to $\mathscr{E}_n' (n = 1, 2, \dots)$, then by the projection property of conditional expectations

$$E(M_n|\mathscr{E}'_{n-1}) = E[E(M_n|\mathscr{E}_{n-1})|\mathscr{E}'_{n-1}] = E(M_{n-1}|\mathscr{E}'_{n-1}) = M_{n-1},$$

so M_n, $n = 1, 2, \ldots$ is an $\mathscr{E}'_n$-martingale.

For any sequence a_n, let Δa_n denote the sequence of differences $a_n - a_{n-1}$, $n = 2, 3, \ldots$ ($\Delta a_1 = a_1$). The condition (A.7) is obviously equivalent to

$$E[\Delta M_n|\mathscr{E}_{n-1}] = 0$$

with probability one. The adapted sequence ΔM_n is called a martingale difference sequence. A simple example of a martingale is the sequence of partial sums of independent random variables having mean 0. If $x_1, x_2, \ldots$ is any adapted sequence of random variables having finite expectation, then

(A.9) $$x_1 + \sum_{k=2}^{n} [x_k - E(x_k|\mathscr{E}_{k-1})], \qquad n = 1, 2, 3, \ldots$$

is a martingale.

Let M_n, $n = 1, 2, \ldots$ be an $\mathscr{E}_n$-martingale and let ΔM_n be the associated martingale difference sequence. The sequence of partial sums $\sum_1^n (\Delta M_k)^2$ is called the quadratic variation of the martingale M_n, $n = 1, 2, \ldots$; and $\sum_1^n E[(\Delta M_k)^2|\mathscr{E}_{k-1}] = \langle M \rangle_n$, say, is called its predictable quadratic variation. Since all the terms defining $\langle M \rangle_n$ are $\mathscr{E}_{n-1}$-measurable, the predictable quadratic variation is predictable in the sense defined above. Now let h_n, $n = 1, 2, \ldots$ be a predictable sequence. The sequence of partial sums

(A.10) $$J_n = \sum_{k=1}^{n} h_k \Delta M_k, \qquad n = 1, 2, \ldots$$

is called a discrete stochastic integral. Since h_k is $\mathscr{E}_{k-1}$-measurable and consequently is conditionally a constant given the class of events $\mathscr{E}_{k-1}$, it follows that $E[h_k \Delta M_k|\mathscr{E}_{k-1}] = h_k E[\Delta M_k|\mathscr{E}_{k-1}] = 0$ with probability one, so J_n, $n = 1, 2, \ldots$ is itself a martingale. A similar calculation shows that the predictable quadratic variation of J_n is

(A.11) $$\langle J \rangle_n = \sum_{k=1}^{n} h_k^2 \Delta \langle M \rangle_k,$$

and

(A.12) $$E(J_n^2) = E(\langle J \rangle_n).$$

Example A.13. Although the following example plays no role elsewhere in this book, it provides particularly enlightening illustrations of the concepts of this appendix. Suppose that $x_0 = 0$ and for $n = 1, 2, \ldots, x_n$ satisfies the first order autoregressive equation

(A.14) $$x_n = \beta x_{n-1} + u_n,$$

where $u_1, u_2, \ldots$ are independent and identically distributed with mean 0 and variance 1, and $\beta \in [-1, 1]$ is an unknown parameter. If $x_1, \ldots, x_n$ are observed, the least-squares estimator of β (which is also the maximum likelihood estimator if the u's are normally distributed) is easily calculated to be

$$b_n = \sum_1^n x_{k-1}x_k \Big/ \sum_1^n x_{k-1}^2.$$

From (A.14) one obtains

$$b_n - \beta = \sum_1^n x_{k-1}u_k \Big/ \sum_1^n x_{k-1}^2. \tag{A.15}$$

Let $\mathscr{E}_n$ denote the class of events defined by $u_1, \ldots, u_n$. From (A.14) it follows that x_n is $\mathscr{E}_n$-measurable. Hence the numerator of (A.15) is a discrete stochastic integral of the form (A.10), whose predictable quadratic variation equals the denominator of (A.15).

The stochastic integrals of interest in V.5 have martingale differences of the form $\Delta M_n = v_n - a_n$, where v_n is a random variable assuming non-negative integral values and $a_n = E(v_n|\mathscr{E}_{n-1})$. The process $A_n = a_1 + \cdots + a_n$ is called the compensator of the point process $N_n = v_1 + \cdots + v_n$. If the random variables v_k are Bernoulli, then $E[(\Delta M_k)^2|\mathscr{E}_{k-1}] = a_k(1 - a_k)$, and the predictable quadratic variation of the martingale $M_n = N_n - A_n$ is $\sum_1^n a_k(1 - a_k)$. If the conditional distribution of each v_k given $\mathscr{E}_{k-1}$ is Poisson, then $E[(\Delta M_k)^2|\mathscr{E}_{k-1}] = a_k$, so the predictable quadratic variation of $N_n - A_n$ is the compensator A_n itself. In this case (A.11) and (A.12) yield

$$E(J_n^2) = E\left\{\sum_1^n h_k^2 \Delta A_k\right\} = E\left\{\sum_1^n h_k^2 \Delta N_k\right\}. \tag{A.16}$$

(The primary simplifying feature of a continuous time formulation—after one overcomes the technicalities—is that the point process $N(t)$ can simultaneously increase only by unit amounts, as in the Bernoulli case, and yet have compensator $A(t)$ equal to the predictable quadratic variation of $N(t) - A(t)$ as in the Poisson case. The simplest example is, of course, a Poisson process of intensity a, which satisfies $E\{N(dt)|\mathscr{E}_t\} = a\,dt$ and $E\{[N(dt) - a\,dt]^2|\mathscr{E}_t\} = a\,dt(1 - a\,dt) = a\,dt$. Whenever $E\{[N(dt) - A(dt)]^2|\mathscr{E}_t\} = A(dt)$ the equation (A.16) becomes

$$\begin{aligned} E\left\{\left[\int_0^t h(s)\{N(ds) - A(ds)\}\right]^2\right\} &= E\left\{\int_0^t h^2(s)A(ds)\right\} \\ &= E\left\{\int_0^t h^2(s)N(ds)\right\}. \end{aligned} \tag{A.17}$$

Since a martingale is a generalization of a zero mean random walk, one expects that under fairly general conditions martingales obey a central limit theorem and behave asymptotically like a Brownian motion process. However, it turns out to be necessary for the predictable quadratic variation of the martingale to behave asymptotically like a sequence of constants. In order to guarantee that this condition is satisfied, it is helpful to introduce a change in the time scale. This purely probabilistic device is closely related to the idea in III.9 of measuring statistical time by the increase in observed Fisher information (see Example A.20 below).

Given the increasing family of σ-fields $\mathscr{E}_n$, $n = 1, 2, \ldots$ a random variable τ assuming values in the set $\{1, 2, \ldots, \infty\}$ is called a stopping time if $\{\tau \leq n\} \in \mathscr{E}_n$ for all $n = 1, 2, \ldots$.

Lemma A.18. *If M_n is an $\mathscr{E}_n$-martingale and τ is a stopping time, then $M_{\tau \wedge n}$ is also an $\mathscr{E}_n$-martingale* (*cf.* (5.44) *for an application*).

PROOF. From $M_{\tau \wedge n} = M_\tau I_{\{\tau \leq n\}} + M_n I_{\{\tau > n\}}$ and the definition of a stopping time it follows that $M_{\tau \wedge n}$ is $\mathscr{E}_n$-measurable. Moreover,

$$\begin{aligned} E(M_{\tau \wedge (n+1)} | \mathscr{E}_n) &= M_\tau I_{\{\tau \leq n\}} + I_{\{\tau > n\}} E(M_{n+1} | \mathscr{E}_n) \\ &= M_\tau I_{\{\tau \leq n\}} + I_{\{\tau > n\}} M_n \\ &= M_{\tau \wedge n}. \end{aligned}$$

□

The following central limit theorem for martingales is given by Freedman (1971, pp. 90–92). A version suitable for direct application in the continuous time case is due to Robolledo (1980) (see also Jacobsen (1982)).

Theorem A.19. *Let M_n be an $\mathscr{E}_n$-martingale with predictable quadratic variation $\langle M \rangle_n$. Suppose that the martingale differences ΔM_n are uniformly bounded and that*

$$P\left\{\lim_{n \to \infty} \langle M \rangle_n = +\infty\right\} = 1.$$

Let $r > 1$ and define the stopping times

$$\tau(r) = \inf\{n: \langle M \rangle_n > r\}.$$

Then for $k = 1, 2, \ldots,$ and $0 < t_1 < t_2 < \cdots < t_k < \infty$, the joint distribution of $r^{-1/2} M_{\tau(rt_i)}$, $i = 1, 2, \ldots, k$ converges as $r \to \infty$ to that of $W(t_i)$, $i = 1, 2, \ldots, k$, where W is a standard Brownian motion with 0 drift.

Example. Consider again the autoregressive model of Example A.13. Note that if the u's are normally distributed, then the observed Fisher Information about β in the sample $x_1, \ldots, x_n$ is $\sum_1^n x_{k-1}^2$, which is also the predictable quadratic variation of the martingale in the numerator of (A.15). In this case the time change of III.9 and that of Theorem A.19 are exactly the same. Although the martingale differences in the numerator of (A.15) are not uniformly bounded, it is possible to show by a truncation argument that Theorem A.19 is applicable; and as $r \to \infty$ $r^{1/2}(b_{\tau(r)} - \beta)$ converges in distribution to standard normal for all $-1 \leq \beta \leq 1$. Also this convergence is uniform in $\beta \in$

$[-1, 1]$. Without making any time change, it is possible to show that $(\sum_1^n x_{k-1}^2)^{1/2}(b_n - \beta)$ converges in distribution to standard normal for all $\beta \in (-1, 1)$, because the predictable quadratic variation behaves like a constant in the limit. But for $\beta = \pm 1$ it is not difficult to show that

$$\left(\sum_1^n x_{k-1}^2\right)^{1/2}(b_n - \beta) \xrightarrow{\mathscr{L}} [W^2(1) - 1]/2\left\{\int_0^1 W^2(t)\,dt\right\}^{1/2}.$$

See Lai and Siegmund (1983) for a complete discussion.

Example A.20. Assume that $x_1, x_2, \ldots, x_n, \ldots$ are random variables whose joint distribution depends on a real parameter θ. Let $\mathscr{E}_n$ denote the class of events defined by $x_1, \ldots, x_n$, and suppose that $f_k(\cdot|\mathscr{E}_{k-1}; \theta)$ denotes the conditional probability (density) function of x_k given $x_1, \ldots, x_{k-1}$. The log likelihood function of $x_1, \ldots, x_n$ is

$$l_n(\theta) = \sum_{k=1}^n \log f_k(x_k|\mathscr{E}_{k-1}; \theta) = \sum_{k=1}^n u_k(\theta),$$

say, and its derivative with respect to θ is

$$\dot{l}_n(\theta) = \sum_{k=1}^n \dot{u}_k(\theta).$$

Under regularity conditions permitting the interchange of differentiation and summation (integration),

$$E_\theta[\dot{u}_k(\theta)|\mathscr{E}_{k-1}] = \int \frac{\partial}{\partial\theta} \log f_k(\xi|\mathscr{E}_{k-1}; \theta) f_k(\xi|\mathscr{E}_{k-1}; \theta)\,d\xi = 0, \tag{A.21}$$

so $\dot{l}_n(\theta)$ is an $\mathscr{E}_n$-martingale. Its predictable quadratic variation is

$$\langle \dot{l}(\theta) \rangle_n = \sum_1^n E_\theta[\dot{u}_k^2(\theta)|\mathscr{E}_{k-1}], \tag{A.22}$$

which by differentiating (A.21) under the integral can be shown to equal

$$\sum_1^n E_\theta[-\ddot{u}_k(\theta)|\mathscr{E}_{k-1}]. \tag{A.23}$$

The observed Fisher Information is $-\ddot{l}_n(\theta)$. It follows from (A.9) and the equality of (A.22) and (A.23) that

$$\ddot{l}_n(\theta) + \langle \dot{l}(\theta) \rangle_n$$

is a martingale. Since $-\ddot{l}_n$ and $\langle \dot{l}(\theta) \rangle_n$ are equal "on average," it seems plausible that under fairly general conditions they will grow at the same asymptotic rate for large n. When this is so, the time change defined in Theorem A.19 by the predictable quadratic variation $\langle M \rangle_n = \langle \dot{l}(\theta) \rangle_n$ is asymptotically equivalent to one defined by $-\ddot{l}_n(\theta)$ (see V.5—especially (5.59)).

APPENDIX 4

Renewal Theory

The renewal theorem provides an important tool in Chapters VIII and IX. Strictly speaking it is only the simplest version of that theorem, namely the one quoted as Theorem 8.24, which is used. For a textbook discussion with different proofs see Feller (1971) and Breiman (1968). However, the discussions following Example 8.65 and Remark 10.6 allude to stronger versions of the renewal theorem with an estimate for the remainder. For these stronger results the Fourier analytic method introduced by Feller and Orey (1961) and developed by Stone (1965 a,b) seems particularly appropriate. This method is described below. In order to convey the main ideas and avoid some technicalities, only the case of integer-valued random variables is discussed in detail. For the non-arithmetic case the reader is referred to Stone's papers.

Let $z_1, z_2, \ldots$ be independent, identically distributed, integer-valued random variables with distribution function F and positive, finite mean μ. Let $S_n = z_1 + \cdots + z_n$ and

$$f(\lambda) = E[\exp(i\lambda z_1)].$$

Assume that the span of z_1 is 1. This is easily seen to imply that

$$\text{(A.24)} \qquad |f(\lambda)| < 1 \qquad 0 < |\lambda| < 2\pi$$

(cf. Problem 8.1). Our starting point is the inversion formula

$$\text{(A.25)} \quad P\{S_n = k\} = (2\pi)^{-1} \int_{-\pi}^{\pi} \exp(-ik\lambda)[f(\lambda)]^n \, d\lambda \quad (k = 0, \pm 1, \ldots).$$

Proposition A.26. $\mathrm{Re}\{1/[1 - f(\lambda)]\}$ *is integrable over* $[-\pi, \pi]$, *and for* $k = 0, 1, \ldots$

$$\sum_{n=0}^{\infty} [P\{S_n = -k\} + P\{S_n = k\}] = \mu^{-1} + \pi^{-1} \int_{-\pi}^{\pi} \cos k\lambda \operatorname{Re}\{1/[1 - f(\lambda)]\}\, d\lambda.$$

PROOF. Since $\operatorname{Re}\{1/[1 - f(\lambda)]\} = \operatorname{Re}[1 - f(\lambda)]/|1 - f(\lambda)|^2 \sim \operatorname{Re}[1 - f(\lambda)]/\mu^2\lambda^2$ as $\lambda \to 0$, by (A.24) to show $\operatorname{Re}\{1/[1 - f]\}$ integrable it suffices to show $\operatorname{Re}[1 - f(\lambda)]/\lambda^2$ integrable. By Fubini's theorem

$$\begin{aligned}\int_{-\pi}^{\pi} \lambda^{-2} \operatorname{Re}[1 - f(\lambda)]\, d\lambda &= \int_{-\pi}^{\pi} \lambda^{-2} \int_{-\infty}^{\infty} (1 - \cos \lambda x) F(dx)\, d\lambda \\ &= \int_{-\infty}^{\infty} \int_{-\pi}^{\pi} (\lambda x)^{-2}(1 - \cos \lambda x)|x|\, d\lambda |x| F(dx) \\ &\le \int_{-\infty}^{\infty} t^{-2}(1 - \cos t)\, dt \int_{-\infty}^{\infty} |x| F(dx) < \infty.\end{aligned}$$

Let $0 < r < 1$. By (A.25)

$$\sum_{n=0}^{\infty} r^n[P\{S_n = -k\} + P\{S_n = k\}] = \pi^{-1} \int_{-\pi}^{\pi} \cos k\lambda \operatorname{Re}\{1/[1 - rf(\lambda)]\}\, d\lambda. \tag{A.27}$$

As $r \to 1$ the left hand side of (A.27) converges to $\sum_{n=0}^{\infty} [P\{S_n = -k\} + P\{S_n = k\}]$. From the fact that $|f(\lambda)|$ is bounded away from 1 for $\varepsilon \le \lambda \le \pi$ and the integrability of $\operatorname{Re}[1/(1 - f)]$ it follows that

$$\lim_{\varepsilon \to 0} \lim_{r \to 1} \int_{\varepsilon \le |\lambda| \le \pi} \cos k\lambda \operatorname{Re}\{1/[1 - rf(\lambda)]\}\, d\lambda = \int_{-\pi}^{\pi} \cos k\lambda \operatorname{Re}\{1/[1 - f(\lambda)]\}\, d\lambda.$$

Hence by (A.27), to complete the proof it suffices to show that

$$\lim_{\varepsilon \to 0} \lim_{r \to 1} \pi^{-1} \int_{|\lambda| < \varepsilon} \operatorname{Re}\{1/(1 - rf)\}\, d\lambda = \mu^{-1}. \tag{A.28}$$ □

By algebra one finds that

$$\operatorname{Re}\left(\frac{1}{1 - rf}\right) = \frac{(1 - r) + r \operatorname{Re}(1 - f)}{|1 - rf|^2}. \tag{A.29}$$

For $r > \frac{1}{2}$, $|1 - rf|^2 \ge |1 - f|^2/4$ (cf. the proof of Remark 8.53). Hence the integral of the second term in (A.29) is majorized by $\int_{|\lambda| \le \varepsilon} \operatorname{Re}\{1/(1 - f)\}\, d\lambda$, which does not involve r and converges to 0 as $\varepsilon \to 0$. By the change of variable $\omega = \lambda/(1 - r)$ and some algebra, the integral of the first term in (A.29) becomes

$$\begin{aligned}\int_{-\varepsilon/(1-r)}^{\varepsilon/(1-r)} &[1 + r^2(1 - r)^{-2}[\operatorname{Re}^2\{1 - f[\omega(1 - r)]\} \\ &+ \operatorname{Im}^2\{f[\omega(1 - r)]\}] + 2r(1 - r)^{-1} \operatorname{Re}\{1 - f[\omega(1 - r)]\}]^{-1}\, d\omega.\end{aligned} \tag{A.30}$$

From the Taylor expansions $\operatorname{Re}[1 - f(\lambda)] = o(\lambda)$ and $\operatorname{Im} f(\lambda) = \mu\lambda + o(\lambda)$ as $\lambda \to 0$, it follows that for each fixed ω the integrand in (A.30) converges to

$1/(1 + \mu^2\omega^2)$ as $r \to 1$. Moreover, if the integrand is multiplied and divided by $1 + \mu^2\omega^2$, it is easily seen that the factor converging to one is uniformly bounded. Hence by the dominated convergence theorem, as $r \to 1$, (A.30) converges to

$$\int_{-\infty}^{\infty} [1 + \mu^2\omega^2]^{-1}\, d\omega = \pi/\mu,$$

which proves (A.28) and hence the proposition.

The renewal theorem is a simple corollary.

Theorem A.31. $\lim_{k\to\infty} \sum_{n=0}^{\infty} P\{S_n = k\} = \mu^{-1}$.

Proof. From Proposition A.26 and the Riemann–Lebesgue lemma it follows that

$$\text{(A.32)} \qquad \lim_{k\to\infty} \sum_{n=0}^{\infty} [P\{S_n = -k\} + P\{S_n = k\}] = \mu^{-1}.$$

For $k > 0$, by the Markov property of the random walk S_n, $n = 1, 2, \ldots$

$$\begin{aligned}\sum_{n=0}^{\infty} P\{S_n = -k\} &= E\left(\sum_{n=0}^{\infty} I_{\{S_n=-k\}}\right)\\ &= P\{S_n = -k \text{ for some } n \geq 1\} E\left(\sum_{n=0}^{\infty} I_{\{S_n=0\}}\right)\\ &= P\{S_n = -k \text{ for some } n \geq 1\}\left(\sum_{n=0}^{\infty} P\{S_n = 0\}\right).\end{aligned}$$

From Proposition A.26 with $k = 0$ it follows that $\sum_{n=0}^{\infty} P\{S_n = 0\} < \infty$; and by the strong law of large numbers

$$P\{S_n = -k \text{ for some } n \geq 1\} \leq P\left\{\inf_{n\geq 1} S_n \leq -k\right\} \to 0$$

as $k \to \infty$. Hence $\sum_{n=0}^{\infty} P\{S_n = -k\} \to 0$ as $k \to \infty$, which together with (A.32) completes the proof. □

For the common probability models of statistical inference, e.g. for models which can be embedded in exponential families, one can reasonably make stronger hypotheses about the tails of the distribution F and draw some conclusions about the speed of convergence in Theorem A.31. To be specific assume that

$$\text{(A.33)} \qquad \mu_2 = Ez_1^2 < \infty \text{ and } \tilde{f}(u) = E\exp(uz_1) < \infty$$

for some positive u, say $u \leq u_0$. Then considered as a function of a complex variable $w = u + iv$,, $\tilde{f}$ is analytic in $0 < \operatorname{Re} w < u_0$, continuous in $0 \leq \operatorname{Re} w < u_0$, and for real λ, $\tilde{f}(i\lambda) = f(\lambda)$.

It is easy to see that $(\sin k\lambda)\operatorname{Im} f(\lambda)/|1 - f(\lambda)|^2$ is bounded at $\lambda = 0$, so trivial modifications in the proof of Proposition A.26 yield for $k \geq 0$

$$\sum_{n=0}^{\infty} P\{S_n = k\} = (2\mu)^{-1} + (2\pi)^{-1} \int_{-\pi}^{\pi} \operatorname{Re}\{e^{-ik\lambda}/[1 - f(\lambda)]\}\, d\lambda.$$

The special case $f(\lambda) = e^{i\lambda}$ gives

$$1 = \tfrac{1}{2} + (2\pi)^{-1} \int_{-\pi}^{\pi} \operatorname{Re}\{e^{-ik\lambda}/(1 - e^{i\lambda})\}\, d\lambda,$$

so by subtraction

$$\text{(A.34)} \qquad \begin{aligned} &\sum_{n=0}^{\infty} P\{S_n = k\} - \mu^{-1} \\ &\quad = (2\pi)^{-1} \int_{-\pi}^{\pi} [e^{-ik\lambda}\{1/[1 - \tilde{f}(i\lambda)] - 1/\mu(1 - e^{i\lambda})\}]\, d\lambda. \end{aligned}$$

Obviously $\tilde{f}(w + 2\pi i) = \tilde{f}(w)$ for $0 \leq \operatorname{Re} w < u_0$, and $\tilde{f}(i\lambda) \neq 1$ unless $\lambda/2\pi$ is an integer (Problem 8.1). These observations combined with the Taylor series expansion

$$\tilde{f}(w) = 1 + \mu w + \mu_2 w^2/2 + o(w^2) \qquad (w \to 0, \operatorname{Re} w \geq 0)$$

show that there exists $u_1 \in (0, u_0)$ such that in $0 < \operatorname{Re} w < u_1$, $|\operatorname{Im} w| \leq \pi$, $\tilde{f}(w) \neq 1$ and hence

$$\text{(A.35)} \qquad e^{-kw}\{1/[1 - \tilde{f}(w)] - 1/\mu(1 - e^{w})\}$$

is analytic. Since (A.35) is continuous in $0 \leq \operatorname{Re} w < u_1$, it follows from (A.34), the periodicity of $\tilde{f}$, and Cauchy's theorem that for arbitrary $u \in (0, \mu_1)$

$$\text{(A.36)} \qquad \begin{aligned} &\sum_{n=0}^{\infty} P\{S_n = k\} - \mu^{-1} \\ &\quad = (2\pi)^{-1} e^{-ku} \int_{-\pi}^{\pi} e^{-ki\lambda}\{1/[1 - \tilde{f}(u + i\lambda)] - 1/\mu(1 - e^{u+i\lambda})\}\, d\lambda. \end{aligned}$$

Since the integrand in (A.36) is bounded, this proves the following result.

Theorem A.37. *If $Ee^{u_0 z_1} < \infty$ for some $u_0 > 0$, there exists $u \in (0, u_0)$ such that as $k \to \infty$*

$$\sum_{n=0}^{\infty} P\{S_n = k\} - \mu = 0(e^{-ku}).$$

If the probability P can be embedded in an exponential family (so (A.33) is satisfied for some positive u), Theorem A.37 implies that the error in many of the approximations of VIII.3–VIII.5 is exponentially small. For example, from Problems 8.1 and 8.9, and from Theorem A.37 it follows that the $o(b^{-1})$

in (8.84) is actually $o(e^{-bu})$ for some $u > 0$. The exponential rate of convergence helps to explain the extremely good numerical accuracy of these approximations.

In order to obtain similar refinements for the results of X.2, 3, e.g. to verify the claim made in Remark 10.6 that the error in (10.7) is actually $o(\Delta^2)$, an additional argument is required. It suffices to show, in the notation of Chapter X, that the non-vanishing of $1 - E_\mu \exp(wS_{\tau_+})$ in $0 \leq \operatorname{Re} w \leq u_1$ except at $w = 0$, should hold for all μ in some neighborhood of 0 with a u_1 which does not depend on μ. This is an easy consequence of the dominated convergence theorem. The details have been omitted.

For a version of Theorem A.37 when z_1 is non-arithmetic—more precisely, strongly non-arithmetic—see Stone (1965b).

Bibliographical Notes

Chapter II

Most of the material in Sections 1–6 was developed by A. Wald during the 1940s and is described in Wald (1947a). An interesting alternative optimality property of the sequential probability ratio test is discussed by Berk (1975).

The Monte Carlo trick of Remark 2.26 is usually called importance sampling (cf. Hammersley and Handscomb, 1964, p. 57). For a systematic discussion of its use in sequential analysis, see Siegmund (1975a). A useful application is suggested in Remark 4.45.

Cusum tests were first suggested by Page (1954). An outstanding paper in the large literature of this subject is Lorden (1971), who shows that as $B \to \infty$, subject to the constraint (2.45), $\sup_\nu \operatorname{ess\,sup} E\{(\tau - \nu + 1)^+ | x_1, \ldots, x_{\nu-1}\}$ is asymptotically minimized by the stopping rule (2.47). For a systematic although now somewhat dated discussion of cusum procedures, see van Dobben de Bruyn (1968). Shiryayev (1963) and Roberts (1966) have introduced an interesting competitor (cf. Problem 2.14). In order to study this procedure in detail one would require some additional mathematical technique, which unfortunately cannot be developed here. Although understanding of the Shiryayev–Roberts procedure is still rudimentary, it appears to be particularly useful for multiparameter problems and problems where the initial value of the parameter is unknown (cf. Pollak, 1985a, 1985b, and Pollak and Siegmund, 1985).

Chapter III

The classical approach to the probability calculations in Sections 3, 6, and 7 is via the reflection principle (e.g. Karlin and Taylor, 1975, pp. 345 ff.). For a very general discussion see Anderson (1960). The likelihood ratio method used here is due to Siegmund and Yuh (1982).

Until recently there has been very little discussion of attained significance levels and confidence intervals for sequential tests. Berk and Brown (1978) gave more than one definition of attained level, including the one adopted here. Independent discussions leading to essentially the same definition are given by Siegmund and Gregory (1979) and Fairbanks and Madsen (1982). Armitage (1958) studied numerically confidence intervals for a Bernoulli parameter p following sequential tests. Siegmund (1978) considered a normal mean, while Bryant and Schmee (1979) discussed the scale parameter of an exponential distribution, again by numerical methods. A completely different approach has been taken by Wijsman (1981).

Recursive numerical integration to obtain the operating characteristics of sequential tests has been used by a number of authors, notably Armitage *et al.* (1969) and Aroian and Robison (1969). The first systematic discussion of group sequential tests (using numerical methods) is Pocock (1977).

The tests studied in Section 7 are discussed by Anderson (1960). Theorem 3.57 is new. Ferebee (1982) and Daniels (1982) (cf. IV.7) propose competitors. The approximate tests of Section 9 were suggested by Whitehead (1978), who was motivated by a related proposal of Cox (1963). For development of a precise asymptotic theory to justify Brownian approximations in similar contexts, see Hall and Loynes (1977) and Sen (1981).

Problem 3.3 is concerned with a particularly simple truncated sequential probability ratio test, for which approximations along the lines of (3.36) and (3.37) are suggested. It is also possible to obtain exact results analogous to Corollaries 3.43 and 3.44 (see Feller (1968)). Samuel-Cahn and Wax (1984) describe an interesting medical application.

Chapter IV

For a more complete discussion of tests of power one see Robbins (1970) or Robbins and Siegmund (1969).

The derivation of what are essentially repeated significance tests as asymptotic Bayes tests is due to Schwarz (1962). As noted in Example 4.10 Schwarz's viewpoint is different than that presented here, although one of the modifications for early termination to accept H_0 which is suggested in Problem 4.1 leads back to Schwarz's boundaries.

The first systematic studies of repeated significance tests, by numerical methods, are due to Armitage *et al.* (1969) and McPherson and Armitage (1971). See also Armitage (1975). Group tests are discussed by Pocock (1977). The theoretical developments with which Sections 2 and 3 are concerned begin with Woodroofe (1976) and Lai and Siegmund (1977), who essentially derive (4.40). Subsequent developments are found in Siegmund (1977), Woodroofe,

(1978), Lai and Siegmund (1979), Siegmund and Gregory (1979), and Siegmund (1982). Many of these results are summarized by Woodroofe (1982). The method of proof of Theorem 4.14 and the use of this result to give unified derivations of (4.17), (4.18), and (4.20) are new. Propositon 4.29 is due to Siegmund and Gregory (1979), but the terms of order $1/b$ in Theorem 4.27 and the approximation (4.42) are new.

Modified repeated significance tests are suggested independently by Haybittle (1971), Peto *et al.* (1976), and Siegmund (1978), and are studied by Siegmund and Gregory (1979). The use of truncated sequential probability ratio tests as in Table 4.9 is sometimes attributed to O'Brien and Fleming (1979), but for earlier discussions see Miller (1970) and Samuel-Cahn (1974).

The discussion in Section 7 is new, but is motivated by Daniels (1982), who gives an analytic treatment of (4.73) based on the method of images for solving the heat equation. The method of Section 8 is taken from Lai and Siegmund (1977), but the suggestion to apply it outside the domain of exponential families is new.

Chapter V

For the most part Section 3 is taken from Siegmund (1980), who also considered the case of unknown σ. The approximations (5.13)–(5.15) are new.

Section 4 expands on results of Siegmund (1977). See also Woodroofe (1978, 1979) and Hu (1985).

Sellke and Siegmund (1983) give a complete proof of Theorem 5.59 Slud (1984) has obtained similar results. A more satisfactory formulation (as indicated in the text) is found in Sellke (1985), but the proof is very complicated. For a different approach to sequential analysis of survival data, which controls the significance level but ignores the power function almost completely, see Tsiatis (1981), Slud and Wei (1982), Tsiatis (1982), and Harrington *et al.* (1982). A thorough and informative Monte Carlo experiment is reported by Gail *et al.* (1982).

Chapter VI

A recent discussion of sequential randomization tests has been given by Hall (1983). The relation (6.10) is due to Blackwell and Hodges (1958).

Theorem 6.19 is a reformulation of a result of Robbins and Siegmund (1974) (cf. also Siegmund, 1983). The discussion in Section 4 is based on Hayre (1979). For a different approach to the problems of Sections 3 and 4, see Bather (1980, 1981).

Chapter VII

Theorem 7.10 is due to Anscombe (1953). A more rigorous argument, which forms the basis for the discussion given here, is due to Woodroofe (1977). An early inquiry in this direction is Stein (1949).

The results of Section 3 come from Robbins and Siegmund (1972) and Siegmund (1982). In particular, the latter reference contains a proof of Theorem 7.18.

Chapter VIII

Sections 1–4 are standard random walk and renewal theory. Most of these results can be found in Feller (1971, Chapters XI and XVIII), although they are occasionally expressed in a somewhat different language. The first person to notice the asymptotic independence in Theorem 8.34 appears to have been Stam (1968). The proof given here comes from Siegmund (1975b). Theorem 8.51 is due to Woodroofe (1979). The essential ingredients in the proof of Remark 8.53 were supplied by S. Lalley.

The results of Sections 5 and 6 are taken from Siegmund (1975c) and (1982) respectively.

Chapter IX

The general theoretical discussion of Section 2 follows Lai and Siegmund (1977, 1979), as simplified in some of its technical aspects by Hagwood and Woodroofe (1982).

Theorem 9.54 and Corollary 9.55 are due to Siegmund (1982). The argument given here is new. Theorem 9.69 is new. Except for minor modifications the discussion of Section 4 comes from Woodroofe (1982).

The method of proof of Theorem 9.88 originated with Woodroofe (1976a). See Woodroofe (1982) for systematic generalizations and applications to repeated likelihood ratio tests in exponential families.

Repeated likelihood ratio tests in multiparameter exponential families and for nonparametric statistics have been developed in a beautiful series of papers by Woodroofe (1978, 1979, 1982, 1983) and Lalley (1983). Hu (1985) has adapted the method of proof of Theorem 9.54 to multiparameter exponential families—especially the sequential t-test of V.4. (Theorem 3 of Woodroofe, 1978, appears to contain a slight error, which is corrected in Theorem 2 of Lalley, 1983.)

Chapter X

For the most part the results of Sections 1 and 2 come from Siegmund (1979). The exceptions are Theorem 10.16 and Proposition 10.37, which are stated without proof in Siegmund (1985). Theorem 10.41 has predecessors in Siegmund (1979) and Siegmund and Yuh (1982). In the present form it is taken from Siegmund (1985). Theorem 10.55 comes from Siegmund (1979); in the special case of symmetric random variables it is due to Spitzer (1960). The suggestion to begin a proof with (10.58) is due to the author; the present proof in all other particulars (including the refinement of Problem 10.7) is due to M. Hogan. Hogan (1984) makes a first attempt at extending corrected diffusion approximations to problems with nonlinear boundaries. This appears to be a much more difficult project than the large deviation type of approximations that are the subject of Chapters VIII and IX.

Chapter XI

Theorem 11.1 and its method of proof are new. The special case $c = b$ has been discussed by Dirkse (1975) (with a heuristic argument), Siegmund (1977), DeLong (1981) (again heuristically), and Jennen (1985). However, the expressions given by the first three of these authors are incorrect. Miller and Siegmund (1982) give a review of this history.

Proposition 11.13 with some of the constants in the form of a definite integral was obtained by Siegmund (1980) in the two dimensional case. The result given here is new. Theorem 11.30, Problem 11.1, and Problem 11.4 give new results.

References

Anderson, T. W. (1960). A modification of the sequential probability ratiotest to reduce the sample size, *Ann. Math. Statist. 31*, 165–197.

Anscombe, F. J. (1946). Linear sequential rectifying inspection for controlling fraction defective, *J. Roy. Statist. Soc. Suppl. 8*, 216–222.

Anscombe, F. J. (1952). Large sample theory of sequential estimation, *Proc. Cambridge Philos. Soc. 48*, 600–607.

Anscombe, F. J. (1953). Sequential estimation, *J. Roy. Statist. Soc. Ser. B 15*, 1–21.

Anscombe, F. J. (1963). Sequential medical trials. *J. Amer. Statist. Assoc. 58*, 365–383.

Armitage, P. (1958). Numerical studies in the sequential estimation of a binomial parameter, *Biometrika 45*, 1–15.

Armitage, P. (1975). *Sequential Medical Trials*, 2nd ed., Oxford: Blackwell.

Armitage, P., McPherson, C. K., and Rowe, B. C. (1969). Repeated significance tests on accumulating data, *J. Roy. Statist. Soc., Ser. A 132*, 235–244.

Aroian, L. A. (1963). Exact truncated sequential tests for the exponential density function, *Ninth National Symposium on Reliability and Quality Control*, Institute of Radio Engineers, 1 East 79th Street, New York, pp. 7–13.

Aroian, L. A. and Robison, D. E. (1969). Direct methods for exact truncated sequential tests of the mean of a normal distribution, *Technometrics, 11*, 661–675.

Arrow, K. J., Blackwell, D., and Girshick, M. A. (1949). Bayes and minimax solutions of sequential decision problems, *Econometrica 17*, 213–244.

Asmussen, S. (1984). Approximations for the probability of ruin within finite time, *Scand. Actuarial. J.*, 31–57.

Barndorff-Nielsen, E. and Cox, D. R. (1984). The effect of sampling rules on likelihood statistics, *Internat. Statist. Rev. 52*, 309–326.

Barraclough, E. D. and Page, E. S. (1959). Tables for Wald tests for the mean of a normal distribution, *Biometrika*, *46*, 169–177.

Bather, J. A. (1980). Randomized allocation of treatments in sequential trials, *Adv. Appl. Probab.*, *12*, 174–182.

Bather, J. A. (1981). Randomized allocation of treatments in sequential experiments, *J. Roy. Statist. Soc. B 43*, 165–292.

Baum, L. and Katz, M. (1965). Convergence rates in the law of large numbers, *Trans. Amer. Math. Soc. 120*, 108–123.

Berger, J. (1980). *Statistical Decision Theory*, Springer-Verlag, New York-Heidelberg-Berlin.

Berk, R. (1975). Locally most powerful sequential tests, *Ann. Statist.*, *3*, 373–381.

Berk, R. and Brown, L. (1978). Sequential Bahadur efficiency, *Ann. Statist. 6*, 567–581.

BHAT (β-blocker heart attack trial research group). (1981). β-blocker heart attack trial—design features, *Controlled Clin. Trials*, *2*, 275–285.

BHAT (β-blocker heart attack trial research group). (1982). A randomized trial of propranolol in patients with acute myocardial infarction, *J. Amer. Med. Assoc. 147*, 1707–1714.

Billingsley, P. (1968). *Convergence of Probability Measure*, John Wiley and Sons, New York.

Blackwell, D. and Hodges, J. L. (1957). Design for the control of selection bias, *Ann. Math. Statist. 28*, 449–460.

Borovkov, A. A. (1962). New limit theorems in boundary problems for sums of independent terms, *Selected Translations in Mathematical Statistics and probability 5*, 315–372.

Breiman, L. (1968). *Probability*, Addison-Wesley, Reading, Massachusetts.

Brillinger, D. (1962). A note on the rate of convergence of a mean, *Biometrika*, *49*, 574–576.

Brown, A., Mohamed, S. D., Montgomery, R. D., Armitage, P., and Lawrence, D. R. (1960). Value of a large dose of antitoxin in clinical tetanus, *Lancet 2*, 227–230.

Bryant, C. and Schmee, J. (1979). Confidence limits on MTBF for sequential test plans of MIL-STd 781, *Technometrics*, *21*, 33–39.

Butler, D. A. and Lieberman, G. J. (1981). An early accept modification to the test plans of military standard 781C, *Naval Research Logistics Quarterly*, *28*, 221–229.

Chernoff, H. (1972). *Sequential Analysis and Optimal Design*, Society for Industrial and Applied Mathematics, Philadelphia.

Chernoff, H. and Roy, S. N. (1965). A Bayes sequential sampling inspection plan, *Ann. Math. Statist.*, *36*, 1387–1407.

Chow, Y. S., and Robbins, H. (1965). On the asymptotic theory of fixed width sequential confidence intervals for the mean, *Ann. Math. Statist. 36*, 457–462.

Chow, Y. S., Robbins, H., and Siegmund, D. (1971). *Great Expectations: The Theory of Optimal Stopping*, Houghton-Mifflin, Boston.

Cornfield, J. (1966). A Bayesian test of some classical hypotheses—with applications to sequential clinical trials, *Jour. Amer. Statist. Assoc. 61*, 577–594.

Cox, D. R. (1952). A note on the sequential estimation of means, *Proc. Cambridge Phil. Soc.*, *48*, 447–450.

Cox, D. R. (1963). Large sample sequential tests for statistical hypotheses, *Sankhyā, Ser. A*, *25*, 5–12.

Cox, D. R. (1972). Regression models and life tables (with discussion), *J. Roy. Statist. Soc. B 34*, 187–220.

Cox, D. R. (1975). Partial likelihood, *Biometrika 62*, 269–276.

Cox, D. R., and Miller, H. D. (1965). *The Theory of Stochastic Processes*, Methuen, London.

Cramér, H. (1946). *Mathematical Methods of Statistics*, Princeton University Press, Princeton.

Crane, M. A. and Lemoine, A. J. (1977). *An Introduction to the Regenerative Method for Simulation Analysis*, Springer-Verlag, New York-Heidelberg-Berlin.

Cuzick, J. (1981). Boundary crossing probabilities for stationary Gaussian processes and Brownian motion. *Trans. Amer. Math. Soc. 263*, 469–492.

Cuzick, J. (1981). Boundary crossing probabilities for Brownian motion and partial sums, unpublished manuscript.

Daniels, H. (1982). Sequential tests constructed from images, *Ann. Statist.*, *10*, 394–400.
Dantzig, G. (1940). On the non-existence of tests of "Student's" hypothesis having power function independent or σ, *Ann. Math. Statist. 11*, 186–192.
DeLong, D. M. (1981). Crossing probabilities for a square root boundary by a Bessel process, *Commun. Statist.–Theor. Math.*, *A10*, 2197–2213.
DeMets, D., Hardy, R., Friedman, L., and Lan, K. K. G. (1984). Statistical aspects of early termination in the beta-blocker attack trial, *Controlled Clinical Trials*, *4*.
Dirkse, J. (1975). An absorption probability for the Ornstein-Uhlenbeck process, *J. Appl. Probab.*, *12*, 595–599.
van Dobben de Bruyn, C. S. (1968). *Cumulative Sum Tests*, London, Griffin.
Doeblin, W. (1938). Sur deux problémes de M. Kolmogoroff concernant les chaines dénombrables, *Bull. Soc. Math. France*, *66*, 218–220.
Efron, B. (1971). Forcing a sequential experiment to be balanced, *Biometrika*, *58*, 403–417.
Efron, B. and Hinkley, D. (1978). Assessing the accuracy of the maximum likelihood estimator: observed versus expected Fisher information, *Biometrika*, *65*, 457–487.
Epstein, B., Patterson, A. A., and Qualls, C. R. (1963). The exact analysis of sequential life tests with particular application to AGREE plans, *Proc. Aerospace Reliability and Maintainability Conference*, 284–311.
Erdös, P. and Kac, M. (1946). On certain limit theorems of the theory of probability, *Bull. Amer. Math. Soc.*, *52*, 292–302.
Fairbanks, K. and Madsen, R. (1982). *P* values for tests using a repeated significance test design, *Biometrika*, *69*, 69–74.
Feller, W. (1940). Statistical aspects of ESP, *Jour. Parapsychology*, *4*, 271–298.
Feller, W. (1968). *An Introduction to Probability Theory and Its Applications*, Vol. I, John Wiley and Sons, New York.
Feller, W. (1972). *An Introduction to Probability Theory and Its Applications*, Vol. II, John Wiley and Sons, New York.
Feller, W. and Orey, S. (1961). A renewal theorem, *J. Math. and Mech.*, *10*, 619–624.
Ferebee, B. (1980). Unbiased sequential estimation of the drift of a Wiener process, University of Heidelberg preprint.
Ferebee, B. (1982). Tests with parabolic boundary for the drift of a Wiener process, *Ann. Statist.*, *10*, 882–894.
Ferguson, T. (1967). *Mathematical Statistics: A Decision Theoretic Approach*, Academic Press, New York.
Flehinger, B., Louis, T. A., Robbins, H., and Singer, B. (1972). Reducing the number of inferior treatments in clinical trials. *Proc. Nat. Acad. Sci. USA*, *69*, 2993–2994.
Fogel, M., Knauer, C., Andres, L., Mahal, A., Stein, D. E. T., Kemeny, M., Rinki, M., Walker, J., Siegmund, D., and Gregory, P. (1982). Continuous vasopressin in active upper gastrointestinal bleeding, *Ann. Internal Medicine*, *96*, 565–569.
Freedman, D. (1971). *Brownian Motion and Diffusion*, Holden Day, San Francisco.
Freireich, E., Gehan, E., Frei, E., Schroeder, C., Wolman, I., Anbar, R., Burgert, E., Mills, S., Pinkel, D., Selawry, O., Moon, J., Gendel, B., Spurr, C., Storrs, R., Haurani, F., Hoogstraten, B., and Lee, S. (1963). The effect of 6-Mercaptopurive on the duration of steroid-induced remissions in acute leukemia, *Blood 21*, 699–716.
Gail, M., DeMets, D., and Slud, E. (1981). Simulation studies on increments of the two-sample logrank score test for survival time data, with application to group sequential boundaries, in *Survival Analysis*, J. Crowley and R. A. Johnson eds., Institute of Mathematical Statistics, Hayward, 287–301.
Gehan, E. (1965). A generalized Wilcoxon test for comparing arbitrarily singly censored samples, *Biometrika*, *52*, 203–223.

Ghosh, B. K. (1970). *Sequential Tests of Statistical Hypotheses*, Addison-Wesley, Reading, Massachusetts.
Grambsch, P. (1982). Sequential sampling based on observed Fisher information to guarantee the accuracy of the maximum likelihood estimator, submitted to *Ann. Statist.*, *11*, 68–77.
Haggstrom, G. W. (1979). Sequential tests for exponential populations and Possion processes. Rand Corporation Technical Report.
Hagwood, C. and Woodroofe, M. (1982). On the expansion for expected sample size in nonlinear renewal theory, *Ann. Probab.*, *10*, 844–848.
Hald, A. (1981). *Statistical Theory of Sampling Inspection by Attributes*, Academic Press, New York.
Hall, W. J. and Loynes, R. M. (1977). Weak convergence of processes related to likelihood ratios, *Ann. Statist.*, *5*, 330–341.
Hall, W. J. (1983). Some sequential tests for matched pairs: a sequential permutation test, *Contributions to Statistics: Essays in Honor of Norman L. Johnson*, P. K. Sen, ed., North Holland, Amsterdam, 211–228.
Hammersley, J. and Handscomb, D. (1964). *Monte Carlo Methods*, Methuen, London.
Harrington, D. P., Fleming, T. R., and Green, S. J. (1982). Procedures for serial testing in censored survival data, in *Survival Analysis*, J. Crowley and R. A. Johnson, eds., Institute of Mathematical Statistics, Hayward, 269–286.
Haybittle, J. L. (1971). Repeated assessments of results in clinical trials of cancer treatment, *British J. of Radiology*, *44*, 793–797.
Hayre, L. S. (1979). Two population sequential tests with three hypotheses, *Biometrika*, *66*, 465–474.
Hoeffding, W. (1960). Lower bounds for the expected sample size and average risk of a sequential procedure, *Ann. Math. Statist.*, *31*, 352–368.
Hogan, M. (1984). Problems in boundary crossings for random walks, Ph. D. thesis, Stanford University.
Hu, Inchi (1985). On repeated significance tests. Stanford University dissertation.
Ito, K., and McKean, H. P., Jr. (1965). *Diffusion Processes and Their Sample Paths*, Springer-Verlag, New York-Heidelberg-Berlin.
Jacobsen, M. (1982). *Statistical Analysis of Counting Processes*, Springer-Verlag, New York-Heidelberg-Berlin.
Jennen, C. (1985). Second order approximations to the density, mean, and variance of Brownian first exit times, *Ann. Probab. 13*, 126–144.
Jennen, C. and Lerche, R. (1981). First exit densities of Brownian motion through one-sided moving boundaries, *Z. Wahrsch. verw. Gebiete*, *55*, 133–148.
Jennen, C. and Lerche, H. R. (1982). Asymptotic densities of stopping times associated with tests of power one, *Z. Wahrsch. verw. Gebiete*, *61*, 501–511.
Jennison, C., Johnstone, I. M., and Turnbull, B. W. (1981). Asymptotically optimal procedures for sequential adaptive selection of the best of several normal means, *Proc. Third Purdue Symposium on Statistical Decision Theory and Related Topics*, S. S. Gupta and J. Berger, eds., 55–86.
Jones, D. and Whitehead, J. (1979). Sequential forms of the log rank and modified wilcoxon test for censored data, *Biometrika*, *66*, 105–113.
Jones, H. L. (1952). Formulas for group sequential sampling of attributes, Ann. Math. *Statist.*, *23*, 72–87.
Karlin, S. and Taylor, H. (1975). *A First Course in Stochastic Processes*, Academic Press, New York.
Kemp, K. W. (1958). Formulae for calculating the operating characteristic and the average sample number of some sequential tests. *J. Roy. Statist. Soc. B*, *20*, 379–386.
Kennet, R. and Pollak, M. (1981). On sequential detection of a shift in the probability of a rare event, *jour. Amer. Statist. Assoc.*, *78*, 389–395.

Kiefer, J. (1959). k-sample analogues of the Kolmogorov-Smirnov and Cramér-von Mises tests, *Ann. Math. Statist. 30*, 420–447.
Lai, T. L. and Siegmund, D. (1977). A nonlinear renewal theory with applications to sequential analysis I, *Ann. Statist.*, *5*, 946–954.
Lai, T. L. and Siegmund, D. (1979). A nonlinear renewal theory with applications to sequential analysis II, *Ann. Statist.*, *7*, 60–76.
Lai, T. L. and Siegmund, D. (1983). Fixed accuracy estimation of an autoregressive parameter, *Ann. Statist.*, *11*, 478–485.
Lalley, S. (1982). Non-linear renewal theory for lattice random walks, *Commun. Statist.–Sequential Analysis*, *1*, 193–205.
Lalley, S. (1983). Repeated likelihood ratio tests for curved exponential families, *Z. Wahrsch. verw. Gebiete*, *62*, 293–321.
Lalley, S. (1984). Limit theorems for first passage times in linear and non-linear renewal theory, *Adv. Appl. Probab.*, *16*.
Lerche, H. R. (1984). On the optimality of open-ended sequential tests with parabolic boundaries, *Proceedings of the Berkeley Conference in Honor of Jerzy Neyman and Jack Kiefer*, L. LeCam and R. Olshen eds., Wadsworth, Belmont.
Lerche, H. R. (1984). The shape of Bayes tests of power one, MSRI preprint.
Levin, B. and Kline, J. (1984). The cusum test of homogeneity, with an application to spontaneous abortion epidemiology, Columbia University, preprint.
Loève, M. (1963). *Probability Theory*, 3rd ed., Van Nostrand, Princeton.
Lorden, G. (1971). Procedures for reacting to a change in distribution, *Ann. Math. Statist.*, *42*, 1897–1908.
Lorden, G. (1976). 2-SPRT's and the modified Kiefer-Weiss problem of minimizing an expected sample size, *Ann. Statist.*, *4*, 281–291.
Lorden, G. and Eisenberger, I. (1974). Detection of failure rate increases, *Technometrics*, *15*, 167–175.
Louis, T. (1975). Optimal allocation in sequential tests comparing the means of two Gaussian populations, *Biometrika*, *62*, 359–369.
McPherson, C. K. and Armitage, P. (1971). Repeated significance tests on accumulating data when the null hypothesis is not true, *J. Roy. Statist. Soc.*, *Ser. A*, *134*, 15–26.
Mardia, K. V. (1972). *Statistics of Directional Data*, Academic Press, New York.
Miller, R. G. (1970). Sequential rank tests—one sample case, *Proc. Sixth Berk. Symp. Math. Statist. and Prob.*, *1*, University of California Press, Berkeley, 97–108.
Miller, R. G., Jr. and Siegmund, D. (1982). Maximally selected chi square statistic, *Biometrics*, *38*, 1011–1016.
Morgan, M. E., MacLeod, P., Anderson, E. O., and Bliss, C. I. (1951). *A sequential procedure for grading milk by microscopic counts*, Conn. (Storrs) Agric. Exp. Sta. bull, *276*, 35 pp.
Oakland, G. B. (1950). An application of sequential analysis to whitefish sampling, *Biometrics*, *6*, 59–67.
O'Brien, P. C. and Fleming, T. R. (1979). A multiple testing procedure for clinical trials, *Biometrics*, *35*, 549–556.
Page, E. S. (1954). Continuous inspection schemes, *Biometrika*, *41*, 100–115.
Peto, R. and Peto, J. (1972). Asymptotically efficient rank invariant test procedures (with discussion), *J. Roy. Statist. Soc. A*, *135*, 185–206.
Peto, R., Pike, M. C., Armitage, P., Breslow, N. E., Cox, D. R., Howard, S. V., Mantel, N., McPherson, K., Peto, J., Smith, P. G. (1976). Design and analysis of randomized clinical trials requiring prolonged observation of each patient, *Br. J. Cancer*, *34*, 585–612.
Petrov, V. V. (1972). *Sums of Independent Random Variables*, Springer-Verlag, New York-Heidelberg-Berlin.
Pocock, S. (1977). Group sequential methods in the design of an analysis of clinical

trials, *Biometrika*, *64*, 191–200.

Pocock, S. (1982). Interim analysis for randomized clinical trials: the group sequential approach, *Biometrics*, *38*, 153–162.

Pollak, M. (1985a). Optimal detection of a change in distribution, *Ann. Statist. 13*, 206–227.

Pollak, M. (1985b). Average run lengths of an optimal method of detecting a change in distribution, to appear in *Ann. Statist.*, *13*.

Pollak, M. and Siegmund, D. (1975). Approximations to the expected sample size of certain sequential tests, *Ann. Statist.*, *3*, 1267–1282.

Pollak, M. and Siegmund, D. (1985). A diffusion process and its application to detecting a change in the drift of Brownian motion, to appear in *Biometrika*.

Prentice, R. L. (1978). Linear rank tests with right censored data, *Biometrika*, *65*, 167–179.

Rebolledo, R. (1980). Central limit theorem for local martingales, *Z. Wahrsch. verw. Gebiete*, *51*, 269–286.

Robbins, H. (1952). Some aspects of the sequential design of experiments, *Bull. Amer. Math. Soc.*, *58*, 527–535.

Robbins, H. (1970). Statistical methods related to the law of the iterated algorithm, *Ann. Math. Statist.*, *41*, 1397–1409.

Robbins, H. (1974). A sequential test for two binomial populations, *Proc. Nat. Acad. Sci. USA*, *71*, 4435–4436.

Robbins, H. and Siegmund, D. (1969). Confidence sequences and interminable tests, *Bull. Int. Statist. Inst.*, *43*, 379–387.

Robbins, H. and Siegmund, D. (1970). Boundary crossing probabilities for the Wiener process and sample sums, *Ann. Math. Statist.*, *41*, 1410–1429.

Robbins, H. and Siegmund, D. (1971). A convergence theorem for non-negative almost supermartingales and some applications, *Optimizing Methods in Statistics*, J. S. Rustagi, ed., Academic Press, New York, 233–257.

Robbins, H. and Siegmund, D. (1974). Sequential estimation of p in Bernoulli trials, in *Studies in Probability and Statistics*, E. J. Willims ed., Jersusalem Academic Press, 103–107.

Robbins, H. and Siegmund, D. (1974). Sequential tests involving two populations, *Jour. Amer. Statist. Assoc.*, *69*, 132–139.

Roberts, S. W. (1966). A comparison of some control chart procedures, *Technometrics*, *8*, 411–430.

Samuel-Cahn, E. (1974). Repeated significance tests II, for hypotheses about the normal distribution, *Commun. Statist.*, *3*, 711–733.

Samuel-Cahn, E. and Wax, Y. (1984). A sequential "gambler's-ruin" test for $p_1 = p_2$, with a medical application, and incorporation of data accumulated after stopping, Hebrew University, preprint.

Schwarz, G. (1962). Asymptotic shapes of Bayes sequential testing regions, *Ann. Math. Statist.*, *33*, 224–236.

Sellke, T. (1984). Evolution of the partial likelihood over time, Purdue University Technical Report.

Sellke, T. and Siegmund, D. (1983). Sequential analysis of the proportional hazards model, *Biometrika*, *70*, 315–326.

Sen, P. K. (1981). *Sequential Nonparametrics*, John Wiley and Sons, New York.

Shiryayev, A. N. (1963). On optimal methods in earliest detection problems, *Theory of Probab. and Its Appl.*, *8*, 26–51.

Siegmund, D. (1975a). Importance sampling in the Monte Carlo study of sequential tests, *Ann. Statist.*, *4*, 673–684.

Siegmund, D. (1975b). The time until ruin in collective risk theory, *Mitt. Verein. Schweiz. Versich.-Math.*, *75*, 157–166.

Siegmund, D. (1975c). Error probabilities and average sample number of the sequen-

tial probability ratio test, *J. Roy. Statist. Soc. Ser. B*, *37*, 394–401.
Siegmund, D. (1977). Repeated significance tests for a normal mean, *Biometrika*, *64*, 177–189.
Siegmund, D. (1978). Estimation following sequential tests, *Biometrika*, *65*, 341–349.
Siegmund, D. (1979). Corrected diffusion approximations in certain random walk problems, *Adv. Appl. Probab.*, *11*, 701–719.
Siegmund, D. (1980). Sequential χ^2 and F tests and the related confidence intervals, *Biometrika*, *67*, 389–402.
Siegmund, D. (1982). Large deviations for boundary crossing probabilities, *Ann. Probab.*, *10*, 581–588.
Siegmund, D. (1983). Allocation rules for clinical trials, *Mathematical Learning Models: Theory and Algorithms*, U. Herkenrath, D. Karlin, and W. Vogel, eds., Springer-Verlag, New York-Heidelberg-Berlin, pp. 203–212.
Siegmund, D. (1985). Corrected diffusion approximations and their applications, *Proc. Berkeley Conference in Honor of Jerzy Neyman and Jack Kiefer*, L. LeCam and R. Olshen, eds., Wadsworth, Belmont.
Siegmund, D. (1985). Boundary crossing probabilities and statistical applications, submitted to *Ann. Statist.*
Siegmund, D. and Gregory, P. (1979). A sequential clinical trial for testing $p_1 = p_2$, *Ann. Statist.*, *8*, 1219–1228.
Siegmund, D. and Yuh, Yih-Shyh (1982). Brownian approximations to first passage probabilities, *Z. Wahrsch. verw. Gebiete*, *59*, 239–248.
Slud, E. (1984). Sequential linear rank tests for two-sample censored survival data, *Ann. Statist.*, *12*, 551–571.
Slud, E. and Wei, L. J. (1982). Two-sample repeated significance tests based on the modified Wilcoxon statistics, *Jour. Amer. Statist. Assoc.*, *77*, 862–868.
Spitzer, F. (1960). The Wiener-Hopf equation whose kernel is a probability density II, *Duke Math. Jour.*, *27*, 363–372.
Stam, A. J. (1968). Two theorems in r-dimensional theory, *Z. Wahrsch. verw. Gebiete*, *10*, 81–86.
Starr, N. (1966). On the performance of a sequential procedure for the fixed width interval estimation of the mean, *Ann. Math. Statist.*, *37*, 36–50.
Starr, N. and Woodroofe, M. (1968). Remarks on a stopping time, *Proc. Nat. Acad. Ser.*, *61*, 1215–1218.
Stein, C. (1945). A two sample test for a linear hypothesis whose power is independent of the variance, *Ann. Math. Statist.*, *16*, 243–258.
Stein, C. (1949). Some problems in sequential estimation, *Econometrica*, *17*, 77–78.
Stone, C. J. (1965a). Characteristic functions and renewal theory, *Trans. Amer. Math. Soc.*, *120*, 327–342.
Stone, C. J. (1965b). On moment generating functions and renewal theory, *Ann. Math. Statist.*, *36*, 1298–1301.
Switzer, P. (1983). A two-sample sequential test for shift with one sample size fixed in advance, in *Recent Advances in Statistics*, Rizvi, Rustagi, and Siegmund, eds., Academic Press, New York, pp. 95–114.
Takahashi, H. and Woodroofe, M. (1981). Asymptotic expansions in non-linear renewal theory, *Commun. Statist.–Theory. Meth.*, *21*, 2112–2135.
Tsiatis, A. (1981). The asymptotic joint distribution of the efficient scores test for the proportional hazards model calculated over time, *Biometrika*, *68*, 311–315.
Tsiatis, A. (1982). Group sequential methods for survival analysis with staggered entry, in *Survival Analysis*, J. Crowley and R. A. Johnson, eds., Institute of Mathematical Statistics, Hayward, 257–268.
Vaishnava, H., Goyal, R. V., Neogy, C. N., and Mather, G. P. (1966). A controlled trial of autiserum in the treatment of tetnus, *Lancet*, *2*, 1371–1373.
Vakhil, B. J., Tulpole, T. H., Armitage, P., and Laurence, D. R. (1968). A comparison

of the value of 200,000 I.U. of tetanus antitoxin (horse) with 10,000 I.U. in the treatment of tetanus, *Clin. Pharmac. Ther.*, *9*, 465–471.

Wald, A. (1947a), *Sequential Analysis*, John Wiley and Sons, New York.

Wald, A. (1947b), Foundations of a general theory of sequential decision functions, *Econometrica*, *15*, 279–313.

Wald, A. and Wolfowitz, J. (1948). Optimum character of the sequential probability ratiotest, *Ann. Math. Statist.*, *19*, 326–339.

Wei, L. J. (1977). A class of designs for sequential clinical trials, *Jour. Amer. Statist. Assoc.*, *72*, 382–386.

Wei, L. J. (1978). The adaptive biased coin design for sequential experiments, *Ann. Statist.*, *6*, 92–100.

Whitehead, J. (1978). Large sample sequential methods with application to the analysis of 2 × 2 contingency tables, *Biometrika*, *65*, 351–356.

Whitehead, J. (1983). *The Design and Analysis of Sequential Clinical Trials*, Ellis Horwood, Chicester.

Whittle, P. (1982, 1983). *Optimization Over Time*, Vol. I and II, John Wiley and Sons, New York.

Wijsman, R. (1981). Confidence sets based on sequential tests, *Commun. Statist.–Theor. Meth. A*, *10*, 2137–2147.

Wilson, D. W., Griffiths, K., Kemp. K. W., Nix, A. B. J., Rowlands, R. J. (1979). Internal quality control of radioimmunoassays: monitoring of error, *Jour. Endocr.*, *80*, 365–372.

Woodall, R. C. and Kurkjian, B. M. (1962). Exact operating characteristics for sequential life tests in the exponential case, *Ann. Math. Statist.*, *33*, 1403–1412.

Woodroofe, M. (1976a). Frequentist properties of Bayesian sequential tests, *Biometrika*, *63*, 101–110.

Woodroofe, M. (1976b). A renewal theorem for curved boundaries and moments of first passage times, *Ann. Probab.*, *4*, 67–80.

Woodroofe, M. (1977). Second order approximations for sequential point and interval estimation, *Ann. Statist.*, *5*, 984–995.

Woodroofe, M. (1978). Large deviations of the likelihood ratio statistic with applications to sequential testing, *Ann. Statist.*, *6*, 72–84.

Woodroofe, M. (1979). Repeated likelihood ratio tests, *Biometrika*, *66*, 453–463.

Woodroofe, M. (1982). *Nonlinear Renewal Theory in Sequential Analysis*, Society for Industrial and Applied Mathematics, Philadelphia.

Woodroofe, M. (1983). On sequential rank tests, in *Recent Advances in Statistics*, Rizvi, Rustagi, and Siegmund, eds., Academic Press, New York, pp. 115–142.

Woodroofe, M. and Takahashi, H. (1982). Asymptotic expansions for the error probabilities of some repeated significance tests, *Ann. Statist.*, *10*, 895–908. Correction (1985) *13*.

Yuh, Yih-Shyh (1982). Second order corrections for Brownian motion approximations to first passage probabilities, *Adv. Appl. Probab.*, *14*, 566–581.

Index

Lecture Notes in Statistics

Vol. 1: R. A. Fisher: An Appreciation. Edited by S. E. Fienberg and D. V. Hinkley. xi, 208 pages, 1980.

Vol. 2: Mathematical Statistics and Probability Theory. Proceedings 1978. Edited by W. Klonecki, A. Kozek, and J. Rosiński. xxiv, 373 pages, 1980.

Vol. 3: B. D. Spencer, Benefit-Cost Analysis of Data Used to Allocate Funds. viii, 296 pages, 1980.

Vol. 4: E. A. van Doorn, Stochastic Monotonicity and Queueing Applications of Birth-Death Processes. vi, 118 pages, 1981.

Vol. 5: T. Rolski, Stationary Random Processes Associated with Point Processes. vi, 139 pages, 1981.

Vol. 6: S. S. Gupta and D.-Y. Huang, Multiple Statistical Decision Theory: Recent Developments. viii, 104 pages, 1981.

Vol. 7: M. Akahira and K. Takeuchi, Asymptotic Efficiency of Statistical Estimators. viii, 242 pages, 1981.

Vol. 8: The First Pannonian Symposium on Mathematical Statistics. Edited by P. Révész, L. Schmetterer, and V. M. Zolotarev. vi, 308 pages, 1981.

Vol. 9: B. Jørgensen, Statistical Properties of the Generalized Inverse Gaussian Distribution. vi, 188 pages, 1981.

Vol. 10: A. A. McIntosh, Fitting Linear Models: An Application of Conjugate Gradient Algorithms. vi, 200 pages, 1982.

Vol. 11: D. F. Nicholls and B. G. Quinn, Random Coefficient Autoregressive Models: An Introduction. v, 154 pages, 1982.

Vol. 12: M. Jacobsen, Statistical Analysis of Counting Processes. vii, 226 pages, 1982.

Vol. 13: J. Pfanzagl (with the assistance of W. Wefelmeyer), Contributions to a General Asymptotic Statistical Theory. vii, 315 pages, 1982.

Vol. 14: GLIM 82: Proceedings of the International Conference on Generalised Linear Models. Edited by R. Gilchrist. v, 188 pages, 1982.

Vol. 15: K. R. W. Brewer and M. Hanif, Sampling with Unequal Probabilities. ix, 164 pages, 1983.

Vol. 16: Specifying Statistical Models: From Parametric to Non-Parametric, Using Bayesian or Non-Bayesian Approaches. Edited by J. P. Florens, M. Mouchart, J. P. Raoult, L. Simar, and A. F. M. Smith. xi, 204 pages, 1983.

Vol. 17: I. V. Basawa and D. J. Scott, Asymptotic Optimal Inference for Non-Ergodic Models. ix, 170 pages, 1983.

Lecture Notes in Statistics

Vol. 18: W. Britton, Conjugate Duality and the Exponential Fourier Spectrum. v, 226 pages, 1983.

Vol. 19: L. Fernholz, von Mises Calculus For Statistical Functionals. viii, 124 pages, 1983.

Vol. 20: Mathematical Learning Models—Theory and Algorithms: Proceedings of a Conference. Edited by U. Herkenrath, D. Kalin, W. Vogel. xiv, 226 pages, 1983.

Vol. 21: H. Tong, Threshold Models in Non-linear Time Series Analysis. x, 323 pages, 1983.

Vol. 22: S. Johansen, Functional Relations, Random Coefficients and Nonlinear Regression with Application to Kinetic Data. viii, 126 pages. 1984

Vol. 23: D. G. Saphire, Estimation of Victimization Prevalence Using Data from the National Crime Survey. v, 165 pages. 1984.

Vol. 24: S. Rao, M.M. Gabr, An Introduction to Bispectral Analysis and Bilinear Time Series Models. ix, 280 pages, 1984

Vol. 25: Times Series Analysis of Irregularly Observed Data. Proceedings, 1983. Edited by Parzen. vii, 363 pages, 1984.

Vol. 26: Robust and Nonlinear Time Series Analysis. ix, 296 pages, 1984.

Vol. 27: Infinitely Divisible Statistical Experiments. vi, 166 pages, 1984.

Vol. 28: Differential-Geometrical Methods in Statistics. v, 296 pages, 1984.